Animal Signals and Communication

Volume 7

More information about this series at http://www.springer.com/series/8824

Thierry Aubin • Nicolas Mathevon
Editors

Coding Strategies in Vertebrate Acoustic Communication

Springer

Editors
Thierry Aubin
Equipe Communications Acoustiques
Neuro-PSI, CNRS UMR9197
University Paris-Saclay
Orsay, France

Nicolas Mathevon
Equipe de Neuro-Ethologie Sensorielle,
ENES/CRNL, CNRS UMR5292, INSERM
UMRS1028
University of Lyon
Saint-Etienne, France

ISSN 2197-7305 ISSN 2197-7313 (electronic)
Animal Signals and Communication
ISBN 978-3-030-39199-7 ISBN 978-3-030-39200-0 (eBook)
https://doi.org/10.1007/978-3-030-39200-0

This Springer imprint is published by the registered company Springer Nature Switzerland AG.
The registered company address is: Gewerbestrasse 11, 6330 Cham, Switzerland

Occasio facit furem

Acknowledgments

We are grateful to the series editors, Vincent Janik and Peter McGregor, for their invitation to prepare this book and for their comments on an earlier version. We want to express our deepest thanks to all the colleagues, collaborators, and friends who accepted to write a chapter. Last but not least, we warmly thank Patricia Mathevon for her help during the editing process.

Thierry Aubin
Nicolas Mathevon

Contents

Chapter 1
Acoustic Coding Strategies Through the Lens of the Mathematical Theory of Communication

Nicolas Mathevon and Thierry Aubin

Abstract The Mathematical Theory of Communication predicts how the amount of information of a signal is transmitted from an emitter to a receiver after propagation through the environment. This theory can be applied to explain the principles of animal communication and can be, in the acoustic domain, a strong framework to explore crucial questions on communication strategies such as which code for which environment, which code for which social life, how the information is decoded at the receiver's level, how physiological mechanisms constrain the information coding. Such an approach encompasses all aspects of the acoustic communication process, including its dynamic dimensions.

In 1949, Claude Shannon & Warren Weaver published their seminal book *The Mathematical Theory of Communication* in which they define the chain of events supporting the transmission of information: an emitter codes a message into a signal, which propagates through a transmission channel to a receiver, who decodes it to formalize a message (Shannon and Weaver 1949). Primarily devoted to engineers involved in technologies supporting human communication, the Mathematical Theory of Communication aimed at predicting the amount of information transferred in a message. By its ability to encompass all aspects of a communication chain, this theory goes far beyond the technical aspects of human communication and was rapidly adopted by other fields. As stated by Weaver himself in the first part of *The Mathematical Theory of Communication*: "This is a theory so general that one does not need to say what kinds of symbols are being considered—whether written letters

N. Mathevon
Equipe de Neuro-Ethologie Sensorielle, ENES/CRNL, CNRS UMR5292, INSERM UMRS1028, University of Lyon, Saint-Etienne, France
e-mail: mathevon@univ-st-etienne.fr

T. Aubin (✉)
Equipe Communications Acoustiques, Neuro-PSI, CNRS UMR9197, University Paris-Saclay, Orsay, France
e-mail: thierry.aubin@u-psud.fr

T. Aubin, N. Mathevon (eds.), *Coding Strategies in Vertebrate Acoustic Communication*, Animal Signals and Communication 7,
https://doi.org/10.1007/978-3-030-39200-0_1

or words, or musical notes, or spoken words, or symphonic music, or pictures. The theory is deep enough so that the relationships it reveals indiscriminately apply to all these and to other forms of communication." Animals routinely produce, acquire, process, and store information, and information is central to biological systems at every scale (Maynard Smith 2000): organisms are information processing units and the Mathematical Theory of Communication is thus the prerequisite for biologists to understand the principles of animal communication. It will be the philosophy of this book.

We will focus on one subphylum, the vertebrates, and one mode of communication, the acoustic, in which a sound signal is coded by an emitter, then propagated through a channel—air, water, or solid—and finally decoded by one or several receivers. This process corresponds to the transmission of acoustic information. At the receipt of information and as a consequence of the signal emitted, the receivers generally react by modifying their behaviors and consequently by sending back new information. This exchange of information constitutes a communication process. To understand this process, it is primordial to study each step of the transmission chain. For example, if one undertakes to study individual vocal recognition between members of a given species, it is not sufficient to identify through signal analysis the idiosyncratic acoustic parameters likely to carry the vocal signatures. It has indeed been demonstrated that some of these parameters may not necessarily be used by individuals to vocally identify the individual identity of their conspecific. To detect what is really used by animals, it is mandatory to question them through experimental protocols such as playback experiments aimed at testing the importance of each individual acoustic feature. Identifying acoustic coding strategies thus relies on the experimental approach.

According to the Mathematical Theory of Communication, an emitter sends a signal with a finite amount of information which transmits through the channel and the receiver collects only a part of this amount. The loss of information is due to the noise (in the Theory of Communication sense of the word) that takes place at different levels of the chain: at the coding level (e.g., bad motor control during vocal production), at the channel transmission level (the "channel capacity," i.e., the maximum rate at which information can be reliably transmitted over a channel), and at the decoding level (e.g., through the filtering of received signal by sensory organs, errors of meaning interpretation during cognitive processing of information). As a consequence, the emitted and the received messages will differ. In a study investigating how a tropical bird's song transmits information, we showed that the efficiency of a sound communication system results from a coding/decoding process well-tuned to the acoustic properties of the environment (Mathevon et al. 2008; Aubin et al. 2014). Using sound analysis combined to propagation and playback experiments, we demonstrated that the white-browed warbler *Basileuterus leucoblepharus* extracts various information from a received song such as the species and individual identities of the emitter as well as its location in the environment. Strikingly, we found that species information is encoded in acoustic features resistant to propagation changes while individual identity is supported by features that degrade quickly. In their chapter, Ole Larsen et al. provide a thorough review of these communication strategies that allow "public" or "private" signaling. They describe how signals are subjected to attenuation and various other modifications during propagation through the environment that decrease the reliability of information transfer, and how senders can evolve signals within these propagation constraints to match the function of the signal. A sender may choose acoustic behaviors that will help to increase the active space of its signal by overcoming the limitations of the environment, e.g., by increasing the intensity of signal, switching to specific frequency bands, or choosing a singing post. On the other hand, for private signaling, the sender may make its signals more subject to propagation constraints. The efficiency of these strategies depends ultimately on the level of masking background sounds, including sounds generated by physical processes, like wind, rain, and sounds produced by other species. Abiotic and/or biotic environmental noises jam the signal and modify its spectral and temporal characteristics, limiting the ability of receivers to detect it or to discriminate between signals. This is particularly obvious during the dawn chorus of birds whose origin and function are discussed by Diego Gill and Diego Llusia. During the dawn chorus, singing birds are themselves responsible for adding noise to the transmission channel and thus, according to the Mathematical Theory of Communication, for diminishing the volume of coded information which can be transmitted.

At the level of the emitter, the challenge is to transmit to receivers precise and reliable information through an environment that can be constraining for acoustic signals and despite the biomechanical constraints related to anatomo-physiological factors that may affect voice production. In his chapter, Julien Meyer shows how humans shift from "normal" speech to shouting, whistling, and drumming to secure the information when the emitter–receiver distance increases. Indeed, normal speech does not project beyond circa 30 m and the three other registers help to circumvent these constraints. Shouted speech is performed by increasing energy power of the signal which then can transmit information up to a few hundred meters. Whistled speech relies on the whistlers' selection of salient features of a given language and enables people to communicate over 2 km. Drummed speech—which consists in using a musical instrument (drums) to produce sounds mimicking salient cues of spoken languages can extend this range up to 20 km! Whistling and drumming represent simpler coding strategies which keep the same encoded messages as their spoken equivalent. These coding strategies fit perfectly with one prediction of the Mathematical Theory of Communication: the information coding strategy has to be adapted to the capacity of the transmission channel. As stated by Weaver (1949): "The best transmitter, in fact, is that which codes the message in such a way that the signal has just those optimum statistical characteristics which are best suited to the channel to be used."

Which information can be coded in acoustic signals? During animal communication, a part of the information carried by acoustic signals is borne by "static" cues that can be markers of individual idiosyncratic characteristics (e.g., body size, sex, age, identity, etc.). Focusing on terrestrial mammals, Benjamin Charlton and his co-authors illustrate how static vocal cues supporting identity, sex, or body size information depend on the biomechanical constraints applied to the vocal tract of the emitter. Using the source–filter framework, they show that two distinct acoustic features of mammals' voice—the fundamental frequency (driven by the larynx) and the formants (linked to the vocal tract)—have the potential to independently code for this "static" information. Besides its "static" components, part of the information carried by animal acoustic signals is dynamic and related to the current emotional and physiological states of the emitter. This other informative facet of acoustic signals is of major importance because it allows senders to modulate—voluntarily or not—the biological meaning of their signaling. In her chapter, Elodie Briefer focuses on the coding of emotions in acoustic vocalizations. Although the expression of emotions by animals and humans has been of interest since a long time (Darwin 1872), it is now scrutinized in the context of animal welfare and Briefer's review pinpoints the recent increase of our knowledge in this domain. She firstly reports that animals code their emotional state by varying the different dimensions of their vocalizations (signal intensity, spectral content, and temporal dynamics). She then suggests that coding emotional states results from a complex combination of features predicted by motivation-structural rules (the motivational state of the emitter), emotion-dimension rules (valence and arousal), and characteristics of the social links between the emitter and receiver. Finally, she suggests that vocal expression of emotional arousal has been conserved throughout evolution in mammals and maybe in birds.

Besides its sound transmission properties, the environment can impact communication through seasonal-induced modulation of signal production mechanisms. Manfred Gahr's chapter focuses on seasonal singing activity in songbirds, examining the relationships between seasons, hormonal levels, and neural control of song production. Gahr firstly emphasizes that song production is not restricted to males in many songbirds, and that seasonal impact varies among species. He then examines the hormonal systems that support song production. Finally, Gahr discusses the evidence for neural mechanisms of hormone-dependent seasonal song structure. His review emphasizes the urge for developing field studies of female and male singing behavior, hormone production as well as molecular approach of hormones' roles to fully understand the proximate mechanisms of seasonal singing.

At the receiver's end, the challenge is to extract information from signals degraded by transmission through the environment: information has to be decoded and this process ultimately lays on neuronal activity. In her chapter, Solveig Mouterde investigates both sides of the information transmission chain, examining how the "individual identity" information is coded in a songbird's call at the emitter's level, how these vocal signatures are degraded along with sound propagation through the environment, and how the relevant information is received and processed at the receiver's neuronal level. Importantly, Mouterde's chapter underlines the importance of looking at the whole picture, i.e., the whole chain of transmission of information, from information coding in the original sound signal, propagation-induced degradation, to how receivers deal with decoding this altered information at the auditory cortex level. By promoting a quantitative approach of information transfer, Mouterde's chapter provides a nice demonstration of how the Mathematical Theory of Communication can help in fully deciphering a communication strategy.

In the real world, the exchange of information occurs in a network rather than in a "one emitter–one receiver" dyad, with a social environment implicating simultaneously several signalers and receivers. Although this aspect was largely ignored by the Mathematical Theory of Communication, this theory easily extends to communication networks (McGregor 2005). Communication is the glue that holds animal groups or societies together and, in general, sociality goes hand in hand with sophisticated communication systems. In her chapter, Isabelle Charrier investigates mother–offspring acoustic recognition in pinnipeds, a group of mammals with a large diversity of social structures, from solitary to highly colonial species. Through playback experiments, she demonstrates that species with high selective pressures for mother–pup recognition show the most reliable recognition systems, with high vocal stereotypy, a rapid onset of vocal recognition and a multi-parametric vocal signature mainly based on amplitude and frequency modulation features. Through the lens of the Mathematical Theory of Communication, the work of Charrier demonstrates that when noise and risk of confusion between different individuals are significant, the emitted individual information is secured through adapted coding strategies such as redundancy. Moreover, the pinnipeds' data reported by Charrier underline the importance of comparative and large-scale studies of communication systems throughout clades of animals which experience different types and levels of constraints.

Experimental field research that combines sound analysis with elegant playback experiments is a prominent tool to understand how coding of information in vocalizations is related to constraints imposed by animal's social organization and ecology. While this approach can be used to compare information coding strategies in different species as in Charrier's chapter, it can also serve to finely decipher the dynamics of an acoustic communication network within a given species. In her chapter, Caroline Casey reports such approach with another pinniped species, the Northern elephant seal *Mirounga angustirostris*. This species constitutes a remarkable biological model to understand how the complexity of social relations can interfere with information coding in acoustic signals. Casey and her collaborators firstly analyzed the information contained in the calls males and then set up playback experiments with both natural and synthetic signals. By showing that elephant seal males may vary their behavioral response to other male's calls depending on their past experience with the emitter, Casey provides evidence for the importance of learning in the ability of individuals to use this information.

A network environment provides opportunities to multiple receivers to eavesdrop on signals exchanged (interception of communication). Eavesdropping occurs in a situation in which one or more observers (eavesdroppers) extract information from a signaling interaction between others. For example, in numerous bird species, females sample males' song to asses various male quality traits such as age, dominance rank, paternal ability, parasitic load, etc. This situation is developed in the chapter by Nina Bircher and Marc Naguib. Songbirds have been from many years a choice model to investigate questions revolving around acoustic communication. However, most of the effort has been put on how males code for information in their songs and how same-sex competitors interpret it. Deciphering how females decode information related to "quality," motivation, resource holding potential or personality of individual males' signals is thus mandatory to get a whole picture of males' communication strategies.

Social eavesdropping among communication networks occurs not only in intraspecific communication but also between species. Some signals, such as alarm calls, are particularly subject to heterospecific eavesdrop, as illustrated in the chapter by Robert Magrath and his co-authors. Alarm signals are of widespread ecological importance because many birds and mammals give alarm calls when they detect predators or other threats, and have thus been used as classic models for understanding signal design and the evolution of communication. Magrath et al. firstly consider the information conveyed by alarm calls and how it is encoded. They then propose different scenarios that could explain the evolutionary history of information coding in these particular signals. They tell how social eavesdropping by other species can lead to interspecific communication, deception, or suppression of information. They also consider the potential mechanisms involved in the ability of social eavesdropping the alarm calls of a different species, and emphasize the role of learning. Magrath et al.'s chapter however underlines that we still know little about the evolutionary history of alarm coding and about the combined importance of acoustic structure and learning in the development of responses to heterospecific alarm calls.

The presence of social eavesdroppers within a communication network can increase the costs associated with signaling. For the emitter, a possible mechanism for balancing costs and benefits is to engage more than one of the receiver's sensory channels (Smith et al. 2011). These multimodal signals increase the complexity of the communication process, especially when this process is interspecific. In the final chapter of this book, Alexis Billings and Daniel Blumstein emphasize that the use of multimodal signals can be explained by two main hypothesis: the multiple messages hypothesis and the backup signals hypothesis. Multimodal signaling encompasses two different coding strategies: either the addition of another communication channel to acoustics will serve to increase the information content, or it will allow increasing the robustness of information transmission. Both strategies had been suggested by the Mathematical Theory of Communication. Moreover, Billings and Blumstein's chapter goes far beyond intraspecific multimodal communication by developing a framework to understand interspecific multimodal signaling systems. They underline that, while conspecifics usually share similar sensory systems and thresholds, different species may not necessarily have the same sensory systems, the same sensory sensitivity, the same cognitive abilities, or the same information processing abilities. They suggest that interspecific multimodal communication is accomplished through the coevolution of senders and receivers or through sensory exploitation. Their chapter emphasizes that investigations on coding strategies in acoustic communication should now be integrated in a more general framework encompassing other communication channels and not be restricted to interspecific interactions [in this perspective see the recent study on birds-of-paradise by Ligon et al. (2018)]. Here the Mathematical Theory of Communication could be of great help. An important step will be to quantify the information brought by each transmission channel, and to calculate the global information that emerges from this multisensory communication.

Overall, the present book encompasses all aspects of the communication chain, as defined by the Mathematical Theory of Communication. While the past years have witnessed divergences about the nature of communication systems—and especially on the definition of what is information (Bergstrom and Rosvall 2011; Rendall et al. 2009; Sterner 2014; Stegmann 2017), all chapters will emphasize the strength of Shannon and Weaver's approach. Although the Mathematical Theory of Communication is often erroneously called a mathematical theory of information, it is however true that information is one of its core concepts. Minimizing the role of information thus brings a serious risk of misunderstanding the basic principles of animal communication (Seyfarth et al. 2010; Stegmann 2013). While we acknowledge that the term "information" can be ambiguous, since it is often used in the metaphorical sense of *meaning* (Rendall and Owren 2013), this does not justify the abandonment of the term as long as it is correctly defined following the Mathematical Theory of Communication, i.e., as *uncertainty reduction*. Thus, Weaver (1949) states that "this word information in communication theory relates not so much to what you *do* say, as to what you *could* say." By reducing uncertainty, information helps an individual to adapt to its environment, and although information has no universal meaning since it depends on who receives it, it is an embedded characteristic of any signal (van Baalen 2013). According to the Mathematical Theory of Communication, the amount of information can be calculated in *bits*. Calculating this amount of information can help making predictions about the efficiency of a communication system. For instance, Garcia et al. (*in prep*) calculated the amount of information related to species identity in the drumming signal of woodpeckers using measured acoustic parameters from recorded sounds, and found through playback experiments that this mathematical calculation predicts well the performance of birds to discriminate between species. Information is also a powerful concept to compare between different signals. Seminal studies on swallows by Beecher (1982, 1989) used information calculation to demonstrate that the degree of individuality in calls depend on the species' degree of coloniality. This path has been followed by numerous studies that used information calculations to predict the number of individuals that can be potentially discriminated on the basis of their calls [e.g., in penguins, Aubin and Jouventin (2002) and Searby et al. (2004); in hyenas, Mathevon et al. (2010)]. Lengagne et al. (1999) showed that wind limits the amount of information related to individual identity in penguins' calls and further demonstrated that these birds increase the number of calls emitted and the number of syllables per call, using redundancy to maintain the efficiency of communication as the Mathematical Theory of Communication would have predicted. Recently, Elie and Theunissen used information theory to conduct a deep exploration of the call repertoire of zebra finches *Taeniopygia guttata* in order to establish an acoustical and functional classification of this bird's vocalizations: their study is a nice illustration of how a non-supervised approach following the Mathematical Theory of Communication can help interpreting behavioral observations (Elie and Theunissen 2016). Besides, these authors showed that the information theory is of primary interest to explore the neural correlates of acoustic signals at the receiver's level (Elie and Theunissen 2019). They also investigated the information related to individual identity in the zebra finch calls: they found that distinct signatures differentiate zebra finch individuals for each call type, and that birds memorize these multiple signatures (Elie and Theunissen 2018). All these studies (along with many others) demonstrate the utility of an approach based on information in the sense of the Mathematical Theory of Communication. Above all, this approach is a powerful tool to make hypothesis that can be tested through observations or experiments. Besides, and conversely to what is sometimes argued, the Mathematical Theory of Communication is not a reductionist theory that would for instance ignore the psychological dimension of emitters and receivers. It represents a framework which welcomes all aspects of the communication process, including its dynamic dimensions. As stated by Weaver (1949): "The word communication will be used here in a very broad sense to include all of the procedures by which one mind may affect another." He even proposes to use "a broader definition of communication ... which could include the procedures by means of which one mechanism affects another mechanism." A few paragraphs later, he insists on the fact that the Theory does not restrict to the engineering of a communication system (what could represent, in the context of animal acoustic communication, the design of sound production organs, signal features, and sensory systems), but contains "most if not all of the philosophical content of the general problem of communication" including the "capacity of the audience." In the context of animal acoustic communication, we assume that this encompasses all the physiological and psychological aspects of both the emitters and receivers. Communicative intentions and mental state attributions are part of the story: when a human being shifts from normal speech to whistling speech, he or she does so with the intention of communicating at long distance, demonstrating that he or she has integrated the channel constraints. Our growing knowledge about animal intentional cognitive abilities provides objective reasons to think that some species are able to behave similarly. For instance, the choice of a song post by a bird to optimize the active space of its vocalizations could result from the cognitive processes supporting efficient communication behavior. Besides, it is often said that the semantic and the pragmatic levels of communication are not concerned by the Mathematical Theory of Communication which would be interested only in the engineering problem of sending information through a transmission channel (Sterner 2014). Yet, Weaver clearly stated that among the three levels of communication problems ("Level A: the technical problem," "Level B: the semantic problem," and "Level C: the effectiveness problem"), "any limitations discovered in the theory at level A necessarily apply to levels B and C," and that level A "overlaps the other levels more than one could possibly naively suspect": "The theory of Level A is, at least to a significant degree, also a theory of levels B and C." Recent philosophical advances are now recognizing the strength of the Mathematical Theory of Communication on all these aspects (Lean 2014). However, there is still one issue that the Mathematical Theory of Communication does not deal with: it is the information quality. In other words, the theory does not distinguish between relevant and irrelevant information (since it was built to only deal with relevant information).

While the field of behavioral ecology usually focuses on this problem by measuring information in terms of fitness consequences (Donaldson-Matasci et al. 2010), it is not fully satisfying if we want to measure the total amount of information coded in a signal, whatever its fitness consequence. Neurobiologists routinely use this approach when measuring, without any assumptions, the information sent and received by neurons (Reinagel 2000; Mouterde, Chap. 8). Although this question goes beyond the aim of the present book, we think that a full understanding of coding strategies would require such a holistic approach.

What is next? Our knowledge about the acoustic coding strategies developed by animals to communicate has considerably increased during the past 40 years. Although quantitative approaches of communication have been developed in the past, they seem to attract less interest nowadays. Yet, we are still lacking quantitative calculations of the quantity of information coded by emitters and decoded by receivers, especially for acoustic signals having complex spectro-temporal dynamics. These calculations will be even more challenging in the context of multimodal signaling, with signals using in parallel different transmission channels (e.g., acoustics, visual, chemical, etc.). Recent papers advocated for a system approach of animal signaling systems, supporting the idea that we now need to think of animal signals as complex dynamic systems (for details see Hebets et al. 2016; Patricelli and Hebets 2016). To understand how multiple signals using various transmission channels support information coding, transfer, and decoding, we will have to quantify, analyze, and compare sets of communication signals that vary in time and space. We believe that such an approach, driven by the concepts of the Mathematical Theory of Communication, are more than ever needed if we want to have a comprehensive understanding of the mechanisms and evolution of complex animal communication systems.

References

Aubin T, Jouventin P (2002) How to vocally identify kin in a crowd: the penguin model. Adv St Behav 31:243–277

Aubin T, Mathevon N, Da Silva ML (2014) Species identity coding by the song of a rainforest warbler: an adaptation to long-range transmission? Acta Acustica 100:748–758

Beecher MD (1982) Signature systems and kin recognition. Am Zool 22:477–490

Beecher MD (1989) Signalling systems for individual recognition: an information theory approach. Anim Behav 18:465–482

Bergstrom CT, Rosvall M (2011) The transmission sense of information. Biol Philos 26:159–176

Darwin CR (1872) The expression of the emotions in man and animals. John Murray, London

Donaldson-Matasci MC, Bergstrom CT, Lachmann M (2010) The fitness value of information. Oikos 119:219–230

Elie JE, Theunissen FE (2016) The vocal repertoire of the domesticated zebra finch: a data-driven approach to decipher the information-bearing acoustic features of communication signals. Anim Cogn 19:285–315

Elie JE, Theunissen FE (2018) Zebra finches identify individuals using vocal signatures unique to each call type. Nat Commun 9:4026

Elie JE, Theunissen FE (2019) Invariant neural responses for sensory categories revealed by the time-varying information for communication calls. PLoS Comput Biol 15:e1006698
Hebets EA, Barron AB, Balakrishnan CN, Hauber ME, Mason PH, Hoke KL (2016) A systems approach to animal communication. Proc R Soc B 283:20152889
Lean OM (2014) Getting the most out of Shannon information. Biol Philos 29:395–413
Lengagne T, Aubin T, Lauga J, Jouventin P (1999) How do king penguins (*Aptenodytes patagonicus*) apply the mathematical theory of information to communicate in windy conditions? Proc R Soc B 266:1623–1628
Ligon RA, Diaz CD, Morano JL, Troscianko J, Stevens M, Moskeland A et al (2018) Evolution of correlated complexity in the radically different courtship signals of birds-of-paradise. PLoS Biol 16:e2006962
Mathevon N, Aubin T, Vielliard J, da Silva M, Sèbe F, Boscolo D (2008) Singing in the rain forest: how a tropical bird song transfers information. PLoS One 3:e1580
Mathevon N, Koralek A, Weldele M, Glickman SE, Theunissen FE (2010) What the hyena's laugh tells: sex, age, dominance and individual signature in the giggling call of *Crocuta crocuta*. BMC Ecol 10:9
Maynard Smith J (2000) The concept of information in biology. Phil Sci 67:177–194
McGregor PK (2005) Animal communication networks. Cambridge University Press, Cambridge
Patricelli GL, Hebets EA (2016) New dimensions in animal communication: the case for complexity. Curr Opin Behav Sci 12:80–89
Reinagel P (2000) Information theory in the brain. Curr Biol 10:R542–R544
Rendall D, Owren MJ (2013) Communication without meaning or information: abandoning language-based and informational constructs in animal communication theory. In: Stegmann UE (ed) Animal communication theory: information and influence. Cambridge University Press, Cambridge
Rendall D, Owren MJ, Ryan MJ (2009) What do animal signals mean? Anim Behav 78:233–240
Searby A, Jouventin P, Aubin T (2004) Acoustic recognition in macaroni penguins: an original signature system. Anim Behav 67:615–625
Seyfarth RM, Cheney DL, Bergman T, Fischer J, Zuberbühler K, Hammerschmidt K (2010) The central importance of information in studies of animal communication. Anim Behav 80:3–8
Shannon CE, Weaver W (1949) The mathematical theory of communication. University of Illinois Press, Urbana
Smith CL, Taylor A, Evans CE (2011) Tactical multimodal signalling in birds: facultative variation in signal modality reveals sensitivity to social costs. Anim Behav 82(3):521–527
Stegmann UE (2013) A primer on information and influence in animal communication. In: Stegmann UE (ed) Animal communication theory: information and influence. Cambridge University Press, Cambridge, pp 1–39
Stegmann UE (2017) Evolution and information: an overview. In: Joyce R (ed) Routledge handbook of evolution and philosophy. Routledge, London, New York, pp 79–90
Sterner B (2014) The practical value of biological information for research. Philos Sci 81:175–194
Van Baalen M (2013) Biological information: why we need a good measure and the challenges ahead. Interface Focus 3:20130030
Weaver W (1949) Recent contributions to the mathematical theory of communication. In: Shannon CE, Weaver W (eds) The mathematical theory of communication. University of Illinois Press, Urbana, pp 1–28

Chapter 2
To Shout or to Whisper? Strategies for Encoding Public and Private Information in Sound Signals

Ole Naesbye Larsen

Abstract To make sound communication public or private is a question of making the active space of the emitted sound signals large or small. A sender can only encode the sound signals within the parameter space defined by its own anatomy and physiology but, in addition, it may choose acoustic behaviors that will help to increase or decrease the active space. Both signal encoding and behavioral choices are limited by the acoustical properties of the noisy environment that changes the propagating sound signal. To make the sound signals public by increasing the size of the active space, the sender must try to overcome the limitations of the environment, whereas to make the sound signals private the sender could make use of these limitations to reduce the size of the active space. Signal encoding is also limited by the auditory physiology of the receiver whose auditory sensitivity and critical ratio relative to the ambient sound level determine the distance, over which a sound signal can be received. The best documented and most important parameter for changing the size of the active space is the source level of the emitted sound, which may be divided into high-amplitude broadcast sounds and low-amplitude soft sounds. In addition, sound signal frequency parameters may help to improve reception of public sound signals and to reduce reception of private signals, whereas other potential active space regulating parameters such as signal duration only follow predictions for some species.

2.1 Introduction

Our own experience and common sense tell us that there are good reasons for keeping some conversation private whereas other types of vocal communication such as warning signals must be public. A vocalizing male songbird may also face conflicting demands as loud public singing on one hand may give away its position

---

O. N. Larsen (✉)
Department of Biology, University of Southern Denmark, Odense, Denmark
e-mail: onl@biology.sdu.dk

T. Aubin, N. Mathevon (eds.), *Coding Strategies in Vertebrate Acoustic Communication*, Animal Signals and Communication 7,
https://doi.org/10.1007/978-3-030-39200-0_2

to listening predators and parasites but on the other hand, in the fitness context, loud signals are worth the risk by attracting willing females and keeping away conspecific competitors from its territory. Once the male is face to face with a soliciting female or an aggressive competitor, however, loud singing or calling may no more be the best communication strategy.

Classic studies on avian sound communication have focused on adaptations evolved to maximize the area or space, over which a song or call can be heard and interpreted by conspecific listeners above the ambient noise of their habitat (Chappuis 1971; Morton 1975; Marten and Marler 1977; Henwood and Fabrick 1979; Dooling 1982; Brenowitz 1982; Wiley and Richards 1978; Ryan and Brenowitz 1985; Dabelsteen et al. 1993; Klump 1996; Aubin and Jouventin 1998). Especially the "acoustic adaptation hypothesis" proposed by Morton (1975, 1986) has received much attention (Boncoraglio and Saino 2007; Ey and Fischer 2009). It postulates that animals' acoustic signals have been shaped by selection pressures exerted by the acoustic properties of their habitats leading to microevolutionary changes. The acoustic signals of a given species consequently are confined within a certain parameter space or "room for variation" that can only change over evolutionary time (Dabelsteen 1985). Such classic studies have suggested, for instance, that male songbirds' territorial proclamation song, the broadcast song, in general may be detectable by conspecifics located up to two average territory diameters away from the singer (e.g., Brenowitz 1982; Dabelsteen et al. 1993). In addition, it has become clear that focusing on the sender–receiver dyad often gives a wrong impression of sound communication in nature, where acoustic communication is better described by a network of senders and both conspecific and hetero-specific receivers (McGregor and Dabelsteen 1996).

More recently scientists have started to investigate the occurrence and function of unobtrusive low-amplitude vocalizations (McGregor and Dabelsteen 1996; Dabelsteen et al. 1998; Titus 1998; Dabelsteen 2005; Searcy and Yasukawa 2017; Vargas-Castro et al. 2017; Niederhauser et al. 2018). Whereas the former vocalizations definitely are "public," the latter may be loosely categorized as "private." According to the Oxford English Dictionary (Hornby et al. 1970) the adjective "private" is opposed to "public" and in general means "belonging to or for the use of one particular person or group of people only." In the present context, "private" may then be defined as "sound communication involving only a particular pair or group of individuals and dealing with matters that are not to be disclosed to others."

In theory there are at least two ways for a sender to produce a private sound signal: (1) to "encrypt the message" of the sound signal such that, although many neighboring animals will be able to hear the sound signal, only receivers who possess the encryption key can decode its message; (2) to ensure that the message is impossible to detect and decode beyond a certain distance from the sender. The former strategy is problematic for a vocalizing animal, since receivers in many cases will be able to extract information to at least localize the position of, or direction to, the sender, which may be detrimental in case of predator and parasite listeners. However, special sound signals with faint or no directional cues may overcome this problem (Marler 1955). The second strategy is more robust as no listeners beyond a certain distance from the sender will be able to retrieve any information from the sound signal. This certain distance defines the absolute outer boundary of the so-called "active acoustic space" of a sound signal (Marten and Marler 1977; Brenowitz 1982; Wiley and Richards 1982; Ryan and Brenowitz 1985; Lohr et al. 2003).

Who are the receivers? In his famous paper, Peter Marler (1955) proposed to distinguish between intended and unintended receivers. He further proposed that sound signals have evolved not only to facilitate communication with intended receivers but also in many cases to impede information transfer to unintended receivers. First, there are the intended receivers, those with high fitness value to the sender—mate, offspring, group/family—and those who carry low or negative fitness values: known predators and conspecific competitors to resources (for mates, food, or shelter) who should be avoided or scared away. Secondly, there are the unintended receivers, most of which should not be alerted by the vocalization because of negative effects on signaler fitness such as hetero-specific predators and parasites (Zuk and Kulluru 1998) but also concealed conspecific competitors who might use the information obtained by eavesdropping on the communication between the sender and intended receivers to their own advantage in future encounters (McGregor and Dabelsteen 1996). Eavesdropping by unintended receivers, however, does not always have negative fitness value for the sender. Eavesdropping on the alarm calls of other species subjected to the same predators, for instance, may be beneficial to the sender (Marler 1955; Magrath et al. 2015).

If the sender wishes to avoid the unintended receivers, a safe strategy is somehow to reduce the active space. The "Eavesdropping Avoidance Hypothesis" states that animals actively reduce their active space to avoid unintended receivers (McGregor and Dabelsteen 1996; Searcy and Yasukawa 2017; Niederhauser et al. 2018). A small active space excludes most unintended receivers and maximizes communication with closely located intended receivers. As we shall see, animals do in fact under certain circumstances reduce their active space but so far there has been little support for the eavesdropping avoidance hypothesis as a general explanation for this behavior (Searcy and Yasukawa 2017), at least in the predator context (Akçay et al. 2016; Niederhauser et al. 2018).

In this chapter we will investigate the concept of active space and attempt to predict by which signal encoding and behavioral mechanisms a vocalizing bird could vary its active space and thus make its vocalizations public or private. We will elaborate on the properties and constraints of the three components of the communication chain: the sender, the sound signal propagating through the noisy environment, and the receivers, the network of listening individuals. Finally, we will compare the predicted coding strategies with observed coding strategies exemplified in a few case studies on presumed private and public acoustic communication.

The general findings presented here may apply to any taxon of sound communicating animals, but we limit ourselves to birds, since the subject of private–public sound communication has been discussed especially in relation to birds (for recent reviews see Dabelsteen 2005; Mathevon et al. 2008; Akçay et al. 2015; Reichard and Welklin 2015; Zollinger and Brumm 2015; Searcy and Yasukawa 2017). However, a number of similar studies have also been performed in terrestrial mammals (for a recent review see Gustison and Townsend 2015) and a change from high-amplitude public sound signals to low-amplitude private sound signals has also been observed in fishes (e.g., "purring sounds" of croaking gourami (*Trichopsis vittata*) females; Ladich 2007), in frogs (e.g., close-range courtship calls of male golden rocket frogs (*Colostethus beebei*); Bourne et al. 2001), and in insects (e.g., courtship songs of male field crickets (*Gryllus campestris*); e.g., Huber 1955).

2.2 The Active Space of a Propagating Sound Signal

When a bird sender vocalizes in a natural noisy environment, its sound signals propagate away from the source and become softer with distance as the sound energy is distributed over a larger and larger spherical surface and reduced by additional mechanisms. A conspecific receiver located very far from the sender is not able to hear the emitted sound signal, since the signal amplitude here is below the bird's absolute hearing threshold (and below the ambient noise level). At progressively shorter distances from the sender, the signal amplitude may be above the receiver's absolute hearing threshold but the signal-to-noise ratio (SNR) may still be too low for detection by the listener. Only at even shorter distances will both the signal amplitude and SNR be sufficiently large for the listening bird to extract relevant information from the sender's vocalization (Lohr et al. 2003; Dooling et al. 2009; Dooling and Blumenrath 2013; Dooling and Leek 2018).

The active space of a sound signal could then be defined in general as "the physical space surrounding a sender, within which a listening individual can extract relevant information from the sender's sound signals and act adaptively." As suggested by Lohr et al. (2003) it seems reasonable to expect that as a conspecific receiver approaches the sender in the natural habitat, it may be able to extract progressively more information from the sound signals. At the farthest distance, where the sound signal is just *detectable*, the receiver may be able to determine the direction to the sender but not much more. At shorter distances, the receiver may be able to *discriminate* hetero-specific from conspecific calls and discriminate between conspecific song or call types (Lohr et al. 2003) or to determine if the sound signals are produced by two or more different senders that require different responses (Wiley 2013). At progressively shorter distances, the receiver may extract more information from the signal on, for instance, the sender's sex, identity, the messages, and the communication context (*recognition*). At the closest ranges, receivers may be able to engage in what in humans is referred to as *comfortable conversation* (Dooling et al. 2009; Dooling and Blumenrath 2013; Dooling and Leek 2018) and extract even more detailed information, for instance, on the motivational state of the sender. These four broad perceptual categories of detection, discrimination, recognition, and comfortable conversation are based on psychoacoustical measurements in the laboratory on both humans and birds and may also apply to the natural environment, where they may be interpreted as geographical "information zones" surrounding the sender (Lohr et al. 2003; Dooling and Blumenrath 2013; Dooling and Leek 2018).

Hetero-specific receivers with different hearing ranges and sensitivities will experience different extents of the active space (measured, e.g., by its radius if circular) of a given sender's vocalization. A Eurasian sparrowhawk (*Accipiter nisus*), for instance, experiences a much smaller active space of the 8-kHz seet-alarm call of the great tit (*Parus major*) than listening great tits, which can detect the call at much longer distances than the predator (Klump et al. 1986). In general, larger birds like the sparrowhawk are less sensitive to high frequencies (have lower high-frequency cut-off in their audiograms) than smaller birds (Dooling 1992). This means that the extent of the active space is an ambiguous entity depending not only on the sender but also on the individual receiver's hearing physiology.

However, no matter what definition of active space extent is used, by varying its size a sender can potentially vary the number of possible receivers in the acoustic communication network. A very small active space with only one or a few possible receivers defines "private" communication, whereas a large active space with many potential receivers defines "public" communication. The interplay between sender, environment, and receivers constantly and dynamically determines the size and shape of the active space. The extent of the active space may then be modeled by a simple equation:

$$\mathrm{SL} - A = \mathrm{RL} \tag{2.1}$$

where SL is the source level at the sender, defined as the sound pressure level measured in front of the vocalizing bird on its beak's length axis and at a distance of 1 m. *A* is the attenuation of the signal from sender to receiver caused by geometrical and environmental factors. RL is the received level, defined as the sound pressure level of the arriving signal measured at the ears of the receiver. The receiver can only detect the signal, if RL exceeds the thresholds defined by the auditory system of the receiver, which we here refer to as "the required $\mathrm{RL}_{\mathrm{Det}}$" (see Sect. 2.3.4), relative to the ambient noise level in the frequency band of the signal.

2.3 Coding Constraints and Possibilities

How can a sound signal producing animal, a sender, then vary the size of the active space of its songs or calls? Before answering the question, we must realize that there are coding constraints imposed on the sender by its biological characteristics, by properties of the environment, through which the sound signal propagates, and by characteristics of the receiver's auditory system. So, the sender can only maximize its active space within certain limits to make it public. However, theoretically it may also take advantage of these limitations to minimize its sound signals' active space to make it private.

2.3.1 Constraints Imposed by Sender Anatomy and Vocal Physiology

The sender cannot vary the output volume of its vocalizations arbitrarily because it is limited by its body size, anatomy, and physiology. A vocalizing songbird, for instance, cannot increase the sound pressure amplitude beyond a certain limit, even in its peak frequency band as there is an upper limit to the subsyringeal pressure that it can produce by muscular control when forcing an airflow through the specialized valves of the syrinx (e.g., Zollinger et al. 2011; Elemans et al. 2015; Srivastava et al. 2015; Mencio et al. 2017). On the other hand, there also may be a lower limit to the SL that a bird can produce for a given song element, since the syringeal mechanism needs a certain subsyringeal air pressure to produce the self-sustained oscillation that generate the radiated sound (Goller and Larsen 1997; Jensen et al. 2007). However, to my knowledge such quantitative information on a lower limit to SL is not available in the literature.

Besides the sender's ability to produce high subsyringeal pressures within a certain frequency range, its body size sets a lower limit to the sound frequencies that it can efficiently radiate. Generally, sounds with wavelengths much longer than the head and body dimensions cannot be broadcast efficiently (for a discussion of this constraint see, e.g., Larsen and Wahlberg 2017). For many birds therefore, the vocalization frequency interval correlates negatively with body mass (and therefore with body size) on a double logarithmic scale such that larger body mass correlates with lower vocalization frequency (Fletcher 2004). A 58-kg cassowary (*Casuarius casuarius*), for instance, produces intense booming sounds with a fundamental frequency down to about 30 Hz (e.g., Mack and Jones 2003), whereas an 8-g blue-throated hummingbird (*Lampornis clemenciae*) does not produce loud sounds below about 3 kHz but its songs may contain strong harmonic components up to about 30 kHz (Pytte et al. 2004).

Within songbirds, however, there is little evidence for such an unambiguous relation between body size (or body mass) and maximum song amplitude, at least for nightingales (*Luscinia megarhynchos*) and zebra finches (Brumm 2009). However, other song parameters than amplitude may correlate with body size, for instance in tanagers (*Thraupidae*) where 9 out of 10 song parameters, especially frequency parameters, correlated with body mass (Mason and Burns 2015). Generally, birds sing their broadcast songs with higher amplitude when using higher frequency (HF) song elements than low frequency (LF) ones. This has been shown experimentally in European blackbirds (*Turdus merula*) and great tits, which increase their territorial public song amplitude in response to increase in ambient noise by switching to higher frequency song types (Nemeth et al. 2013). On the other hand, birds could theoretically use sounds with wavelengths longer than their head and body dimensions to produce soft private signals, since LF signals would not be emitted efficiently.

The sender's head size also determines the directionality of emitted sounds and can be modified by body postures as observed in the red-winged blackbird (*Agelaius phoeniceus*) (Patricelli et al. 2008). When the head and body dimensions become large relative to the wavelength of the emitted sound, the radiation pattern (and thereby the active space) indicated by iso-pressure curves changes from circular (or omnidirectional) to more elongated (or directional) as the relative sound pressure is reduced behind the bird but remains in its frontal field (Larsen and Dabelsteen 1990; Brumm 2002; Patricelli et al. 2007, 2008). For a given HF sound signal emitted by a nonmoving bird, the active space therefore is directionally elongated but the bird may increase the size of its operational active space by frequently changing its beak direction. There is evidence that some bird species even use elongation of the active space to specifically address known predator receivers, by adjusting the directionality of their public antipredator calls with higher frequencies (Yorzinski and Patricelli 2010).

There are further limitations regarding to what extent birds can vary their active space. In the time domain birds can only vary sound signal envelopes (amplitude modulation) and frequencies (frequency modulation) up to a certain pace—even by superfast muscles, which have an inherent upper contraction "speed limit" of about 250 Hz (Mead et al. 2017). Most of these limitations are either dictated by physical laws or can only be changed over evolutionary time by anatomical or physiological adaptations.

In addition, vocalizing birds have a sometimes very large but always limited signal "library" (MacDougall-Shackleton 1997), i.e., memory of inherited or learned vocalizations, from which to choose calls or song elements with specific characteristics (e.g., LF or HF, narrow frequency band or broadband, simple or complex sounds). Finally, very complex songs require appropriate motor control skills (Podos 1997) that vary between but also within species depending on individual cognitive skills and conditions experienced during individual development (Naguib and Riebel 2014).

2.3.2 *Potential Active Space Regulation by Sender Behavior*

Over the past 20 years much research has been aimed at elucidating the effects of ambient noise on animal sound communication, especially the possible detrimental effects of man-made noise (Slabbekoorn et al. 2018). Territorial songbird songs are public and broadcast with high sound levels that seem energetically optimized (Ward et al. 2004; Zollinger et al. 2011). Studies also suggest that songbirds seek out song posts at specific heights above ground to optimize the transmission range (e.g., Lemon et al. 1981; Dabelsteen et al. 1993). So, it seems reasonable to expect that a territorial bird singing a public song should try to maintain its active space when conditions change, e.g., with increase in ambient sound. Both correlational and experimental studies suggest that birds singing such songs do in fact attempt to maintain the size of their active space by regulating their sound output in response to changes in ambient noise through different mechanisms. For instance, if the ambient noise level increases naturally or experimentally, an avian sender automatically increases its source level.

Part of this increase is caused by a feedback mechanism, the so-called Lombard response that occurs when there is spectral overlap between signal and noise (Manabe et al. 1998; Brumm and Todt 2002; Pytte et al. 2003; Brumm and Zollinger 2011, 2013; Zollinger et al. 2011). This response does not compensate 1:1 for increased noise levels. In many cases, the Lombard response only compensates by an increase of sound pressure amplitude of 0.1–0.7 dB per dB increase in ambient noise level (Osmanski and Dooling 2009). A physiological consequence of increased signal amplitude above the normal level, however, is an increase in the emitted sound frequency pushing the song to frequency bands with less ambient noise and consequently partly compensating for "inefficient" Lombard responses (for a thorough discussion see Zollinger et al. 2012 and the abovementioned example of city-dwelling blackbirds and great tits by Nemeth et al. 2013).

Other sender responses serving to keep the size of the active space of a public song in response to temporary increases in certain frequency bands of ambient noise, includes switching to other song types that are louder and contain higher frequencies than the normal proclamation song (for a comprehensive review of these responses see Brumm and Zollinger 2013). Finally, the sender may help to keep the size of the active space by repeating its song over and over, i.e., making the sound signals highly redundant, which according to signal detection theory (so-called error correcting coding) improves detectability (Lengagne et al. 1999; Price 2013; Wiley 2013).

In the here-and-now, the sound emitting bird can also choose what, when, and where to sing or call, unless an immediate reaction is called upon by, for instance, intruders, group movements, or roosting. By carefully selecting its song posts (or singing during flight) the sender can—at least theoretically—vary (maximize or minimize) its active space as a function of habitat characteristics, such as singing height above ground in forest environments, or in response to meteorological conditions, such as wind direction in open environments (Henwood and Fabrick 1979; Dabelsteen et al. 1993; Mathevon et al. 1996, 2005; Halfwerk et al. 2012, 2018).

The sender may also vary the extent of its active space by choosing time of day and season when sound propagation characteristics differ, for instance, before or after leaf fall in deciduous forests (Blumenrath and Dabelsteen 2004), or in relation to daily variation in ambient sound levels (e.g., Henwood and Fabrick 1979; Brenowitz 1982; Brumm 2004; Fuller et al. 2007; McLaughlin and Kunc 2013), or by trying to avoid overlapping songs of simultaneously vocalizing nearby senders (Brumm and Todt 2004; Goodwin and Podos 2013; Yang et al. 2014).

Such active choices by the sender to optimize its active space are theoretically possible and have been demonstrated convincingly in many cases but mainly for keeping the signals public. However, whether singing birds on a regular basis do make behavioral choices for making their vocalizations private still needs to be determined by many more careful and comparative studies in the field.

2.3.3 *Coding Constraints and Possibilities by Sound Propagation Through the Noisy Environment*

The sound signal radiated from the sender is inevitably subject to change by several mechanisms during propagation through the environment. Therefore, the signal is no longer in its original form when it reaches receivers in the communication network, even after short distance propagation (Richards and Wiley 1980; Wiley and Richards 1978; Dabelsteen et al. 1993).

The most important environmental coding constraint on propagating sound signals is geometric attenuation caused by spherical spreading (see, e.g., Wahlberg and Larsen 2017; Larsen and Radford 2018). Geometric attenuation acts equally on all sound frequencies and is omnipresent. It dominates sound signal attenuation within a radius of up to about 100 m from the sender in most terrestrial environments (Fig. 2.1). Spherical attenuation in general attenuates the emitted sound as a function of distance from the sender and is measured as sound pressure level (or peak pressure) calculated from the expression $20\log_{10}$ (distance/reference distance) decibel (dB) (for derivation of the expression see, e.g., Wahlberg and Larsen 2017). From this expression it is obvious that the received level (RL) at a distance *r* of, for instance, 100 m from a sender is 40 dB lower than the sender's SL (Fig. 2.1). From 100 m to 200 m the spherical attenuation increases by only 6 dB and at further 100-m steps by even less.

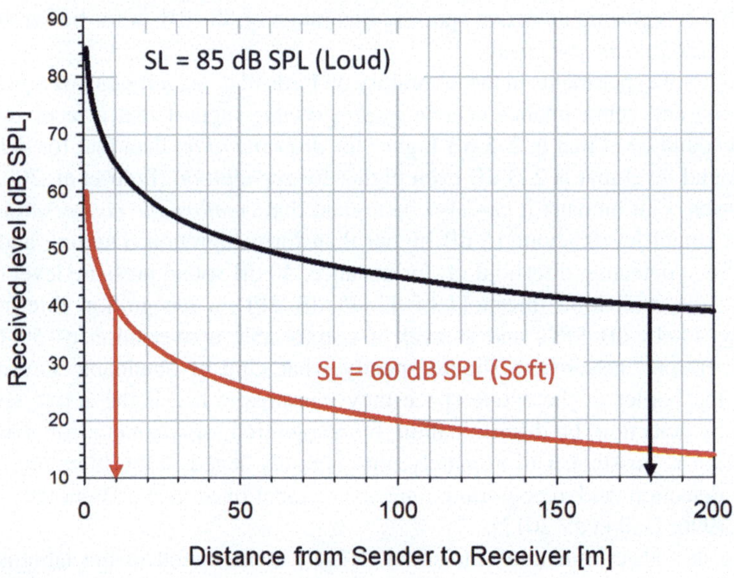

Fig. 2.1 Received level (RL) as a function of distance from a sender emitting songs or calls at two different source levels (SL), when only geometric attenuation from a point source is taken into consideration. If the receiver needs a sound level of 40 dB SPL to be able to detect the vocalization (RL_{Det} = 40 dB SPL), then a source level of 85 dB SPL will give a maximum detection range of 178 m, whereas a source level of 60 dB SPL gives a maximum detection range of only 10 m

Theoretically, a bird can then vary the size of the active space of its calls or songs simply by varying SL of a given sound signal. Rearranging Eq. 2.1, the relationship between SL and RL to a first approximation including only spherical attenuation in *A* is then:

$$\mathrm{RL} = \mathrm{SL} - 20 \log_{10} r/r_{\mathrm{ref}} \tag{2.2}$$

where r is the distance from sender to the ears of the receiver and r_{ref} is the 1-m reference distance at the sender. From this equation it is easy to see that by far the most efficient way for the sender of reducing the active space to avoid unintended receivers is to lower the signal volume, i.e., reducing SL. Using Eq. 2.2 we can get a first impression of the space reduction based on geometric attenuation alone. For instance, if the sender sings or calls out with a SL of 85 dB SPL (Fig. 2.1) and if the receiver's required detection threshold is at $\mathrm{RL}_{\mathrm{Det}} = 40$ dB SPL in the same frequency range, then this will produce an active "detection" space with a radius of about 178 m (that covers an area of 100,000 m^2, if the active space to a first approximation is spherical and consequently circular on the ground). However, if the sender lowers its SL by 25 dB to 60 dB SPL, then the active space radius is reduced to 10 m (area 314 m^2) at ground level. Note that the active space measured as horizontal area is reduced by a factor of about 300. These values of active space radii should be compared with recent comprehensive field studies taking into consideration all relevant factors for active spaces of broadcast song such as the study of Gall et al. (2012) reporting an active space radius range of 88–101 m for brown-headed cowbird (*Molothrus ater*) song.

Behavioral experiments in the laboratory on birds like budgerigars (*Melopsittacus undulatus*) and zebra finches (*Taeniopygia guttata*) suggest that a receiving bird needs a sound level that is 2–5 dB higher for *discrimination* than that for *detection* and a sound level that is 2–3 dB even higher for *recognition* (Lohr et al. 2003). By comparison with humans it can also be argued that *comfortable conversation* may occur at sound levels about 15 dB higher than for recognition (Dooling and Leek 2018). So, a detection threshold of, for instance, 40 dB sound pressure level (SPL), means a discrimination threshold of 42–45 dB SPL, a recognition threshold of typically 44–48 dB SPL, and a level of comfortable conversation of 59–63 dB SPL. Therefore, researchers should specify what kind of obtainable information defines the border of the active space they investigate; i.e., if the active space is defined by detection, by discrimination, by recognition, or when comfortable communication is possible. Unfortunately, this is rarely stated in the literature, where mainly detection and recognition ranges are mentioned (e.g., Brenowitz 1982; Klump 1996; Gall et al. 2012).

Since the "discrimination" threshold ($\mathrm{RL}_{\mathrm{Dis}}$) as measured in the laboratory is 2–5 dB higher than the "detection" threshold, then the "discrimination" space (Fig. 2.2) for a discrimination threshold of, e.g., 43 dB, compared to the detection threshold at 40 dB, reduces the radius of the active space from 178 m to 126 m (and covers an area of about 50,000 m^2) for a source level of 85 dB SPL but to a radius of

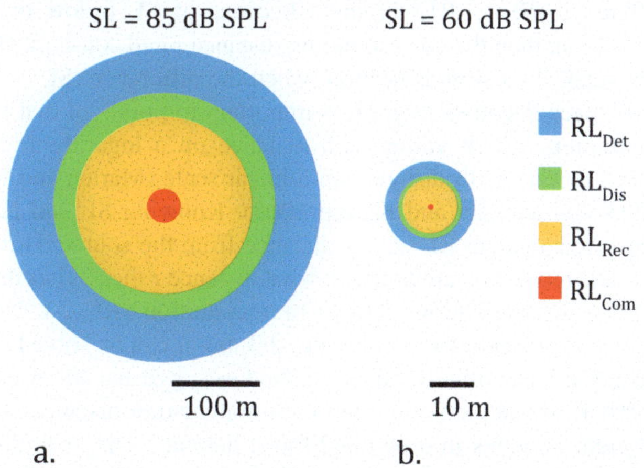

Fig. 2.2 Size of active spaces for two senders with different source levels and taking into consideration only geometric attenuation. The active space can be subdivided in four different "information zones" depending on the sound level required by the receiver for detection (RL_{Det} = 40 dB SPL—blue), the sound level required for discrimination (RL_{Dis} = 43 dB SPL—green), the sound level required for recognition (RL_{Rec} = 45 dB SPL—orange) and the sound level required for comfortable conversation (RL_{Com} = 60 dB SPL—red). (**a**) Sender's source level is 85 dB SPL. (**b**) Sender's source level is 60 dB SPL. Note different distance scales in (**a**) and (**b**)

7 m (area now about 160 m^2) for a source level of 60 dB corresponding to an area reduction by a factor of about 300 (Fig. 2.2).

Geometric attenuation always attenuates a propagating sound wave predictably. At longer distances, however, the measured attenuation typically deviates from the one predicted by geometric attenuation. In most environments the sound pressure level of a song or call measured at a certain distance from the sender is lower than predicted by spherical attenuation. The extra attenuation component is called Excess Attenuation (EA) and is traditionally measured in dB/m (Marten and Marler 1977). Please note, however, that some studies suggest that EA might also be measured on a logarithmic length scale, i.e., in dB/log(distance) (Dabelsteen et al. 1993) or develop in a more complicated fashion (Ręk 2013).

In contrast to spherical attenuation the magnitude of EA is highly frequency (f) dependent and varies with time of day and year (t), with environment (e), and meteorological conditions (m), i.e., it could be expressed as EA(f,t,e,m). After rearranging RL and SL in Eq. 2.2 and setting $r_{ref} = 1$, a second approximation of the relationship between SL and RL at a distance r (in meter) from a point source sender therefore is:

$$\mathrm{SL} - \mathrm{RL} = 20 \log_{10}(r) + r\mathrm{EA}(f, t, e, m) \qquad (2.3)$$

When the difference (SL − RL) is small because of SL is low or because the "required RL" is high, then the corresponding distance *r* and consequently the active space becomes small. In a similar fashion, when the difference (SL − RL) is large, *r* and the active space becomes large. It is important to point out that in contrast to geometric attenuation, which varies with distance on a logarithmic scale, EA is usually supposed to vary with distance on a linear scale (Marten and Marler 1977; Lohr et al. 2003; Langemann and Klump 2005). Knowing SL and EA and using Eq. 2.3 it is easy to calculate RL at any distance from the source. However, in the present context it is more relevant to ask at what distance *r* the RL has dropped to the level required for discrimination (RL_{Dis}) or recognition (RL_{Rec}). Unfortunately, there is no analytical solution for solving Eq. 2.3, but it can be solved computationally, e.g., by using the Newton–Raphson method as suggested and used by Nemeth and Brumm (2010) to calculate maximum communication distances for European blackbird and great tit songs in urban and forest habitats. The magnitude of EA as judged from transmission experiments in different environments and based on values at several distances is of the order of 0.16 dB/m for many types of bird sounds (Morton 1975; Dabelsteen et al. 1993; Holland et al. 1998; Nemeth et al. 2001; Mathevon et al. 2005; Nemeth et al. 2006) but some more recent measurements suggest EA values on the order of 0.3 dB/m (Barker et al. 2009; Sandoval et al. 2015; Piza and Sandoval 2016) or even about 0.8 dB/m (Niederhauser et al. 2018).

Excess attenuation is an umbrella term for a number of habitat-induced effects, the most predictable being frequency dependent atmospheric absorption (e.g., Attenborough 2007). Absorption is directly proportional to distance from the sender and depends not only on sound frequency but also on temperature, atmospheric pressure, and relative humidity (ISO 1993). For instance, at 20 °C and 1 atm, the atmospheric absorption attenuation varies from about 0.0002 dB/m at 100 Hz to about 0.005 dB/m at 1 kHz, whereas from 1 kHz it increases up to about 0.2 dB/m at 10 kHz (e.g., Bass et al. 1995). So, under these atmospheric conditions and at distances of up to 100 m from the sender, the absorption attenuation hardly influences sound at low frequencies (from 0.1 to 1 kHz between 0.02 and 0.5 dB) whereas at higher frequencies the attenuation becomes appreciable (between 1 and 10 kHz up to about 20 dB)—in addition to the 40 dB caused by spherical attenuation at 100 m. In general, the frequency dependence of atmospheric absorption means that sound propagation acts as a low-pass filter (Fig. 2.3), such that LF components of any sound signal dominate when received at distances far from the sender, since HF harmonics attenuate rapidly with distance (Meyer 2015).

Theoretically, a sender can vary its active space further by singing or calling when ambient temperature and humidity produce high EA (more private) or low EA (more public) and/or by using mainly HF or LF sound signals, respectively. Again, more field experiments are needed to test if birds actually do make use of these mechanisms for varying their active space.

Another mechanism of frequency-dependent attenuation that may influence the size and shape of active space is refraction. Phenomena like air turbulence, reflection, and diffraction from the ground and obstacles in the propagation path mainly

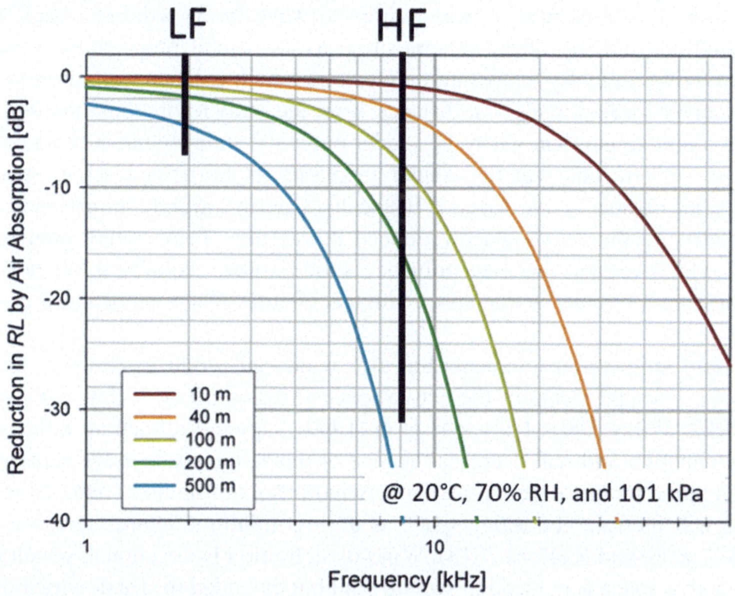

Fig. 2.3 Reduction in received level (*RL*) produced by atmospheric attenuation as a function of frequency at different distances from the sender (point source) showing the lowpass filtering caused by atmospheric absorption. This attenuation adds to the geometric attenuation at the same distances. The level of low frequency (*LF*) sound signals only get a few dB additional attenuation even at 200 m, whereas high frequency (*HF*) sound signals get a marked additional attenuation by atmospheric absorption even at 40 m. The attenuation curves were calculated for 20 °C, 101 kPa, and 70% relative humidity, using absorption values from the Web calculator at the National Physical Laboratory, UK (http://resource.npl.co.uk/acoustics/techguides/absorption/)

distort the envelope of the sound signals by extending their duration from echoes, filtering out certain frequency bands, or introducing variations in their amplitudes (Wiley and Richards 1978; Richards and Wiley 1980; Attenborough 2007; Wahlberg and Larsen 2017; Larsen and Radford 2018). Refraction means that the propagating sound changes direction toward regions of air with lower sound velocity and away from regions with higher sound velocity (Wiley and Richards 1978). The speed of sound (*c*) is mainly determined by temperature and increases with increasing temperature. Daytime air temperature is often highest close to the ground and decreases with height when the air is still, i.e., a temperature gradient is established, the so-called temperature lapse. This means that a wave front of sound propagating parallel to ground will bend skywards. In this situation, a receiver located close to the ground at a given distance from the sender will hear a fainter sound signal than when the air is more turbulent and no temperature gradient can exist. So, under these conditions the active space of the sender is reduced in the horizontal direction but increased in the vertical one. The opposite situation can occur on windless nights when the air temperature becomes lowest close to the ground and increases up to a given height above ground from where it decreases again with height; this is called temperature inversion. Here, upwards directed wave fronts will bend back toward the ground and add by linear superposition to the sound propagating parallel to ground. During a nightly temperature inversion, which extends well into the morning, a receiver located at a given distance from the sender will consequently hear a stronger sound signal than when the sender emits the same signal at noon or in the afternoon. This means that the sender increases its horizontal active space but decreases its vertical active space if it vocalizes during nightly temperature inversions relative to the active space produced in daytime. Temperature gradients are vertical, which means that they influence sound signals equally in all horizontal directions and a circular active space remains circular when temperature gradients change.

Not only temperature gradients but also wind can produce refraction that will change the size and shape of the active space (Morton 1975; Wiley and Richards 1978, 1982; Wiley 1991; Lengagne et al. 1999). Upwind refraction induces additional attenuation and will decrease the size of the active space in the same way as during daytime temperature lapses. Downwind refraction induces reduced attenuation and will increase the active space as during nighttime temperature inversions (see also Larsen and Radford 2018). Wind by definition is directional, which means that the active space is reduced in size upwind but extended in size downwind. So, a circular active space in quiet air becomes "flatter" upwind and more elongated downwind under windy conditions (Henwood and Fabrick 1979).

Finally, the often neglected "ground effect" may comb-filter the propagating sound signal and enhance low frequency sound but attenuate higher frequencies. Including it in calculations of outdoor sound propagation predicts changes in propagating sound signals in open environments quite well (for a general introduction see Wahlberg and Larsen (2017); for a specific study on hooded crows (*Corvus corone cornix*) see Jensen et al. (2008)).

Sound signals virtually always propagate in a "sea" of ambient environmental sound consisting of well-defined localizable biotic sound sources such as roaring mammals or singing birds and an abiotic and biotic background "hum" of non-localizable distributed sources. The latter sources are, for instance, wind-induced rattling of grass in the open, where a light breeze of 4 m/s produces an ambient sound level of about 35 dB(A) (Boersma 1997) or of rattling canopy leaves in closed environments, where the same breeze may produce sound levels of 45–55 dB (A) (Fegeant 1999). Rainy conditions may produce sound levels of 50 dB(A) in forests (Miller 1978) and reduce the active space area of tawny owls (*Strix aluco*) by a factor 69 compared to dry conditions (Lengagne and Slater 2002). Chorusing frogs or insects will also increase the ambient sound level (Brumm and Slabbekoorn 2005). Especially in tropical environments intense insect sounds can totally "drown" or mask all other sounds in a frequency band of typically 5–7 kHz, which may be the reason that bird species in tropical forests sing at lower frequencies than species in open habitats such as grasslands (Ryan and Brenowitz 1985; Weir et al. 2012). In addition to natural biotic and abiotic sound sources there is nowadays in most environments also appreciable sound components generated by human activities that interfere with the abilities of animals to detect important sound signals

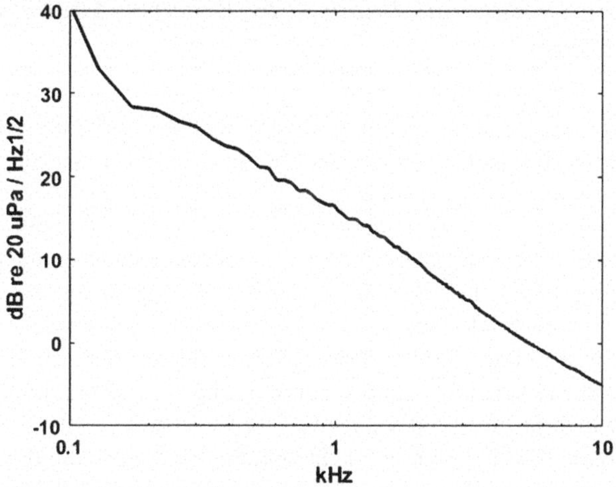

Fig. 2.4 Example of spectrum level of ambient sound as a function of frequency. Ambient sound was recorded close to a puffin rookery on Tjörnes in northern Iceland (Mooney et al. 2019). Note how the spectrum level increases markedly with lower frequencies, especially below 200 Hz and that in a dB-log(frequency) diagram the spectrum level falls off toward higher frequencies, almost as a straight line. So, spectrum levels of ambient sound are very different from the flat (horizontal) spectrum levels of white noise traditionally used in psychoacoustical studies of avian hearing. Courtesy of Adam Smith and Magnus Wahlberg

(Slabbekoorn and Ripmeester 2008; Barber et al. 2010; Francis and Barber 2013; Shannon et al. 2016; Slabbekoorn et al. 2018).

A spectral analysis of ambient sound recorded far from localized sources and where distributed sources prevail often reveals a power spectrum dominated by low frequencies and leveling off toward higher frequencies (Fig. 2.4). This is a direct consequence of the low-pass filtering by atmospheric absorption described above: at the position of the receiver LF sounds are collected and added by linear superposition from a much larger area than HF sounds. So, although LF sound signals propagate much further than HF sounds, since they are not subject to as much low-pass filtering by atmospheric absorption, they may be masked by the omnipresent LF ambient sound, which tends to decrease their active space.

Ambient sound of both biotic and abiotic origin is often referred to as "background noise" or "ambient noise." One should note, however, that the term "noise" is a relative term and used from somebody's perspective: what is noise to one individual may be valuable sound information to another individual. Therefore, "ambient sound" is the neutral term that should be used in most contexts. However, when dealing with the physiological mechanisms producing receiver perception the term "noise" is traditionally used to signify ambient sounds that reduce the receiver's possibility of decoding sound signals carrying relevant information and will consequently be used here.

2.3.4 Coding Constraints Imposed by Receiver Auditory Physiology

The active space of a sound signal is determined not only by sender and the environment, through which it propagates, but also by the properties of the avian receiver's auditory physiology (Gall et al. 2012) that changes only over evolutionary time. In the absolute quiet of a laboratory sound insulated booth the so-called audiogram of a bird can be determined by psychoacoustical or by physiological methods (Dooling 1982). For each of a number of frequencies the lowest sound pressure level, the threshold value, required by the receiver to produce a statistically significant response to indicate that it has detected the test tone, is determined. For birds in general, the result is a U-shaped curve of threshold in the quiet as a function of frequency, which at threshold values below, for instance, 40 dB SPL typically extends from about 300 Hz to about 8 kHz with the lowest threshold of about 10 dB SPL at a best frequency (BF) of about 2 kHz (Dooling and Leek 2018). Big birds tend to have BF at lower frequencies, whereas small birds have BF at higher frequencies and owls are special with much lower thresholds than other birds (cf. the sparrowhawk example in Sect. 2.2).

Knowledge of the audiogram in the quiet, however, will not help much when trying to estimate functional active spaces in the noisy environment of the real world, where sound communication takes place (Klump 1996). First, we need to introduce an important aspect of animal auditory physiology, the so-called critical masking ratio or just critical ratio (CR) (for a comprehensive introduction see, e.g., Klump 1996, Dooling et al. 2000, or Dooling and Leek 2018). In hearing vertebrates, the peripheral auditory system can be perceived as a series of overlapping bandpass filters, each of which lets a certain band of sound frequencies pass for further analysis in the central auditory system but rejects lower and higher frequencies (Fletcher 1940). Whereas the audiogram in the quiet remains constant for an individual bird and species, an audiogram measured in the presence of ambient noise depends on the composition of the noise (frequency, amplitude), which may also fluctuate over time and vary with environment, i.e., the noise level can be expressed as a function $N(f,t,e)$. Audiograms measured in the presence of ambient noise are "masked audiograms" with higher thresholds than in the quiet but are the only ones relevant for sound communication and determination of active spaces.

In the presence of a given noise level, CR can be measured in the laboratory by presenting a pure tone (i.e., a single frequency) at progressively lower sound levels and recording the sound level, for which the receiving bird can no longer detect the tone above the noise, i.e., for which the tone is just masked. The power of the test tone at this threshold (measured in watt) divided by the power (in watt) of $N(f,t,e)$ in the 1-Hz band of the tone is the critical ratio, CR, often expressed in dB (if for instance the power of the test tone at threshold is 10 times larger than that of the noise in the corresponding 1-Hz band, then $\mathrm{CR} = 20\log_{10}(10) = 20$ dB). The noise power in each 1-Hz band is called the spectrum level of the noise at that frequency. It is assumed that the noise power is flat in a frequency band centered at the signal frequency and with a bandwidth of at least one-third of an octave (or 23% of the center frequency).

As pointed out in Fig. 2.4, the spectrum level of the ambient noise in natural environments decreases with frequency. In contrast, the critical ratio of the avian auditory system increases with frequency and should be denoted CR*(f)*. The approximate median CR*(f)* for the 14 bird species tested so far with ambient white noise is 19 dB at 250 Hz, 22 dB at 500 Hz, 24 dB at 1 kHz, 26 dB at 2 kHz, 30 dB at 4 kHz, and 38 dB at 8 kHz (Dooling and Blumenrath 2013; Dooling and Leek 2018). CR*(f)* in many bird species therefore almost, but not quite, increases by about 3 dB/octave (i.e., 3 dB per doubling of frequency) as observed in humans and other animals. So, to detect a sound signal a bird receiver will need a larger signal-to-noise ratio (SNR) of the sound signal if it is composed of high frequencies than of low frequencies.

In contrast to cricket song, birdsong is seldom composed of single frequency tones, but each song varies in a certain frequency band, for instance from 2 to 7 kHz in great tits (McGregor and Krebs 1982), but often has a prominent peak frequency. Fortunately, it turns out that the CR values measured with pure tones by psychoacoustical methods in the laboratory predicts the responses of receiving birds to the peak frequencies of the sender's song quite well (Dooling and Leek 2018). So, if the spectrum level of wind-generated masking noise, *N(f,t,e)*, in a deciduous forest varies from 10 dB re. 20 μPa $Hz^{-1/2}$ at 2 kHz to 5 dB re. 20 μPa $Hz^{-1/2}$ at 4 kHz (Klump 1996), then the "median bird" mentioned above with CR's varying from 26 dB at 2 kHz to 30 dB at 4 kHz will require a received level, RL_{Det}, of more than 36 dB at 2 kHz and more than 35 dB at 4 kHz to detect a song dominated by frequencies in this range. So, in this example the received level required for detection, RL_{Det}, remains rather constant in this frequency range.

In general, the RL (in Eq. 2.3 above) required by a certain bird species for detecting a call or song, RL_{Det}, should then be larger than the sum of the bird's CR*(f)* in the relevant frequency range and the spectrum level of the ambient noise, *N (f,t,e)*, in the same frequency range. If the extra dB values mentioned in Sect. 2.3.3 required for discrimination, recognition, and comfortable conversation applies to birds in general, then for instance $RL_{Rec} = RL_{Det} + 6$ dB (between 4 and 8 dB extra; Lohr et al. 2003) should be used in Eq. 2.3. (For a most educational introduction to calculating communication distances based on realistic source levels, received levels, noise levels, and critical ratios, see Nemeth and Brumm (2010)).

Other properties of the avian auditory system should be taken into consideration when estimating the required RL, for instance, the relation between signal-to-noise ratio at threshold and amplitude (AM) or frequency modulation (FM) of the songs or calls. Unfortunately, to my knowledge, no systematic study of this relation has been reported but Lohr et al. (2003) showed that in budgerigar, zebra finch, and canary (*Serinus canaria*) receivers, high AM of a 2860 Hz pure tone decreased the detection threshold in broadband noise by about 5 dB relative to that of an unmodulated pure tone of 2860 Hz. In contrast, sinusoidal FM did not change the detection threshold. So, an avian sender may conceivably increase the range of its sound signals in ambient noise by changing from unmodulated to AM song and call elements. A 5-dB decrease in threshold may increase the detection range by a factor 1.8, i.e., making the song more public by extending the range from, e.g., 10 to 18 m in Fig. 2.2b. However, we cannot generalize until more systematic studies have been performed. A study on human speech reception (Zeng et al. 2005) found that AM is important for speech recognition in the quiet, but that FM significantly enhances speech recognition in noise. Whether AM and FM have similar roles in avian song recognition remains unknown.

We know much more regarding the relation between auditory threshold and duration of calls or song elements. Traditionally, the peripheral auditory system has been modeled as a leaky-energy detector with a certain integration time, T_i (Plomp and Bouman 1959). In this model, thresholds will remain constant for stimulus tone durations longer than T_i but increase when stimulus duration is progressively shorter than T_i. Such a relation was seemingly found by testing the budgerigar with different duration tones at the best frequency of the audiogram, where T_i was estimated to be 230 ms and where the threshold at a stimulus duration of, for instance, 10 ms was about 15 dB higher than at 230 ms (Dooling and Searcy 1985). This should be compared with average budgerigar contact call durations of about 190 ms (Farabaugh et al. 1998). Later studies examining more frequencies in the European starling (*Sturnus vulgaris*) found a marked frequency dependence of integration times that varied such that at midfrequencies 1–2 kHz T_i was about 550 ms, whereas at lower (500 Hz) and higher (4 kHz) frequencies it was about 200 ms (Klump and Maier 1990). In starlings, the average duration of, for instance, starling motif song is 640 ms (Eens et al. 1989). More recently, careful behavioral studies of threshold level duration (TLD) functions in great tits (Pohl et al. 2013) found time constants of 132 ms in the quiet and 85 ms in noise, which should be compared with an average song element duration of 100 ms (Pohl et al. 2009). So, the relation between duration of song elements and integration times seems rather complicated. Consequently, the validity of the leaky-energy detector model has been questioned and a physiologically more likely probabilistic model has been proposed (for a detailed introduction and discussion, see Pohl et al. 2013). What remains certain though, is the general trend of the threshold level duration (TLD) curves, which for all the mentioned studies show increasing thresholds of 5–15 dB when stimulus sounds are shortened from 1000 ms to 10 ms, which is the duration range of most bird calls and song elements. In the present context this means that the required RL for detection, RL_{Det}, should include not only the CR*(f)* of the bird and *N(f,t,e)* but also the threshold increase depending vocalization duration from the TLD-curves, which may be denoted TLD(Δt). If the sender shortens the element durations, the RL_{Det} increases; this decreases the size of the active space and consequently makes the signals more private. Note however that the call or song element duration at the receiver is often longer than that at the sender, especially in closed environments where reverberation tends to increase the duration with distance (Mathevon et al. 1996). In summary, for detection of a sound signal embedded in the ambient sound of the natural world the receiver's auditory system requires an RL_{Det} that is larger than the sum of (at least) the following parameters:

$$RL_{Det} \geq CR(f) + N(f, t, e) + TLD(\Delta t) \tag{2.4}$$

where CR(*f*) is the critical ratio of the receiver as a function of frequency and *N(f,t,e)* is the spectrum level of the ambient noise as a function of frequency, time, and environment.

2.3.5 Potential Active Space Regulation by Receiver Behavior

The receiver may behaviorally change the extent of the active space of a sound signal emitted from a sender by moving in real time, e.g., in closed environments by flying up in or above vegetation, which may reduce the local EA*(f,t,e,m)* by several dB (Lemon et al. 1981; Dabelsteen et al. 1993; Mathevon et al. 2005). A European blackbird receiver, for instance, located in the forest undergrowth 3 m above ground and close to the border of the active space, where it can just detect the song of a conspecific sender, may gain about 5 dB by flying up to a perch 9 m above ground and thus enter the discrimination (or perhaps even the recognition) zone (Dabelsteen et al. 1993). So, for birds "active spaces" actually are 3D spaces and not just horizontal areas as for many mammals.

Receivers may also extend the range of their active space by different mechanisms of "release from masking." If the receiver is located near a localized masking noise source, such as a vocalizing hetero-specific animal or group of animals, it may experience up 10 dB release from masking, the so-called spatial release from masking, by placing itself such that the angle between a conspecific sender and the noise source is about 90° as has been observed in budgerigars (Dent et al. 1997). Another mechanism for release from masking is the so-called "co-modulation masking release" (CMR). Birds and other animals often (perhaps almost always) experience masking sounds with slow amplitude fluctuations caused by wind gusts or turbulence in the air. If broadband maskers, covering several of the previously mentioned overlapping bandpass filters in the auditory periphery, are coherently and slowly amplitude modulated, then signal detection is improved considerably. CMR in European starlings, for instance, has been studied intensively and release from masking with impressive median values of 28 dB has been reported (for a comprehensive introduction and discussion see Langemann and Klump 2001). These mechanisms of release from masking cannot be used by the sender for encoding public or private information but do emphasize the fact that the extent of the functional active space is much more complicated than predicted by Eq. 2.3.

2.4 Predictions on Encoding of Public and Private Sound Signals

We can now try to answer the introductory question in Sect. 2.3 on how the sender can encode private and public information by varying the size of the active space of its vocalizations. First, we will predict sender actions, each of which would encode for *public* signals, which maximize the active space (Table 2.1). Public calls or song elements should first and foremost be loud, that is, have a high source level (SL). The signals should predominately be concentrated at low frequencies to minimize the effect of atmospheric absorption and other types of excess attenuation (EA(f,t,e,m)) at longer distances. However, signals with frequencies much lower than 1 kHz might be masked by ambient sound, the spectrum level of which in many environments increases at lower frequencies (Fig. 2.4). By concentrating the sound signal energy in a narrow frequency band, the sender would further enhance the active space by increasing the SNR, i.e., (SL – $N(f,t,e)$), in one or only a few auditory bandpass filters of the receiver. High amplitude modulation (AM) of narrow band signals would enhance detection, but AM might be masked by fluctuations caused by reverberation and atmospheric turbulence (Wiley and Richards 1978). Duration of calls or song elements (Δt) should be relatively long, at least as long as the integration time of the receiver's auditory system, if the leaky-energy detector model is used; signal durations longer than a few hundred ms would optimize detectability, since the TLD(Δt) curve would approach a constant low value. Directionality should be close to omnidirectionality to reach as many visually concealed receivers as possible. This would require emitting sounds with wavelengths longer than the body dimensions, such that the sender could be described as a point source (see, e.g., Larsen and Wahlberg 2017), i.e., also argue for using LF signals. In case the bird for other reasons was confined to using directional high-frequency sounds, it could enhance omnidirectionality by frequently turning its head. Each of these parameter choices for encoding the signals mentioned above would help to make the sound signals public. However, different birds may have been adapted by selection to using only a subset of these choices as they are limited by their sound signals' parameter space or "room for

Table 2.1 Overview of possible encoding of signal parameters that a vocalizing bird could use to make calls or song elements public or private by increasing or decreasing the extent of the signal's active space

Parameter of sound signal	Public	Private
Source level (SPL or peak pressure)	High	Low
Frequency (dominant band)	Low	High (or very low)
Bandwidth	Narrow	Broad
Frequency modulation	Low	High (sweeps)
Amplitude modulation	High	Low (or none)
Duration of elements or calls	Long (>about 500 ms)	Short (<about 500 ms)
Directionality	Omnidirectional	Directional

Table 2.2 Overview of possible behavioral choices that a vocalizing bird could make to increase or decrease the extent of its active space by making it easier (public) or more difficult (private) for a receiver to hear and interpret the sender's calls or songs

Behavior	Public	Private
Time of day for calling or singing relative to ambient noise, *N*	When ambient noise is low (morning, evening, or night)	When ambient noise is high (midday)
Time of day for calling or singing relative to excess attenuation, EA	When temperature and humidity minimize EA	When temperature, humidity, and wind maximize EA
Song post choice	High above ground with few intervening objects	Close to ground or intervening objects
Repetition of sound signals	Highly redundant (many repetitions)	Single signal or great variability of signals
Duration of social interactions	Long	Short

variation" (Dabelsteen 1985). So, they could not choose any combination of parameters for enhancing the active space, especially if some combinations changed the signals' information such that, for instance, a blackbird no longer sounded like a blackbird.

In addition to encoding sound parameters for making the signals public, the sender could maximize the range of broadcast signals by some behavioral choices (Table 2.2). The sender should sing or call predominantly when ambient noise, $N(f,t,e)$ was low, and temperature and wind gradients enhanced transmission distance by reducing EA(f,t,e,m). The sender should emit the sound signals from song posts located far away from ground or when flying to avoid attenuation by the ground effect and intervening objects. To make sure that the message got across through the noisy habitat, the sender should finally, like the Eurasian skylark (*Alauda arvensis*), repeat the sound signal over and over, i.e., the sound signal should be highly redundant, since this would increase the receivers' probability of both detecting, discriminating, and recognizing the signal (Price 2013; Wiley 2013). Finally, social communication interactions could be of long duration as this would emphasize, for instance, territory ownership to all receivers.

Secondly, to encode for small *private* active spaces, the sender should use the opposite strategy (Table 2.1). The source level (SL) of calls and song elements should be low, i.e., the signals quiet or soft. As we saw in Fig. 2.2, a source level reduction of just 25 dB would lead to a dramatic reduction in the active space. Soft signals should predominantly contain high frequencies to take advantage of the components of EA(f,t,e,m) (e.g., atmospheric attenuation) that increase with frequency, but they could also be very low frequency, since LF signals cannot be emitted efficiently from small animals and in addition they would be masked by LF ambient noise. In contrast to a narrow band signal of a certain SL, a broadband signal of the same SL would be more private, since it would have its energy smeared out over many auditory bandpass filters of the receiver, which means that the required RL_{Det} might not be reached in any filter (i.e., be below CR(f) + $N(f,t,e)$), unless the receiver came much closer to the sender. Low or no amplitude modulation would further decrease the active space, if the findings regarding AM of Lohr et al. (2003)

apply to birds in general (see Sect. 2.3.4). Calls or song elements with durations much shorter than the integration time of the auditory system should be preferred, since the TLD(Δt) curves add extra dB to RL_{Det} (Eq. 2.4) and thereby reduce the active space further. Highly directional calls could address a specific receiver such as a competitor but be less obvious to potential receivers located in other directions.

Also for soft vocalizations, the sender would have some behavioral choices of when and where to signal (Table 2.2). In the middle of day, the ambient noise level is often rather high and the SNR consequently rather low, which reduces the size of the active space by masking. Signaling when the wind is high will further reduce the active space by increasing the ambient noise in general and enhance attenuation upwind but reduce attenuation downwind. Optimal locations will be close to the intended receivers and on or close to ground. In addition, there is the possibility to combine close quarter sound signals with visual gestures or vibrational signals that work only at close distances. Redundancy of private signals should be low since single messages or great variability in song elements need a high signal-to-noise level to get across to receivers. Finally, vocal interactions should be short duration to reduce probability of catching the attention of unintended receivers.

2.5 Observed Soft Private and Loud Public Calls and Songs

Do vocalizing birds actually encode private and public information in their calls and songs as predicted above? Earlier transmission studies in deciduous forest habitats of European blackbird broadcast song consisting of introductory loud, unmodulated about 2-kHz song elements (whistles) and terminal softer, highly modulated 2–8 kHz elements (twitter), for instance, found that the whistle elements transmitted much further than the twitter elements (Dabelsteen et al. 1993). The structure of the soft twitter elements, sometimes used as "strangled song" for private communication was much in accordance with the predictions listed above including low amplitude, broadband sweeps or harmonics, and high note variability (Dabelsteen et al. 1998). However, this was just one species and we did not take into consideration the auditory physiology of the receiver and overestimated the size of active space of the broadcast song. Later studies, however, have corrected the extent of the active space of blackbird broadcast song (for a comprehensive discussion see Nemeth and Brumm 2010).

2.5.1 Observed Soft and Loud Vocalizations

There is now a substantial literature on low-amplitude private versus high-amplitude public calls and songs. Here, we will use the term "soft" to indicate low-amplitude vocalizations, whereas other terms have been used in the past, such as "strangled song" (Snow 1958), "quiet song" (Dabelsteen et al. 1998), "short-range song" (Titus

1998; Reichard et al. 2013), "whisper song" (Morton 2000; Ishizuka 2009), "quiet call" (Gorissen and Eens 2004), and "low-amplitude song" (Reichard et al. 2011).

The literature mainly discusses the *function* of soft songs or calls, i.e., their correlation with subsequent receiver behavior, where for example the use of soft songs by passerines seems to reliably predict attacks (Hof and Hazlett 2010); for a clear and comprehensive recent review on soft signaling see Reichard and Anderson (2015). In addition, spectrographic parameters of the songs are often reported, whereas quantitative source levels are not. Soft vocalizations have been observed in several types of social interaction, which include, for instance, aggressive interactions (e.g., Searcy and Nowicki 2006; Anderson et al. 2007; Ballentine et al. 2008; Searcy and Beecher 2009; Hof and Hazlett 2010; Akçay et al. 2011, 2015; Zollinger and Brumm 2015; Krieg and Burnett 2017), courtship (e.g., Balsby and Dabelsteen 2005; Reichard et al. 2011, 2013), nest relief when feeding the chicks (e.g., Ishizuka 2009; Elie et al. 2010), and coordination of group movement (e.g., Radford and Ridley 2008). But apart from the coding by change of (mainly qualitatively estimated) amplitude, it is difficult to find enough information to test, if birds in general follow the other six predictions on how to encode their calls and songs with public or private information (Table 2.1).

A recent literature review (Reichard and Welklin 2015) shows that soft vocalizations have been reported from 433 (58%) out of a total of 749 species of North American breeding birds divided in 22 taxonomic orders, out of which 743 species (99%) also sing loud broadcast song. High-amplitude calls are present in twice as many species as high-amplitude songs. Soft calls occur in 2.6 times as many species as soft song, which occur mainly in the context of courtship and territoriality, whereas soft calls mainly occur in the context of contact and courtship but less in aggression and alarm. However, in the literature analyzed, the distinction between loud and soft sound signals was based mainly on qualitative human assessment such as: soft calls are "inaudible at more than 15 m from the bird." Seldom was the evaluation based on quantitative dB measurements.

In about 60% of the species, the reported soft calls or songs were judged to be softer versions of the loud song (Reichard and Welklin 2015). If these numbers are representative, it suggests that 60% of the species code for privacy only by amplitude reduction. But there is a problem: The maximum song amplitude may vary appreciably within a population of conspecific songbirds. In a population of nightingales, for instance, Brumm and Todt (2002) found differences in median vocal sound pressure levels of up to 10 dB between six identified males; differences in median song amplitude of up to 14 dB were reported from 11 zebra finches (Brumm and Slater 2006), Kobayashi and Okanoya (2003) found average amplitude differences of 6 dB among six Bengalese finches (*Lonchura striata*), Brumm and Ritschard (2011) found 9 dB differences in chaffinches, and Nemeth et al. (2012) reported a 12 dB maximum difference in song amplitude between rock sparrows (*Petronia petronia*). So, without quantitative measurements of dB values and without making sure that the same individual sings both a high amplitude and a low amplitude version of the broadcast song, the observed variability in broadcast song amplitude may just be an expression of population variability—not an active choice of reducing the active space to obtain privacy.

In the remaining 40% of species their soft songs were reported to have a structure different from the broadcast songs, such as broader bandwidth and greater note variability (Reichard and Welklin 2015). This is in accordance with predictions of Table 2.1. The soft songs or calls were observed to be emitted in close proximity to receivers and in the contexts of aggression, courtship, contact, and alarm. So, soft songs and calls seem common in vocalizing birds in North America but at this stage we cannot be certain of how common the occurrence is across the whole avian class.

2.5.2 Observed Source Levels of Soft and Loud Vocalizations

Studies of marine mammal sound communication and echolocation always report source levels (SL) or even apparent source levels (ASL) of their experimental animals (e.g., Møhl et al. 2000). ASL is the source level back-calculated from recordings at a distance of the animal without knowledge of the direction of the vocalizing mammal's head or body axis. Back-calculated ASL could also be used for estimating SL in bird species with visually hidden vocalizing individuals. It is therefore rather surprising that quantitative source levels of birdsong are still seldom reported in the literature. This means that for many bird species we are left with only qualitative impressions of their active spaces—and therefore of their possible coding for privacy versus publicity.

Perusing the literature, we find that the broadcast loud songs or calls generally have source levels ranging from 80 to 95 dB SPL (Table 2.3). Sometimes, SL is reported in dB(A), where "(A)" indicates that the sound was high-pass filtered before measuring the sound level, i.e., rejecting the power of low frequencies. Brackenbury (1979) found that the maximum total radiated power varied from 10 mW/(kg body weight) to 870 mW/(kg body weight) in 17 different European songbird species. This corresponded to maximum source levels (SL) ranging from 74 dB SPL in the whitethroat (*Sylvia communis*) to 100 dB SPL in the song thrush (*Turdus philomelos*) (Brackenbury 1979). Later model calculations suggest that avian peak acoustic power is of the order of 200 mW (Fletcher 1988) but this constraint on maximum SL needs further investigation. An outlier regarding avian source levels is the screaming piha (*Lipaugus vociferans*) from Brazilian rain forests with source levels of 110 dB(A) (Nemeth 2004). In contrast, source levels of soft songs and calls are seldom reported but seem to be in the range 50–77 dB SPL (Table 2.4).

Fortunately, there are examples of quantitative studies reporting source levels for both broadcast and soft vocalizations in the same bird species. One study on song sparrows (*Melospiza melodia*; Anderson et al. 2008), found that the SL of their loud broadcast songs ranged from 78 to 85 dB SPL, whereas their soft song SL was reported to vary continuously in the range 55–77 dB SPL with a median difference to their broadcast song levels of 14 dB, corresponding to a reduction of active space radius with a factor five when changing from broadcast to soft vocalizations.

Table 2.3 Broadcast song

Species	SL mean	SL range	References
Song sparrow (*Melospiza melodia*)	80	78–85	Anderson et al. (2008)
Eurasian wren (*Troglodytes troglodytes*)	91	–	Armstrong and Whitehouse (1977)
King penguin (*Aptenodytes patagonicus*)	95	–	Aubin and Jouventin (1998)
Great tit (*Parus major*)	89	–	Blumenrath and Dabelsteen (2004)
Red-winged blackbird (*Agelaius phoeniceus*)	91	88–94	Brenowitz (1982)
Nightingale (*Luscinia megarhynchos*)	85	76–92	Brumm (2004, 2009)
Chaffinch (*Fringilla coelebs*)	82	78–87	Brumm and Ritschard (2011)
Zebra finch (*Taeniopygia guttata*)	80	60–90	Cynx et al. (1998)
Zebra finch (*Taeniopygia guttata*)	72	68–78	Ritschard and Brumm (2011) (in anechoic chamber)
European blackbird (*Turdus merula*)	85	–	Dabelsteen (1981)
Screaming piha (*Lipaugus vociferans*)	110	–	Nemeth (2004)
Red-winged blackbird (*Agelaius phoeniceus*)	83	79–94	Patricelli et al. (2007, 2008)
Corncrake (*Crex crex*)	96	80–101	Ręk and Osiejuk (2011)

Examples of source level (SL) of loud broadcast songs and calls of various bird species measured as sound pressure level (dB SPL, i.e., root-mean-square sound pressure re. 20 μPa)

Table 2.4 Soft song and calls

Species	Mean or median	SL range	References
Song sparrow (*Melospiza melodia*)	65	54–77	Anderson et al. (2008)
House wren (*Troglodytes aedon*)	–	25–33 dB softer than broadcast	Krieg and Burnett (2017)
Corncrake (*Crex crex*)	71	65–76	Ręk and Osiejuk (2011)
Brown-flanked bush warbler (*Cettia fortipes*)	13 dB softer than broadcast	–	Xia et al. (2013)

Examples of source levels (SL) of soft song of various bird species measured as dB SPL. Note that in some studies, soft song levels are given only relative to the SPL of the broadcast song, which has not been quantified

However, the difference between the loudest broadcast song and the softest song was 36 dB corresponding to a space radius reduction of a factor 63. In addition, the authors divided soft songs into "crystallized soft songs," whose structure was much the same as the broadcast songs, and very soft "warbled songs" whose was very different from that of broadcast songs.

Other informative studies concerned corncrakes (*Crex crex*) that belong to quite another order (*Gruiformes*) than songbirds (Ręk and Osiejuk 2011; Ręk 2013). Nevertheless, the corncrakes expressed the same pattern as the song sparrows producing loud calls with median SL of 96 dB SPL (range 80–101 dB SPL) and low-amplitude aggressive gurgling–mewing calls with median SL of 70 dB SPL (range 65–76 dB SPL); i.e., with a median difference of 25 dB between high- and low-amplitude calls corresponding to a reduction of active space radius by a factor 18, when changing from loud to soft vocalizations (cf. Figs. 2.1 and 2.2). When measuring soft call transmission and taking the receiver's presumed CR(*f*) into consideration, this suggests a maximum active space radius of 20–40 m (Ręk 2013).

These two examples raise the question whether the intuitive division by human observers of songs and calls into only two categories, soft and loud, really reflects reality to the birds. Perhaps more amplitude ranges should be included or perhaps the change in amplitude actually is a continuum? In addition, some authors suggest that using peak pressures to calculate active spaces and not only RMS pressures will give a better estimate of the functional active spaces as bird receivers may not need to identify the complete song but only its most intense components.

2.5.3 Other Coding Parameters Observed in Soft and Loud Vocalizations

We predicted that broadcast loud calls and song elements should be of longer duration than soft ones. But in the two quantitative studies just mentioned this is not the case. In song sparrows there was no difference in song element duration (Anderson et al. 2008)—and in the corncrakes the duration of soft calls was actually about four times longer than that of broadcast calls (Ręk 2013). However, Xia et al. (2013) found significantly shorter duration elements in the soft song (40–80 ms) than in the broadcast song (>200 ms) of the brownish-flanked bush warbler (*Cettia fortipes*).

The song sparrow and corncrake examples show two very different frequency ranges of soft vocalizations relative to the broadcast ones. In song sparrows the two versions of soft song had a wider frequency range than broadcast songs—but not much (Anderson et al. 2008). The broadcast song varied in the range 2.0–8.0 kHz, whereas the two versions of the soft song varied in the range 1.7–8.1 kHz (crystallized soft song) and 1.5–9.1 kHz (warbled song). This follows the predictions of Table 2.1.

In corncrakes, on the other hand, the soft call elements were much lower frequency range than the broadcast calls (Ręk 2013); broadcast call frequency range was 0.5–5.0 kHz opposed to soft call ranges of less than 1 kHz (gurgling note) and 1–2 kHz (mewing note). This is the opposite of the predictions of

Table 2.1. This deviation illustrates that other parameters than frequency content and duration should be included to understand detectability of sound signals. Corncrake song is special consisting of sweeps of transients, which are super easy to recognize for humans even at long distances, although the signals' frequency content is broadband. So, this is a good example of a "weird exemption" from most songbird songs, which generally are tonal.

Other studies show marked differences between broadcast and soft songs following all the predictions of Table 2.1. Vargas-Castro (2015), for instance, showed that soft song elements in the white-throated thrush (*Turdus assimilis*) are much more broadband, at higher frequencies, and with much longer frequency sweeps (FM) than broadcast song elements and therefore follow the same pattern as other Turdus species such as the European blackbird (Dabelsteen et al. 1993). Finally, greater variability in soft song elements than in broadcast song seems widespread, as recently documented by Vargas-Castro (2015) and Xia et al. (2013).

2.5.4 *Observed Behavioral Choices of Soft and Loud Calls and Songs*

Behavioral choices for extending the size of the active space by optimizing song post height above ground have been reported for some species, e.g., in rainforest antbirds (Nemeth et al. 2001). However, behavioral choices for reducing the active space seem inadequately reported in the literature to make any strong conclusions about exploitation of environmental properties such as those predicted in Table 2.3 for encoding private information in avian calls or song elements.

2.6 Conclusions

Functionally, birds encode private and public information in their vocalizations. The main mechanism is regulation of the size of the active space. Active space reduction by lowering the voice makes the vocalizations more private and seems common in birds. But we need more quantitative studies to be able to generalize on privacy coding. We also need more quantitative source level measurements of known individuals to investigate whether there are more relevant amplitude categories than just the two, broadcast and soft, perhaps even a continuum. Coding for privacy by frequency parameters can be enhanced either by using broadband high frequency sounds, limited by atmospheric attenuation, or low frequency sounds potentially masked by low frequency ambient noise. Coding of public or private information by sound signal duration follows predictions in some studies but not in other, so to generalize we need more experimental works. Finally, social behavior of vocalizing birds is carefully reported in the literature but behavioral choices concerning where and when to vocalize relative to environment, time, and meteorological conditions are generally not. So, there is still much to do before we can make strong conclusions about the strategies for encoding public and private information in sound signals.

Acknowledgments I am grateful to the editors for inviting me to write this chapter and to Magnus Wahlberg for his comments to improve the manuscript.

References

Akçay Ç, Tom ME, Holmes D, Campbell E, Beecher MD (2011) Sing softly and carry a big stick: signals of aggressive intent in the song sparrow. Anim Behav 82:377–382

Akçay Ç, Anderson RC, Nowicki S, Beecher MD, Searcy WA (2015) Quiet threats: soft song as an aggressive signal in birds. Anim Behav 105:267–274

Akçay Ç, Clay A, Campbell E, Beecher MD (2016) The sparrow and the hawk: aggressive signalling under the risk of predation. Behav Ecol 27(2):601–607

Anderson RC, Nowicki S, Searcy WA (2007) Soft song in song sparrows: response of males and females to an enigmatic signal. Behav Ecol Sociobiol 61:1267–1274

Anderson RC, Searcy WA, Peters S, Nowicki S (2008) Soft song in song sparrows: acoustic structure and implications for signal function. Ethology 114:662–676

Armstrong EA, Whitehouse HLK (1977) Behavioural adaptations of the wren (*Troglodytes troglodytes*). Biol Rev 52:235–294

Attenborough K (2007) Sound propagation in the atmosphere. In: Rossing TD (ed) Handbook of acoustics. Springer, New York, pp 113–147

Aubin T, Jouventin P (1998) Cocktail-party effect in king penguin colonies. Proc R Soc B 265:1665–1673

Ballentine B, Searcy WA, Nowicki S (2008) Reliable aggressive signalling in swamp sparrows. Anim Behav 75:693–703

Balsby TJS, Dabelsteen T (2005) Simulated courtship interactions elicit neighbour intrusions in the whitethroat, *Sylvia communis*. Anim Behav 69(1):161–168

Barber JB, Crooks KR, Fristrup KM (2010) The costs of chronic noise exposure for terrestrial organisms. Trends Ecol Evol 25(3):180–189

Barker NKS, Dabelsteen T, Mennill DJ (2009) Degradation of male and female rufous-and-white wren songs in a tropical forest: effects of sex, perch height, and habitat. Behaviour 146:1093–1122

Bass HE, Sutherland LC, Zuckerwar AJ, Blackstock DT, Hester DM (1995) Atmospheric absorption of sound: further developments. J Acoust Soc Am 97(1):680–683

Blumenrath SH, Dabelsteen T (2004) Degradation of great tit *Parus major* song before and after foliation: implications for vocal communication in deciduous forests. Behaviour 8:935–958

Boersma HF (1997) Characterization of the natural ambient sound environment: measurements in open agricultural grassland. J Acoust Soc Am 101(4):2104–2110

Boncoraglio G, Saino N (2007) Habitat structure and the evolution of bird song: a meta-analysis of the evidence for the acoustic adaptation hypothesis. Funct Ecol 21:134–142

Bourne GR, Collins AC, Holder AM, McCarthy CL (2001) Vocal communication and reproductive behavior of the frog *Colostethus beebei* in Guyana. J Herpetol 35:272–281

Brackenbury JH (1979) Power capabilities of the avian sound-producing system. J Exp Biol 78:163–166

Brenowitz EA (1982) The active space of red-winged blackbird song. J Comp Physiol A 147:511–522

Brumm H (2002) Sound radiation patterns in nightingale (*Luscinia megarhynchos*) songs. J Ornithol 143:468–471

Brumm H (2004) The impact of environmental noise on song amplitude in a territorial bird. J Anim Ecol 73:434–440
Brumm H (2009) Song amplitude and body size in birds. Behav Ecol Sociobiol 63:1157–1165
Brumm H, Ritschard M (2011) Song amplitude affects territorial aggression of male receivers in chaffinches. Behav Ecol 22:310–316
Brumm H, Slabbekoorn H (2005) Acoustic communication in noise. Adv Study Behav 35:151–209
Brumm H, Slater PJB (2006) Animals can vary signal amplitude with receiver distance: evidence from zebra finch song. Anim Behav 71:699–705
Brumm H, Todt D (2002) Noise-dependent song amplitude regulation in a territorial songbird. Anim Behav 63:891–897
Brumm H, Todt D (2004) Male-male vocal interactions and the adjustment of song amplitude in a territorial bird. Anim Behav 67:281–286
Brumm H, Zollinger SA (2011) The evolution of the Lombard effect: 100 years of psychoacoustic research. Behaviour 148:1173–1198
Brumm H, Zollinger SA (2013) Avian vocal production in noise. In: Brumm H (ed) Animal communication and noise. Springer, Berlin
Chappuis C (1971) Un exemple de l'influence du milieu sur les émissions vocales des oiseaux: L'évolution des chants en forêt équatoriale. Terre Vie 118:183–202
Cynx J, Lewis R, Tavel B, Tse H (1998) Amplitude regulation of vocalizations in noise by a songbird, *Taeniopygia guttata*. Anim Behav 56:107–113
Dabelsteen T (1981) The sound pressure level in the dawn song of the blackbird *Turdus merula* and a method for adjusting the level in experimental song to the level in natural song. Z Tierpsychol 56(2):137–149
Dabelsteen T (1985) Messages and meanings of bird song with special reference to the blackbird (*Turdus merula*) and some methodology problems. Biologiske Skrifter 25:173–208
Dabelsteen T (2005) Public, private or anonymous? Facilitating and countering eavesdropping. In: McGregor P (ed) Animal communication networks. Cambridge University Press, Cambridge, NY, pp 38–62
Dabelsteen T, Larsen ON, Pedersen SB (1993) Habitat-induced degradation of sound signals: quantifying the effects of communication sounds and bird location on blur ratio, excess attenuation, and signal-to-noise ratio in blackbird song. J Acoust Soc Am 93:2206–2220
Dabelsteen T, McGregor PK, Lampe HM, Langmore NE, Holland J (1998) Quiet song in song birds: an overlooked phenomenon. Bioacoustics 9:89–105
Dent ML, Larsen ON, Dooling RJ (1997) Free-field binaural unmasking in budgerigars (*Melopsittacus undulatus*). Behav Neurosci 111(3):590–598
Dooling RJ (1982) Auditory perception in birds. In: Kroodsma DE, Miller EH (eds) Acoustic communication in birds, vol 1. Academic, New York, pp 95–130
Dooling RJ (1992) Hearing in birds. In: Webster DB, Fay RR, Popper AN (eds) The evolutionary biology of hearing. Springer, New York, pp 545–559
Dooling RJ, Blumenrath SH (2013) Avian sound perception in noise. In: Brumm H (ed) Animal communication and noise, animal signals and communication, vol 2. Springer, Berlin, pp 229–250
Dooling RJ, Leek MR (2018) Communication masking by man-made noise. In: Slabbekoorn H, Dooling RJ, Fay RR, Popper AN (eds) Effects of anthropogenic noise on animals. Springer, New York, pp 23–46
Dooling RJ, Searcy MH (1985) Temporal integration of acoustic signals by the budgerigar (*Melopsittacus undulatus*). J Acoust Soc Am 77(5):1917–1920
Dooling RJ, Lohr B, Dent ML (2000) Hearing in birds and reptiles. In: Dooling RJ, Fay RR, Popper AN (eds) Comparative hearing: birds and reptiles. Springer, New York, pp 308–359
Dooling RJ, Leek MR, West E (2009) Predicting the effects of masking noise on communication distance in birds. J Acoust Soc Am 125(4):2517
Eens M, Pinxten R, Verheyen RF (1989) Temporal and sequential organization of song bouts in the starling. Ardea 77:75–86

Elemans CP, Rasmussen JH, Herbst CT, Düring DN, Zollinger SA, Brumm H, Srivastava K, Svane N, Ding M, Larsen ON, Sober SJ, Švec JG (2015) Universal mechanisms of sound production and control in birds and mammals. Nat Commun 6:8978

Elie JE, Mariette MM, Soula HA, Griffith SC, Mathevon N, Vignal C (2010) Vocal communication at the nest between mates in wild zebra finches: a private vocal duet? Anim Behav 80:597–605

Ey E, Fischer J (2009) The "acoustic adaptation hypothesis" – a review of the evidence from birds, anurans and mammals. Bioacoustics 19:21–48

Farabaugh SM, Dent ML, Dooling RJ (1998) Hearing and vocalizations of wild-caught Australian budgerigars (*Melopsittacus undulatus*). J Comp Psychol 112:74–81

Fegeant O (1999) Wind-induced vegetation noise. Part II: field measurements. Acta Acust Acust 85 (2):241–249

Fletcher H (1940) Auditory patterns. Rev Mod Phy 12:47–65

Fletcher NH (1988) Bird song – a quantitative acoustic model. J Theor Biol 135:455–481

Fletcher NH (2004) A simple frequency-scaling rule for animal communication. J Acoust Soc Am 115(5):2334–2338

Francis CD, Barber JR (2013) A framework for understanding noise impacts on wildlife: an urgent conservation priority. Front Ecol Environ 11(6):305–313

Fuller RA, Warren PH, Gaston KJ (2007) Daytime noise predicts nocturnal singing in urban robins. Biol Lett 3:368–370

Gall MD, Ronald KL, Bestrom ES, Lucas JR (2012) Effects of habitat and urbanization on the active space of brown-headed cowbird song. J Acoust Soc Am 132(6):4053–4062

Goller F, Larsen ON (1997) In situ biomechanics of the syrinx and sound generation in pigeons. J Exp Biol 200:2165–2176

Goodwin SE, Podos J (2013) Shift of song frequencies in response to masking tones. Anim Behav 85:435–440

Gorissen L, Eens M (2004) Interactive communication between male and female great tits (*Parus major*) during the dawn chorus. Auk 121(1):184–191

Gustison ML, Townsend SW (2015) A survey of the context and structure of high- and low-amplitude calls in mammals. Anim Behav 105:281–288

Halfwerk W, Bot S, Slabbekoorn H (2012) Male great tit song perch selection in response to noise-dependent female feedback. Funct Ecol 26:1339–1347

Halfwerk W, Lohr B, Slabbekoorn H (2018) Impact of man-made sound on birds and their songs. In: Slabbekoorn H, Dooling RJ, Popper AN, Fay RR (eds) Effects of anthropogenic noise on animals, vol 66. Springer, New York, pp 209–242

Henwood K, Fabrick A (1979) A quantitative analysis of the dawn chorus: temporal selection for communicatory optimization. Am Nat 114(2):260–274

Hof D, Hazlett N (2010) Low-amplitude song predicts attack in a North American wood warbler. Anim Behav 80:821–828

Holland J, Dabelsteen T, Pedersen SB, Larsen ON (1998) Degradation of wren *Troglodytes troglodytes* song: implications for information transfer and ranging. J Acoust Soc Am 103 (4):2154–2166

Hornby AS, Gatenby EV, Wakefield H (1970) The advanced learner's dictionary of current English. Oxford University Press, London

Huber F (1955) Sitz und Bedeutung nervöser Zentren für Instinkthandlungen beim Männchen von *Gryllus campestris* L. Z Tierpsychol 12:12–48

Ishizuka T (2009) Whisper song in the grey thrush *Turdus cardis* immediately before and after feeding their young. J Yamashina Instit Ornithol 41:34–41

ISO (1993) Attenuation of sound during propagation outdoors. 1. Calculation of the absorption of sound by the atmosphere. ISO 9613-1. International Organization for Standardization, New York

Jensen KK, Cooper BG, Larsen ON, Goller F (2007) Songbirds use pulse tone register in two voices to generate low-frequency sound. Proc R Soc B 274:2703–2710

Jensen KK, Larsen ON, Attenborough K (2008) Measurements and predictions of hooded crow (*Corvus corone cornix*) call propagation over open field habitats. J Acoust Soc Am 123 (1):507–518
Klump GM (1996) Bird communication in a noisy world. In: Kroodsma DE, Miller EH (eds) Ecology and evolution of acoustic communication in birds. Cornell University Press, London, pp 321–338
Klump GM, Maier EH (1990) Temporal summation in the European starling (*Sturnus vulgaris*). J Comp Psychol 104(1):94–100
Klump GM, Kretzschmar E, Curio E (1986) The hearing of an avian predator and its avian prey. Behav Ecol Sociobiol 18:317–323
Kobayashi K, Okanoya K (2003) Context-dependent song amplitude control in Bengalese finches. Neuroreport 14(3):521–524
Krieg CA, Burnett AD (2017) Female house wrens may use a low-amplitude call as an aggressive signal. Ethology 123:316–327
Ladich F (2007) Females whisper briefly during sex: context- and sex-specific differences in sounds made by croaking gouramis. Anim Behav 73:379–387
Langemann U, Klump GM (2001) Signal detection in amplitude-modulated maskers. I. Behavioural auditory thresholds in a songbird. Eur J Neurosci 13:1025–1032
Langemann U, Klump GM (2005) Perception and acoustic communication networks. In: McGregor PK (ed) Animal communication networks. Cambridge University Press, Cambridge, pp 451–480
Larsen ON, Dabelsteen T (1990) Directionality of blackbird vocalization: implications for vocal communication and its further study. Ornis Scand 21(1):37–45
Larsen ON, Radford C (2018) Acoustic conditions affecting sound communication in air and underwater. In: Slabbekoorn H, Dooling RJ, Fay RR, Popper AN (eds) Effects of anthropogenic noise on animals, vol 66. Springer, New York, pp 109–144
Larsen ON, Wahlberg M (2017) Sound and sound sources. In: Brown CH, Riede T (eds) Comparative bioacoustics: an overview. Bentham Science, Sharjah, pp 3–62
Lemon RE, Struger J, Lechowicz MJ, Norman RF (1981) Song features and singing heights of American warblers: maximization or optimization of distance? J Acoust Soc Am 69:1169–1176
Lengagne T, Slater PJB (2002) The effects of rain on acoustic communication: tawny owls have good reason for calling less in wet weather. Proc R Soc B 269:2121–2125
Lengagne T, Aubin T, Lauga J, Jouventin P (1999) How do king penguins (*Aptenodytes patagonicus*) apply the mathematical theory of information to communicate in windy conditions? Proc R Soc B 266:1623–1628
Lohr B, Wright TF, Dooling RJ (2003) Detection and discrimination of natural calls in masking noise by birds: estimating the active space signal. Anim Behav 65:763–777
MacDougall-Shackleton SA (1997) Sexual selection and the evolution of song repertoires. Curr Ornithol 14:81–124
Mack AL, Jones J (2003) Low-frequency vocalizations by cassowaries (*Casuarius* spp.). Auk 120 (4):1062–1068
Magrath RD, Haff TM, Fallow PM, Radford AN (2015) Eavesdropping on heterospecific alarm calls: from mechanisms to consequences. Biol Rev 90:560–586
Manabe K, Sadr EI, Dooling RJ (1998) Control of vocal intensity in budgerigars (*Melopsittacus undulatus*): differential reinforcement of vocal intensity and the Lombard effect. J Acoust Soc Am 103(2):1190–1198
Marler P (1955) Characteristics of some animal calls. Nature 176:6–8
Marten K, Marler P (1977) Sound transmission and its significance for animal vocalization: I. Temperate habitats. Behav Ecol Sociobiol 2:271–290
Mason NA, Burns KJ (2015) The effect of habitat and body size on the evolution of vocal displays in *Thraupidae* (tanagers), the largest family of songbirds. Biol J Linn Soc 114(3):538–551
Mathevon N, Aubin T, Dabelsteen T (1996) Song degradation during propagation: importance of song post for the wren *Troglodytes troglodytes*. Ethology 102:397–412

Mathevon N, Dabelsteen T, Blumenrath SH (2005) Are high perches in the blackcap *Sylvia atricapilla* song or listening posts? A sound transmission study. J Acoust Soc Am 117 (1):442–449

Mathevon N, Aubin T, Vielliard J, Da Silva ML, Sebe F, Boscolo D (2008) Singing in the RainForest: how a tropical bird's song transmits information. PLoS One 3(2):e1580

McGregor PK, Dabelsteen T (1996) Communication networks. In: Kroodsma DE, Miller EH (eds) Ecology and evolution of acoustic communication in birds. Cornell University Press, Ithaca, pp 409–425

McGregor PK, Krebs JR (1982) Song types in a population of great tits (*Parus major*): their distribution, abundance and acquisition by individuals. Anim Behav 79:126–152

McLaughlin KE, Kunc HP (2013) Experimentally increased noise levels change spatial and singing behaviour. Biol Lett 9:20120771

Mead AF, Osinalde N, Ørtenblad N, Nielsen J, Brewer J, Vellema M, Adam I, Scharff C, Song Y, Frandsen U, Blagoev B, Kratchmarova I, Elemans CPH (2017) Fundamental constraints in synchronous muscle limit superfast motor control in vertebrates. elife 6:e29425. https://doi.org/10.7554/eLife.29425

Mencio C, Huberman B, Goller F (2017) Contributions of rapid neuromuscular transmission to the fine control of acoustic parameters of birdsong. J Neurophysiol 117:637–645

Meyer J (2015) Acoustic adaptation to natural environments. In: Meyer J (ed) Whistled languages: a worldwide inquiry about human whistled speech. Springer, Berlin, pp 91–103

Miller LN (1978) Sound levels of rain and wind in the trees. Noise Control Eng J 11:101–115

Mooney TA, Smith A, Larsen ON, Hansen KA, Wahlberg M, Rasmussen MH (2019) Field-based hearing measurements of two seabird species. J Exp Biol 222:jeb190710. https://doi.org/10.1242/jeb.190710

Morton ES (1975) Ecological sources of selection on avian sounds. Am Nat 109(965):17–34

Morton ES (1986) Predictions from the ranging hypothesis for the evolution of long distance signals in birds. Behaviour 99:65–86

Morton ES (2000) An evolutionary view of the origins and functions of avian vocal communication. Jpn J Ornithol 49:69–78

Møhl B, Wahlberg M, Madsen PT, Miller LA, Surlykke A (2000) Sperm whale clicks: directionality and source level revisited. J Acoust Soc Am 107(1):638–648

Naguib M, Riebel K (2014) Singing in space and time: the biology of birdsong. In: Witzany G (ed) Biocommunication of animals. Springer, Dordrecht

Nemeth E (2004) Measuring the sound pressure level of the song of the screaming piha *Lipaugus vociferans*: one of the loudest birds in the world? Bioacoustics 14:225–228

Nemeth E, Brumm H (2010) Birds and anthropogenic noise: are urban songs adaptive? Am Nat 176 (4):465–475

Nemeth E, Winkler H, Dabelsteen T (2001) Differential degradation of antbird songs in a neotropical rainforest: adaptation to perch height? J Acoust Soc Am 110:3263–3274

Nemeth E, Dabelsteen T, Pedersen SB, Winkler H (2006) Rainforests as concert halls for birds: are reverberations improving sound transmission of long song elements. J Acoust Soc Am 119 (1):620–626

Nemeth E, Kempenaers B, Matessi G, Brumm H (2012) Rock sparrow song reflects male age and reproductive success. PLoS One 7(8):e43259

Nemeth E, Pieretti N, Zollinger SA, Geberzahn N, Partecke J, Miranda AC, Brumm H (2013) Bird song and anthropogenic noise: vocal constraints may explain why birds sing higher-frequency songs in cities. Proc R Soc B 280:20122798

Niederhauser JM, DuBois AL, Searcy WA, Nowicki S, Anderson RC (2018) A test of the eavesdropping avoidance hypothesis as an explanation for the structure of low-amplitude aggressive signals in the song sparrow. Behav Ecol Sociobiol 72:47

Osmanski MS, Dooling RJ (2009) The effects of altered feedback and control of vocal production in budgerigars (*Melopsittacus undulatus*). J Acoust Soc Am 126:911–919

Patricelli GA, Dantzker MS, Bradbury JW (2007) Differences in acoustic directionality among vocalizations of the male red-winged blackbird (*Agelaius phoeniceus*) are related to function in communication. Behav Ecol Sociobiol 61:1099–1110

Patricelli GA, Dantzker MS, Bradbury JW (2008) Acoustic directionality of red-winged blackbird (*Agelaius phoeniceus*) song relates to amplitude and singing behaviours. Anim Behav 76:1389–1401

Piza P, Sandoval L (2016) The differences in transmission properties of two bird calls show relation to their specific functions. J Acoust Soc Am 140(6):4271–4275

Plomp R, Bouman MA (1959) Relation between hearing threshold and duration of tone pulses: J Acoust Soc Am 31:749–758

Podos J (1997) A performance constraint on the evolution of trilled vocalizations in a songbird family (Passeriformes: Emberizidae). Evolution 51:537–551

Pohl NU, Slabbekoorn H, Klump GM, Langemann U (2009) Effects of signal features and environmental noise on signal detection in the great tit, *Parus major*. Anim Behav 78:1293–1300

Pohl NU, Slabbekoorn H, Neubauer H, Heil P, Klump GM, Ulrike Langemann U (2013) Why longer song elements are easier to detect: threshold level-duration functions in the Great Tit and comparison with human data. J Comp Physiol A 199(3):239–252

Price JJ (2013) Why is birdsong so repetitive? Signal detection and the evolution of avian singing modes. Behaviour 150:995–1013

Pytte CL, Rush KM, Ficken MS (2003) Regulation of vocal amplitude by the blue-throated hummingbird, *Lampornis clemenciae*. Anim Behav 66:703–710

Pytte CL, Ficken MS, Moiseff A (2004) Ultrasonic singing by the blue-throated hummingbird: a comparison between production and perception. J Comp Physiol A 190:665–673

Radford AN, Ridley AR (2008) Close calling regulates spacing between foraging competitors in the group-living pied babbler. Anim Behav 75:519–527

Reichard DG, Anderson RC (2015) Why signal softly? The structure, function and evolutionary significance of low-amplitude signals. Anim Behav 105:253–265

Reichard DG, Welklin JF (2015) On the existence and potential functions of low-amplitude vocalizations in North American birds. Auk 132:156–166

Reichard DG, Rice RR, Vanderbilt CC, Ketterson ED (2011) Deciphering information encoded in birdsong: male songbirds with fertile mates respond most strongly to complex, low-amplitude songs used in courtship. Am Nat 178(4):478–487

Reichard DG, Rice RR, Schultz EM, Schrock SE (2013) Low-amplitude songs produced by male dark-eyed juncos (*Junco hyemalis*) differ when sung during intra- and inter-sexual interactions. Behaviour 150:1183–1202

Ręk P (2013) Soft calls and broadcast calls in the corncrake as adaptations to short and long range communication. Behav Process 99:121–129

Ręk P, Osiejuk TS (2011) Nonpasserine bird produces soft calls and pays retaliation cost. Behav Ecol 22:657–662

Richards DG, Wiley RH (1980) Reverberations and amplitude fluctuations in the propagation of sound in a forest: implications for animal communication. Am Nat 115(3):381–399

Ritschard M, Brumm H (2011) Effects of vocal learning, phonetics and inheritance on song amplitude in zebra finches. Anim Behav 82:1415–1422

Ryan MJ, Brenowitz EA (1985) The role of body size, phylogeny, and ambient noise in the evolution of bird song. Am Nat 126(1):87–100

Sandoval L, Dabelsteen T, Mennill DJ (2015) Transmission characteristics of solo songs and duets in a neotropical thicket habitat specialist bird. Bioacoustics 24(3):289–306

Searcy WA, Beecher MD (2009) Song as an aggressive signal in songbirds. Anim Behav 78:1281–1292

Searcy WA, Nowicki S (2006) Signal interception and use of soft song in aggressive interactions. Ethology 112:865–872

Searcy WA, Yasukawa K (2017) Eavesdropping and cue denial in avian acoustic signals. Anim Behav 124:273–282

Shannon G, McKenna MF, Angeloni LM, Crooks KR, Fristrup KM, Brown E, Warner KA, Nelson MD, White C, Briggs J, McFarland S, Wittemyer G (2016) A synthesis of two decades of research documenting the effects of noise on wildlife. Biol Rev 91:982–1005
Slabbekoorn H, Ripmeester EAP (2008) Birdsong and anthropogenic noise: implications and applications for conservation. Mol Ecol 17:72–83
Slabbekoorn H, Dooling RJ, Popper AR, Fay RR (eds) (2018) Effects of anthropogenic noise on animals. Springer, New York, p 309
Snow DW (1958) A study of blackbirds. Allen and Unwin, London
Srivastava KH, Elemans CPH, Sober SJ (2015) Multifunctional and context-dependent control of vocal acoustics by individual muscles. J Neurosci 35(42):14183–14194
Titus RC (1998) Short-range and long-range songs: use of two acoustically distinct song classes by dark-eyed juncos. Auk 115(2):386–393
Vargas-Castro LE (2015) Spatial pattern of syllable sharing in white-throated thrushes: implications for song learning and dispersal behaviours. Behaviour 152(6):775–795
Vargas-Castro LE, Sandoval L, Searcy WA (2017) Eavesdropping avoidance and sound propagation: the acoustic structure of soft song. Anim Behav 134:113–121
Wahlberg M, Larsen ON (2017) Propagation of sound. In: Brown CH, Riede T (eds) Comparative bioacoustics: an overview. Bentham Science, Sharjah, pp 63–121
Ward S, Lampe HM, Slater PJB (2004) Singing is not energetically demanding for pied flycatchers, *Ficedula hypoleuca*. Behav Ecol 15(3):477–484
Weir JT, Wheatcroft DJ, Price TD (2012) The role of ecological constraint in driving the evolution of avian song frequency across a latitudinal gradient. Evolution 66:2773–2783
Wiley RH (1991) Associations of song properties with habitats for territorial oscine birds of Eastern North-America. Am Nat 138:973–993
Wiley RH (2013) Signal detection, noise, and the evolution of communication. In: Brumm H (ed) Animal communication and noise, animal signals and communication, vol 2. Springer, Berlin, pp 7–30
Wiley RH, Richards DG (1978) Physical constraints on acoustic communication in the atmosphere: implications for the evolution of animal vocalizations. Behav Ecol Sociobiol 3:69–94
Wiley RH, Richards DG (1982) Adaptations for acoustic communication in birds: sound transmission and signal detection. In: Kroodsma DE, Miller EH (eds) Acoustic communication in birds. Academic, New York, pp 132–181
Xia C, Liu J, Alström P, Wu Q, Zhang Y (2013) Is the soft song of the brownish-flanked bush warbler an aggressive signal? Ethology 119:653–661
Yang XJ, Ma XR, Slabbekoorn H (2014) Timing vocal behaviour: experimental evidence for song overlap avoidance in Eurasian wrens. Behav Process 103:84–90
Yorzinski JL, Patricelli GL (2010) Birds adjust acoustic directionality to beam their antipredator calls to predators and conspecifics. Proc R Soc B 277:923–932
Zeng FG, Nie K, Stickney GS, Kong YY, Vongphoe M, Bhargave A, Wei C, Cao K (2005) Speech recognition with amplitude and frequency modulations. Proc Natl Acad Sci USA 102(7):2293–2298
Zollinger SA, Brumm H (2015) Why birds sing loud songs and why they sometimes don't. Anim Behav 105:289–295
Zollinger SA, Goller F, Brumm H (2011) Metabolic and respiratory costs of increasing song amplitude in zebra finches. PLoS One 6(9):e23198
Zollinger SA, Podos J, Nemeth E, Goller F, Brumm H (2012) On the relationship between, and measurement of, amplitude and frequency in birdsong. Anim Behav 84(4):e1–e9
Zuk M, Kolluru GR (1998) Exploitation of sexual signals by predators and parasitoids. Q Rev Biol 73:415–438

Chapter 3
The Bird Dawn Chorus Revisited

Diego Gil and Diego Llusia

Thou hearest the Nightingale begin the Song of Spring:
The Lark, sitting upon his earthy bed, just as the morn
Appears, listens silent, then, springing from the waving corn-field, loud
He leads the Choir of Day—trill! trill! trill! trill!
William Blake (Milton: a Poem, 1810)

Abstract The bird dawn chorus has fascinated humans since ancient times, but still today numerous questions remain unclear. This chapter will explore this puzzling phenomenon, a communal display that likely involves the highest level of sound complexity found among animal signals. Covering from the first descriptive studies to recent multidisciplinary approaches, we review the physiological, behavioural and environmental factors affecting dawn chorus. In addition, we provide a critical assessment of the supporting evidence for the functional hypotheses proposed so far to disentangle its proximal and ultimate causes. We find that, despite the latest empirical and theoretical studies, there is still a good degree of confusion, and that four out of the nine hypotheses proposed so far in the literature have not been empirically tested. We show that most of these hypotheses are not incompatible with each other, and that their explanatory value changes depending on the species and

D. Gil (✉)
Departamento de Ecología Evolutiva, Museo Nacional de Ciencias Naturales (CSIC), Madrid, Spain
e-mail: diego.gil@csic.es

D. Llusia
Departamento de Ecología, Terrestrial Ecology Group, Universidad Autónoma de Madrid (UAM), Madrid, Spain

Centro de Investigación en Biodiversidad y Cambio Global (CIBC-UAM), Universidad Autónoma de Madrid, Madrid, Spain

Departamento de Ecologia, Laboratório de Herpetologia e Comportamento Animal, Instituto de Ciências Biológicas, Universidade Federal de Goiás (UFG), Goiânia, GO, Brazil
e-mail: diego.llusia@uam.es

T. Aubin, N. Mathevon (eds.), *Coding Strategies in Vertebrate Acoustic Communication*, Animal Signals and Communication 7,
https://doi.org/10.1007/978-3-030-39200-0_3

the season. We argue that, at any rate, a single explanation may not be a reasonable expectation. The best-supported hypotheses for early singing provide three complementary lines of explanation: (1) singing at dawn has a relatively low energetic cost, most likely because it does not interfere with feeding; (2) is optimal to manipulate female mating or settle territory boundaries; and (3) may promote a handicap mechanism that prevents dishonest signalling. Thus, it follows that a combination of hypotheses based on both an optimality standpoint and costliness assumptions is needed to understand the phenomenon. We provide a series of specific suggestions for further research to refine our knowledge of this intriguing aspect of animal behaviour.

3.1 Introduction

The bird dawn chorus has fascinated humans and aroused their interest since ancient times. The use of roosters as clocks is likely as old as chicken domestication itself, which recent studies trace back to 3000 BT (Peters et al. 2016). The classical Greek and Latin literature, as well as the Bible or later the writings of the Romantics, is rife with references to bird song being used as an announcement of sunrise (Armstrong 1963; Doggett 1974). Romeo famously leaves Juliet's chamber in a hurry when he realises that it was a lark, and not a nightingale, the bird that woke him up (Shakespeare 1597). Today the bird dawn chorus keeps intriguing us, and numerous questions about this puzzling behavioural phenomenon are still unclear.

It is impossible to envisage a review on this question without building on previous efforts, such as those of Kacelnik and Krebs (1983), Mace (1987b) and more recently Staicer et al. (1996), which have been seminal in shaping modern research on the dawn chorus. Staicer et al. (1996) summarised 12 different hypotheses on the dawn chorus, classified in three groups (intrinsic, social and environmental), and examined them in relation to empirical support. Now, more than 20 years since this key review, we wish devote this chapter to revisiting the status of the different hypotheses in the light of the studies published in the last decades. As the approach and terminology in Staicer et al. (1996) has been widely applied, here we proceed using a similar classification of hypotheses (Table 3.1).

Starting by briefly describing the general features of the bird dawn chorus, this chapter outlines a series of issues that appear when we try to disentangle its proximal and ultimate causes. On this overall basis, we move on to discuss, one by one, the main factors affecting this phenomenon and the explanations that have been proposed to explain why birds sing at this particular time. To do so, we use the following structure: (1) we report the patterns, variations and known facts about the bird dawn chorus in relation to physiological, behavioural or environmental factors; (2) we present functional hypotheses that are derived from these observations; and (3) we critically assess the supporting evidence for such hypotheses and its

Table 3.1 List of the nine hypotheses proposed to explain the function of dawn chorus examined in this review, with summary of evidence for each of them and possible tests and studies still needed to provide evidence

Hypotheses	Why at dawn?	Proposed by	Evidence	Research agenda: untested predictions
1: Circadian hormone cycles	Dawn chorus explained by a peak of hormones (hormonal epiphenomenon) or an unknown common driver of activity and physiology (basic metabolism) that stimulates singing, irrespective of adaptive functions	Staicer et al. (1996)	Some experimental support for a role of melatonin. Highly unlikely that a non-functional hypothesis could explain the wide variation in patterns found in nature	Experimental manipulations of melatonin should lead to changes in song timing. Also differences between individuals and populations should be mirrored by differences in melatonin
2: Self-stimulation of androgen production	Singing at dawn would prepare males for their daily social interactions by auto-stimulating their own gonadal production	Staicer et al. (1996)	Untested and highly speculative	According to the hypothesis, modifying song dawn intensity should lead to parallel changes in aggressive behaviour and physiology later in the day
3: Manipulation of female mating behaviour around the fertilisation window	Dawn chorus is directed to females in a window period of fertility at dawn when males can pursue extra-pair fertilizations, influence maternal investment before egg laying, and protect their own offspring paternity	Mace (1987a)	Some direct experimental evidence for song timing affecting fertilisation success. Also correlative evidence in some species showing a link of early chorus with fertile periods. A highly species-specific function	Further experimental tests altering sleeping patterns are needed. Also there is scope for comparative evidence comparing populations or species with different rates of extra-pair paternity. Further work in timing of extra-pair forays in more species would also be useful
4: Territory defence and social dynamics	Dawn singing as territory and status signalling to counteract increasing social instability after the night, when resettling territory borders and social relationships would be favoured	Kacelnik and Krebs (1983)	Some experimental support in several temperate songbirds. Territorial intrusions more likely at dawn in some species	Experimental tests manipulating song timing and testing its effects on territory ownership or instability are still required. There is little information on intrusion rates and sneaking behaviour (detailed radio-tracking of males at dawn

(continued)

Table 3.1 (continued)

Hypotheses	Why at dawn?	Proposed by	Evidence	Research agenda: untested predictions
				likely fruitful) in a sizeable number of species
5: Early singing as a handicap	Early singing is a handicap that enables honest signalling of individual differences in quality and condition, since singing earlier at dawn than average is costly	Montgomerie (1985)	No experimental tests, but wide indirect evidence from correlative studies.	Many possibilities to test whether experimental manipulations of condition, parasite loads or immune challenges lead to changes in song timing. Also necessary to analyse condition dependency of timing vs. other traits
6: Inefficient foraging at dawn	Dawn chorus takes advantage of a time window when singing does not interfere with feeding, since foraging at dawn has lower profitability than later in the day (unfavourable temperature and light levels)	Kacelnik (1979)	No direct tests, but some indirect evidence that foraging at dawn is not profitable. SDP models support the evolution of dawn chorus under these nutritional constraints	Male–female comparisons would be useful to check if feeding profitability is constrained: females should not forage at the time that males sing, and if they do they should show lower profitability. Comparisons between different feeding guilds can also be useful. For instance, differences in the effect of light in foraging of insectivorous vs. granivorous species should be explained by differences in dawn song timing
7: Unpredictable overnight conditions	Surplus of energy stored for unpredictable overnight conditions is free to be used at dawn for singing	McNamara et al. (1987)	Formal SDP models support the evolution of dawn chorus assuming these nutritional constraints. There is some experimental support for some predictions	It is necessary to obtain more evidence of individual singing responses to food predictability. It would also be interesting to take a comparative approach and test

				whether choruses increase in length with decreasing predictability of overnight energy requirements
8: Low predation at dawn	Early singing is favoured due to the lower risk of predation at dawn	Staicer et al. (1996)	Not formally tested and some negative indirect evidence	It is necessary to know if predation is really lower at dawn than later by means of correlative data and experimental evidence. Comparisons between populations subjected to different predation risks can also be useful
9: Better sound transmission conditions at dawn	Dawn chorus benefits from optimal sound transmission at dawn because atmospheric conditions are more stable and birds can maximize the active space reached by their acoustic signals	Henwood and Fabrick (1979)	Equivocal empirical support, despite some indirect evidence	Replication of classic acoustic transmission experiments is desirable. It would also help to estimate daily changes in active space and to test whether individual birds have a better perception at dawn than later on in the day

primary predictions, pointing out remaining questions and proposing further studies to test them.

An important omission of our review is female song. A series of studies have emphasised that female song is more widespread than traditionally assumed (Langmore 1998), actually being the evolutionarily ancestral state in birds (Odom et al. 2014). However, research on female song is still underrepresented in the literature (Riebel et al. 2005) and we have found almost no reference to female song in the dawn chorus literature. We believe that very interesting insights may be gained by comparing male and female patterns of singing with respect to dawn in those species that show female song. But in the remainder of the chapter we will deal with male song only.

3.2 The Bird Dawn Chorus

The dawn chorus is a daily period of high singing activity shown by most passerine and some non-passerine species in the breeding season. This burst of singing typically starts for the earliest species around nautical twilight (sun at 12° under the horizon), and is followed in a rather orderly fashion by other species until sometime around sunrise (Thomas et al. 2002). Although the time of the first song is a species-specific trait (Berg et al. 2006), during this period many species end up overlapping their songs, leading to a more or less cacophonic chorus (Fig. 3.1) that characterises the soundscapes of different avian communities (Farina et al. 2015). As a result, probably no other daily period involves larger levels of sound complexity among animal group displays (Burt and Vehrencamp 2005). Depending on latitude, season and species composition, choruses can be more or less intense and prolonged, but they occur in most breeding bird habitats (Staicer et al. 1996).

The first studies on the dawn chorus in the twentieth century were rather descriptive, and dealt primarily with the order in which species sang (the bird-clock), and with the effects of weather and moonlight on singing time (Allard 1930; Allen 1913; Wright 1912). Further studies started to carefully measure the specific light level that triggered the first song of each species, reaching more conclusive evidence for a species-specific light singing level (Leopold and Eynon 1961; Scheer 1952). However, we have to wait until the 1980s to find the first considerations of the dawn chorus from the perspective of behavioural ecology. In that sense, the pioneering study by Alex Kacelnik and John Krebs (1983) is noteworthy, combining physiological, ecological and behavioural perspectives. Since then, dawn song has been examined from a multitude of angles, including physiology (Shimmura and Yoshimura 2013), sexual selection (Dolan et al. 2007), reproduction (Kempenaers et al. 2010), energy allocation (Hutchinson et al. 1993), noise pollution (Gil et al. 2015) and soundscapes (Farina et al. 2015). Thus, a comprehensive review of this topic needs to cover a very wide range of subjects.

From the point of view of its features, the dawn chorus has also been studied using a relatively broad range of measures. Undoubtedly, dawn chorus timing, and

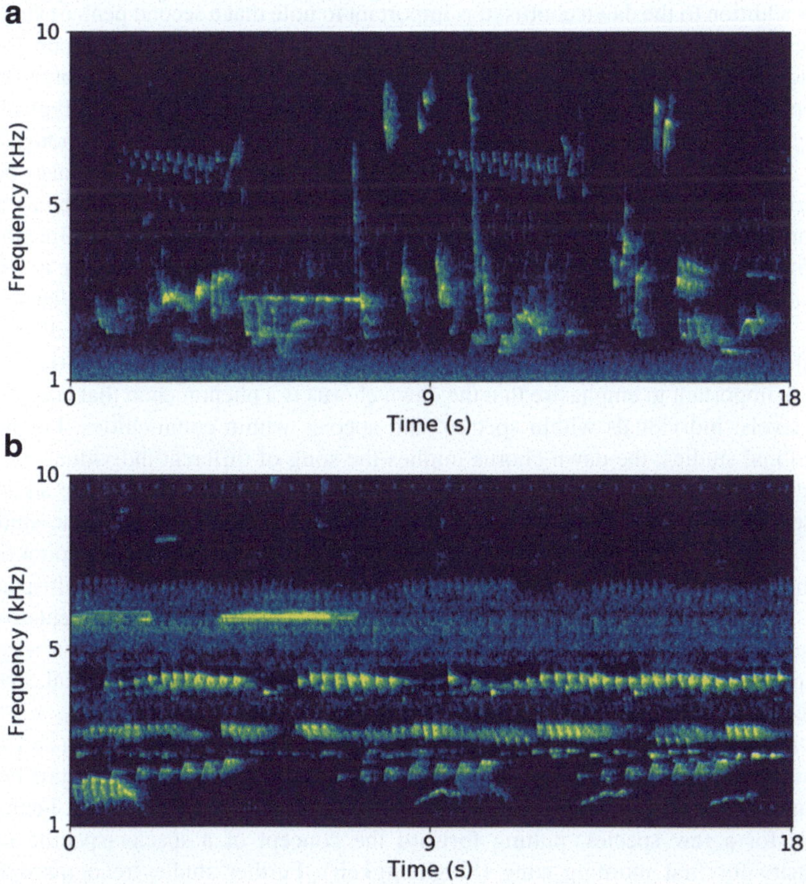

Fig. 3.1 Spectrograms of a temperate and a tropical avian community: (**a**) Forêt domaniale de la Commanderie (Île-de-France, France: 48° 17′ N, 2° 40′ E, 85 m.a.s.l; April 2014) and (**b**) Ducke Reserve (Manaus, Brazil: 2° 57′ S, 59° 55′ W, 92 m.a.s.l; October 2009). The following avian species can be detected: (**a**) *Turdus merula, Turdus philomelos, Erithacus rubecula, Periparus ater* and *Regulus regulus*; (**b**) *Attila spadiceus, Momotus momota, Ramphotrigon ruficauda, Dendrocolaptes certhia* and *Dendrexetastes rufigula* (in addition to at least one anuran: Dendropsophus reticulatus). From unpublished recordings by Diego Llusia (France) and Thiago Bicudo (Manaus). Figure created in Python v.3.7.4. Listen to sound clips at: http://sono.behavecol.es/en/landscapes

particularly the time at which each species or individual starts to sing within the chorus (i.e. the time of the first song), has been one of the most frequently employed parameters (Allard 1930; Gil et al. 2015; Leopold and Eynon 1961; Poesel et al. 2006; Scheer 1952). However, other researchers have focused on additional aspects, such as dawn-related variations in song repertoire (Allard 1930; Allen 1913; Demko et al. 2013; Kroodsma et al. 1989; Nelson and Croner 1991; Staicer et al. 1996), song rate (Liu and Kroodsma 2007; Pärt 1991) or song characteristics (Halfwerk et al. 2011; Lambrechts and Dhondt 1986).

In addition to the dawn chorus, it is important to note that a second peak of singing at the end of the day, the dusk chorus, is also common (Cuthill and Macdonald 1990; Møller 1988; Morse 1989), although less intense and dramatic (for a noteworthy exception see McCabe 1951). This double-peaked circadian pattern is typical of many activity patterns across diverse vertebrate taxa (Rusak 2013). Although we only consider birds in this review, circadian cycles of signalling have also been described in other animal groups, including crickets, crabs, fish, amphibians and mammals (Au et al. 2000; D'Spain and Batchelor 2006; Green 1990; Greenfield 2015; Ord 2008). These similarities suggest two alternatives: (1) that there is widespread selection for acoustic communication to exploit optimal times; (2) that selection focus on synchronization, allowing comparisons between signallers by ritualizing the displays at the same time (West-Eberhard 1979; Zahavi 1980).

It is important to emphasise that the dawn chorus is a phenomenon that comprises two levels: individuals within species, and species within communities. For most functional studies, the dawn chorus implies the song of different individuals of the same species at dawn. In that sense, the multispecies chorus would be the additive consequence of selective pressures acting within species. Although some studies have analysed interactions between species that may reduce song-overlapping time (Planqué and Slabbekoorn 2008), no study that we know of proposes a multispecies function of dawn chorusing in general. Nevertheless, some studies tentatively suggest that some species within the chorus may converge in timing as a way of competing among them or of extracting public information (Phelps et al. 2006; Tobias et al. 2014).

Differences among species in the time of the first song at dawn are strong and well known (Fig. 3.2), and were the main object of early descriptive studies (Allard 1930; Allen 1913). Later on, Scheer (1952) transformed specific times into light intensity levels for a few species, putting forward the concept of a species-specific light intensity for first morning song (Singhelligkeit). Further studies recognised that these values varied with seasonal and meteorological factors, although repeatability in the order of species singing was still high (Leopold and Eynon 1961).

So, what is the cause of these interspecific differences? Armstrong was the first author to recognise that eye size could explain dawn song timing, claiming that one could get a good idea of what time a bird starts to sing by looking at its eyes (Armstrong 1963). The first test of this hypothesis was conducted with European species, showing that species with larger eyes start to sing earlier than species with smaller eyes (Thomas et al. 2002). They also found that body size was an additional predictor, with larger species starting to sing later than smaller species, although this effect was less important than that of eye size. Further studies in other avian communities have confirmed this relationship between eye size and dawn song time (Berg et al. 2006; Chen et al. 2015; Gil et al. 2015). A study conducted in an Ecuadorian rainforest showed that, in addition to eye size and body size, dawn song timing was also explained by feeding stratum: birds in the well-lit canopy sang earlier than birds foraging in the dark understory (Berg et al. 2006). An important consideration in these studies is the need to control for phylogeny, since dawn timing is a phylogenetically conserved trait, and a bias may appear because of the particular

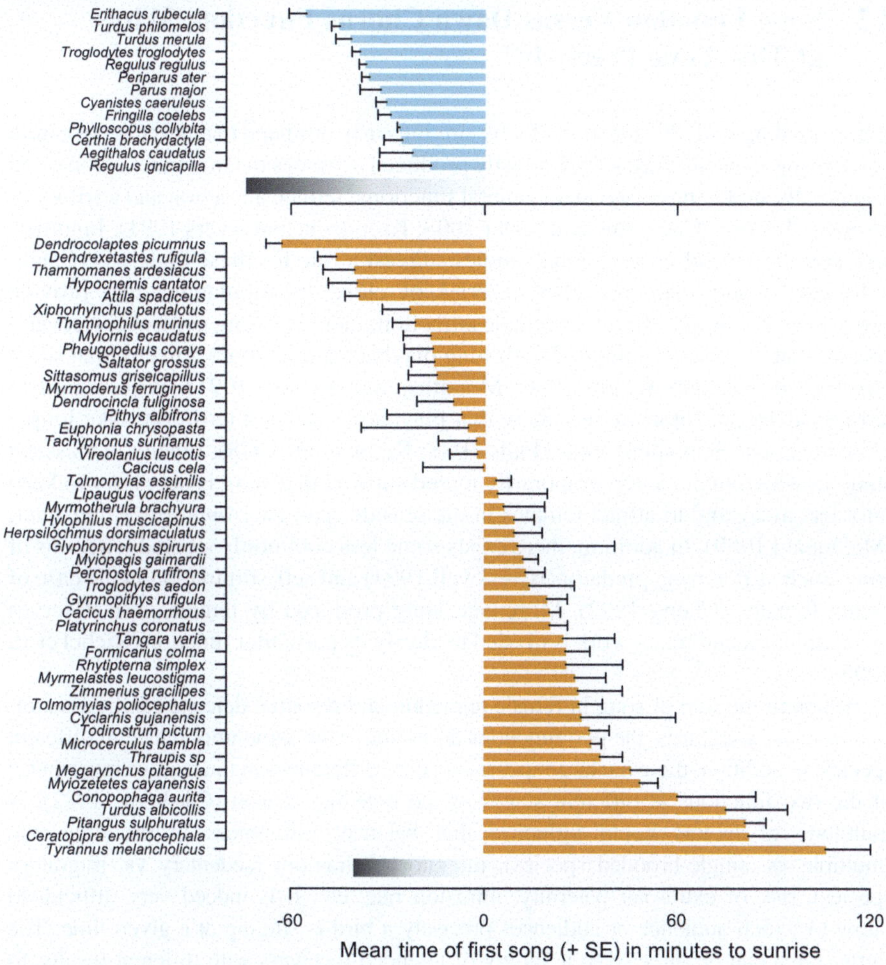

Fig. 3.2 Mean time of the first song for a temperate (blue) and a tropical (orange) avian community (same areas as in Fig. 3.1). Grey scales represents dawn period, from nautical twilight to sunrise, lasting for 70 min in Île-de-France and 45 min in Manaus. Unpublished data by Diego Llusia, Thierry Aubin and Jérôme Sueur (France) and Diego Gil and Marconi Campos-Cerqueira (Brazil). Figure created in R v.3.5.2

sampling of species of a given study (for instance suboscines sing earlier than oscines: Berg et al. 2006).

Larger eye size implies larger pupils and corneas and an increased number of photoreceptors, leading to improved visual acuity, a larger visual field width, and the possibility of perceiving objects at lower light intensities (Martin 1993; Thomas et al. 2002). Larger eye size has evolved in connection to nocturnal habits, capture of mobile prey, use of dark habitats and increased flight distances (Garamszegi et al. 2002; Martínez-Ortega et al. 2014; Møller and Erritzøe 2010).

3.3 Song Function Versus Dawn Chorus Function: Why at This Time Precisely?

Before dealing with the question of why singing early, we need to answer the question of why singing at all. Bird song is mostly produced by males in temperate regions, and is generally said to have two main general functions: female attraction and territory or resource defence (Catchpole and Slater 2008; Kroodsma and Byers 1991). Innumerable correlative and observational studies suggest a role for these functions (Catchpole 1982), and there are also a series of clever manipulations that provide experimental support. Thus, for female attraction, there is strong evidence in several species that nest boxes equipped with song playback attract more females than silent boxes (Eriksson and Wallin 1986; Mountjoy and Lemon 1991). With regard to territory defence, some studies show that playback-defended territories take longer to be occupied than silent areas (Falls 1988; Krebs et al. 1978). Finally, a classical study in which males were temporarily muted showed that song is necessary to keep intruders away and to attract females in the seaside sparrow (*Ammospiza maritima*) (McDonald 1989). In addition, there exists some less commonly studied functions of song, such as deterring predators (Cresswell 1994) and self-stimulation in the case of laying females (Cheng 1992). Moreover, song produced by females also deserves more attention and much work is needed to clarify its particular function (Riebel et al. 2005).

Although the role of song in female attraction and resource defence can be generalised to many songbirds, the two functions do not have the same importance in different species. In addition, there are large species-specific differences in the relative importance of the two functions at different stages of the breeding season (Catchpole 1982). A multitude of factors could influence this balance: differences in brooding rate (multiple vs. single-brooded species), migratory behaviour (sedentary vs. migratory species), risk of extra-pair paternity, intrusion rate, etc. It is indeed very difficult to know to which audience or audiences precisely a bird is singing at a given time. The same signal can be directed to a variety of intended receivers with different results: to attract potential mates; to its already acquired mate, who may modify her investment in response to the song (Gil et al. 2004); to its neighbours, who may use information on his competitive abilities to update the territorial status quo (Sexton et al. 2007); or to floaters, who may use song information to decide future settlement patterns (Amrhein and Erne 2006; Amrhein et al. 2004a). Deciding which is the most relevant receiver at a given time is a difficulty that permeates most research on the dawn chorus, since adaptive hypotheses about this trait are, by definition, intimately related to its function.

We must stress here that many studies attempt to explain the function of the dawn chorus without addressing the key question: why singing at this precise time? In other words, it is not enough to know that the dawn song is directed to the female or to male neighbours; we must also explain why more singing occurs at this time than later or earlier in the day. In this chapter, we will consider in turn all hypotheses that have been proposed to explain dawn song in birds, even some for which the evidence is still scant or negative (see Table 3.1 for a list of hypotheses).

3.4 Factors Affecting the Dawn Chorus and Related Functional Hypotheses

The dawn chorus has been the subject of a large number of disciplines, from physiology of circadian rhythms to ecology of predation, energetics, visual perception and extra-pair paternity. We have classified the information gathered by these different areas in the following sections, with the aim of providing a wide view of the levels of variation shown by this phenomenon. To allow the reader to follow this information more easily, we have associated each factor with the hypotheses that most closely derive from that subject area.

3.4.1 Hormones and Other Internal Drivers: Proximal Causation of the Dawn Chorus

Most of the temporal organization of the behaviour and physiology of animals is affected by circadian rhythms, i.e. daily cycles of endogenous origin (Gwinner and Brandstatter 2001). These rhythms are selectively responsive to specific environmental factors, mainly to light and ambient temperature. In the case of birds, a complex system that includes the retina, the pineal gland and the hypothalamus is responsible for circadian organization (Cassone and Westneat 2012; Gwinner and Brandstatter 2001). The key messenger within this system is melatonin, a hormone that controls circadian pacemaking of behaviour and physiology. Secreted by the pineal gland, melatonin has a primary role in regulating sleep and daily rhythms (Bentley 2001; Haldar and Singh 2001; Rodríguez et al. 2001). Studies in several species have established that daily cycles of singing are regulated by a central clock system, and that neither light nor sound stimuli are always necessary to trigger the response (Shimmura and Yoshimura 2013; Wang et al. 2012). In zebra finches (*Taeniopygia guttata*), pinealectomy stops circadian rhythms of singing in constant environmental conditions, but these can be restored by melatonin implants, showing a role of this hormone in regulating circadian singing cycles (Wang et al. 2012).

Data show that circadian period length is highly heritable in great tits (*Parus major*), suggesting that it might be an object of selection in response to ecological pressures (Helm and Visser 2010). External factors, such as light, sound and singing by conspecifics (see sections below) can interact with melatonin secretion and strongly modify singing patterns. For instance, in the case of roosters (*Gallus gallus domesticus*), both light levels and listening to other roosters increase the probability and rate of crowing (Shimmura and Yoshimura 2013). It is not known whether modifications in timing are brought about by a change in the responsiveness of the organism to internal hormone secretions, or by a modification of hormone levels (for a similar consideration in the case of testosterone see Hau 2007). Studies showing that urban life leads to reduction of melatonin levels in Eurasian blackbirds (*Turdus merula*) suggest that the second mechanism is a likely possibility (Dominoni et al. 2013a, b).

In addition to the effect of melatonin, testosterone is also seasonally related to the timing and performance of singing behaviour. For instance, studies in several species show a seasonal match between the dawn chorus and a yearly peak of testosterone levels (Foerster et al. 2002; Van Duyse et al. 2003). Besides, male great tits whose androgen receptors had been experimentally blocked were less likely to show a dawn song than control birds (Van Duyse et al. 2005), and a correlative study with Eurasian blue tits (*Cyanistes caeruleus*) found a positive relationship between male testosterone levels and dawn singing output before breeding (Foerster et al. 2002). However, these correlations between hormone levels and behaviour do not demonstrate that testosterone is a direct modulator of singing. Moreover, other experiments have failed to observe an effect of testosterone implants on dawn singing characteristics (Kunc et al. 2006; Shimmura and Yoshimura 2013).

Hypothesis 1: Circadian Hormone Cycles
The possibility that dawn singing is an epiphenomenon of circadian fluctuations in hormones was proposed by Staicer et al. (1996). Since androgens play an important role in both the development and the expression of singing behaviour in birds and other taxa (Alward et al. 2018), testosterone was the first hormone that was proposed as candidate for this role. Although there is ample evidence showing that testosterone and singing behaviour are connected in songbirds (see above), the issue here is not whether testosterone promotes singing, but rather whether circadian variation in testosterone levels is responsible for the dawn chorus.

There is not much information on how androgen levels vary through the day, but the existing data do not support this hypothesis at all. Some studies have shown that levels of testosterone and luteinizing hormone (which stimulates testosterone production) are higher at night and decrease in the early morning (Laucht et al. 2011; Needham et al. 2017; Ottinger and Follett 1987; Rattner et al. 1982). This evidence would imply that the daily peak of testosterone does not coincide with the dawn chorus. Furthermore, a study in Japanese quail (*Coturnix japonica*) showed that normal daily cycles of crowing were preserved regardless of whether males had been castrated or not (Wada 1986). In addition, the dawn chorus of the silver-beaked tanager (*Ramphocelus carbo*) in Amazonia shows a strong seasonality that is not connected to an increase in testosterone levels (Quispe et al. 2016).

A more promising candidate for circadian regulation of behaviour is melatonin (Hau et al. 2017). As mentioned above, there is experimental evidence for a role of this hormone in affecting circadian and seasonal singing patterns (Cassone et al. 2008; Wang et al. 2012). However, documenting circadian rhythms of melatonin is extremely difficult given the rapid fluctuations in concentration that occur during the day and night, as well as the individual variation in cycling patterns (Seltmann et al. 2016). A common melatonin curve typically involves basal levels during the day with a sudden peak soon before sleep, a maintained plateau for some hours and a dramatic decrease some hours before dawn (Seltmann et al. 2016). Thus, melatonin can explain why birds are awake before dawn, but this does not tell us much about why they are singing at this time. Although it is possible that species-specific differences in dawn timing are connected to sleep cycles and melatonin secretion patterns, little data are available on this respect. In addition, research shows that links between melatonin and behaviour are extraordinarily plastic (Daan and Aschoff 1975). Thus, it would seem naive to assume that the dawn chorus is a mere epiphenomenon of variation in this hormone. On the contrary, research suggests that selection for early activation may drive changes in melatonin secretion (Dominoni et al. 2013a).

An even more basal explanation of dawn and dusk choruses in birds could be based on the fact that a whole suite of physiological and activity patterns show a bimodal peak (early and late in the day), even in complete isolation from external stimuli (Aschoff 1966; Refinetti 2016). Basic physiological processes related to thermoregulation, metabolism or food intake are often described by a bimodal pattern (Aschoff and Meyer-Lohmann 1955; Brandt and Cresswell 2009). These circadian rhythms are similar to those found in bird singing and other signalling routines (Cuthill and Macdonald 1990; Ord 2008). In that sense, it could be argued that the most parsimonious explanation for the dawn chorus should be sought with reference to basic organic functions, given the whole suite of physiological and behavioural components that are affected in the same manner (Aschoff 1966). However, it seems unlikely that the diversity of patterns observed in dawn chorusing could be simply explained as non-functional consequences of circadian cycles (Hau 2007). The patterns that we describe in this chapter show large individual and species differences, suggesting that ecological factors largely modulate this behaviour.

Hypothesis 2: Self-stimulation of Androgen Production
This hypothesis was proposed in a previous review (Staicer et al. 1996) and, to our knowledge, it has not yet been tested. According to this hypothesis, by singing at dawn males would auto-stimulate their own gonadal hormone production, and this would be advantageous later on in their daily social interactions. This suggestion is based on studies that show that vocalizing influences ovulation patterns in female African collared doves (*Streptopelia roseogrisea*) (Cheng 1992), or the bird's own androgen levels in budgerigars (*Melopsittacus undulatus*) (Brockway 1967). However, a role for this function in dawn singing remains highly speculative.

3.4.2 *Breeding Cycle*

A good way to understand the function of song is to examine the link between breeding stages and singing behaviour (Catchpole 1973). Following this idea, analyses on how dawn singing varies with breeding stage have led to hypotheses about its possible functions (e.g. mate attraction, mate guarding, territory defence, etc.).

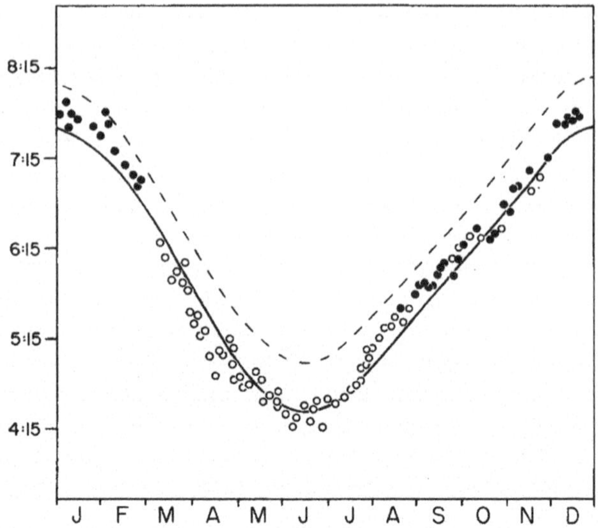

Fig. 3.3 Changes throughout the year in the time of the first song (white dots) or the first call (black dots) in the song sparrow (*Melospiza melodia*) in relation to sunrise (dashed curve) and civil twilight (solid curve). Reprinted from Fisler (1962), by permission of Oxford University Press

Surprisingly there is no comparative analysis examining how song timing with respect to sunrise changes throughout the breeding season for different species (Fig. 3.3). Available data show that in temperate areas, most species sing gradually earlier with respect to sunrise as the reproductive season starts, and tend to sing later when reproduction wanes (Allard 1930; Da Silva et al. 2014; Hasan 2011; Keast 1994a; Lein 2007; Leopold and Eynon 1961; Scheer 1952). This pattern is by no means general and there are strong species- and population-specific differences. In a study conducted in Norway, Amrhein et al. (2008) compared the singing behaviour of blue and great tits across the season. They found that both species sang later relative to sunrise as the season progressed and that, whereas single-brooded blue tits stopped dawn singing after the laying period, great tits continued singing, possibly preparing for a second brood (Amrhein et al. 2008). In contrast, other populations of blue tits sing increasingly early as the season progresses (Da Silva et al. 2014; Poesel et al. 2001).

These data suggest that earlier than average dawn singing is characteristic of the first part of the reproductive season, which includes the pairing stage and the female fertility period. Singing early at the time of pairing is to be expected if song earliness conveys reliable information about male quality (see Hypothesis 5), since it would pay males to show their capacity at this time. However, the period of female fertility is a key phase of the reproductive season that has been specifically linked to the dawn chorus.

Hypothesis 3: Manipulation of Female Mating Behaviour Around the Fertilisation Window

The fertile period of female birds typically lasts from some days before to a few days after laying of the first egg (Birkhead and Møller 1998). Within this period, some authors have assumed that females would be particularly fertile at dawn, right after

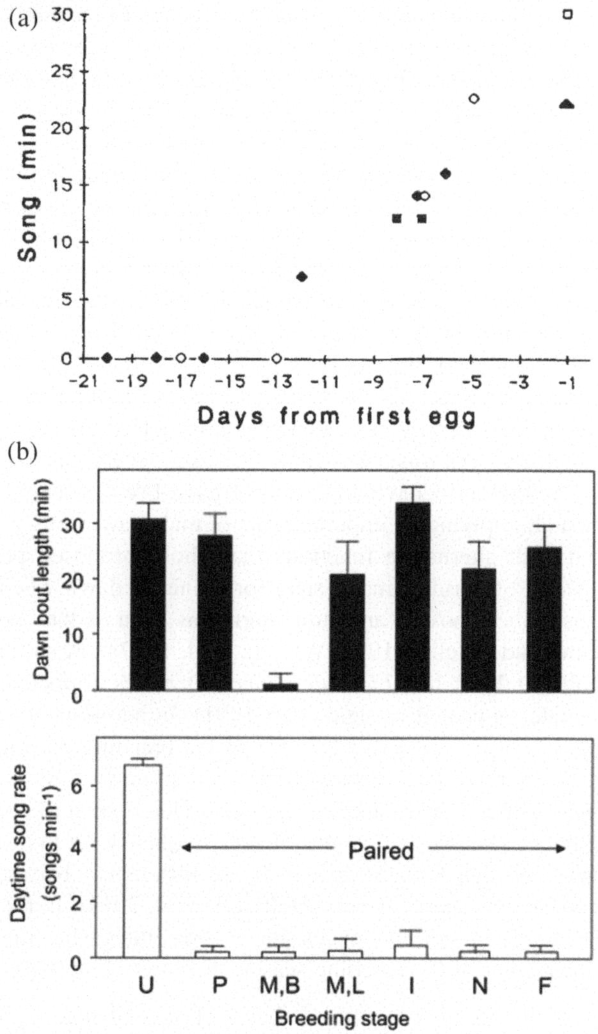

Fig. 3.4 Contrasting patterns in dawn song in relation to female fertility: (**a**) song rate showing an increase around the fertility period in the great tit (*Parus major*). Reprinted from Mace (1987a); (**b**) dawn song, but not day-time song, was almost absent when a male's female was fertile in the chipping sparrow (*Spizella passerine*; M, B = mating or copulations, nest-building) in contrast to unpaired males (U), just paired (P), egg-laying (L), incubation (I), nestling (N) or fledgling (F) periods. Reprinted from Liu and Kroodsma (2007), by permission of Oxford University Press

laying (Birkhead and Møller 1992; Sheldon and Burke 1994). Ruth Mace (1987a) showed that the duration of male song at dawn in the great tit was closely connected to fertility of the male's own female. In her study, the dawn chorus appeared only before laying, increased in duration at the female's fertile period and shortened progressively until stopping altogether a week after the last egg was laid (Fig. 3.4a). This coincidence in time (both the period of the season and the time of the day) led Mace to suggest that the function of dawn singing would be to: a) protect paternity at a time when sneaker males could approach fertile females, and b) to increase female receptivity to copulations by the singing male.

A similar hypothesis was later proposed to explain high singing rates of paired males in general (Møller 1991). These two hypotheses assume that song is an honest signal of male quality (Gil and Gahr 2002), and that a high rate of song at dawn is disproportionally costly for low quality males (see hypothesis 5 further down for evidence of dawn song acting as a handicap).

Additional evidence for a link between female fertility and dawn song has been found in many other species, including the yellowhammer (*Emberiza citrinella*) (Møller 1988), two species of Ficedula flycatchers (Pärt 1991; Vabishchevich 2011), the Eastern phoebe (*Sayornis phoebe*) (Bruni and Foote 2014), the Eastern kingbird (*Tyrannus tyrannus*) (Sexton et al. 2007), the Eurasian blackbird (Cuthill and Macdonald 1990), the dusky warbler *Phylloscopus fuscatus* (Forstmeier and Balsby 2002), and the European starling *Sturnus vulgaris* (Pinxten and Eens 1998). In these cases, fertility is linked with increases in the duration or advances in singing at dawn, but males generally show a dawn chorus before and after female fertility has stopped, suggesting that this behaviour is not exclusively directed to the male's own partner.

Additional behavioural patterns have been observed during the female fertility period, implying an important role of male dawn song at this time for some species, but with alternative functions than those proposed previously. Thus, in several species of Paridae, males sang for longer at dawn when their females were experimentally removed, and stop singing as soon as they appeared again (Mace 1986; Otter and Ratcliffe 1993; Welling et al. 1997a). A similar pattern was found in the collared flycatcher (*Ficedula albicollis*): males stopped dawn singing as soon as their females appeared in sight (Pärt 1991). In addition, song characteristics or singing behaviour at this particular time of the breeding season show important modifications in some species, supporting a role of dawn song in female guarding and extra-pair mating. For instance, a study in a Dutch great tit population showed that males produced songs with slightly lower-frequency on the day before the first egg was laid, and that females appear to use differences between males in these songs to choose extra-pair partners (Halfwerk et al. 2011). In the blacked-capped chickadee (*Poecile atricapillus*), males sing at lower rates when their females are fertile than at other times of the breeding season (Foote et al. 2008b).

However, this hypothesis has met several hurdles that limit its value. Firstly, although further studies on the great tit have also found a link of early dawn song with female fertility, none has documented a complete cessation of dawn song after laying (Amrhein et al. 2008; Naguib et al. 2016), suggesting that the extreme dependency of dawn song with fertility shown by Mace was fortuitous. In addition, there are strong differences between species in whether males sing or not while their females are fertile (Gil et al. 1999b). Examples of birds in which their dawn chorus is completely unrelated to female fertility include the American robin (*Turdus migratorius*) (Slagsvold 1996), the common nightingale (*Luscinia megarhynchos*) (Kunc et al. 2005) and two different Phylloscopus warblers (Gil et al. 1999b; Rodrigues 1996). In an even more contrasting fashion, chipping sparrow males (*Spizella passerina*) do not sing at all when females are fertile (Liu 2004), whereas they sing for the remainder of the season (Fig. 3.4b).

A second hurdle to the hypothesis is the disputed evidence for a fertility window after laying (Birkhead et al. 1996). Birkhead et al. (1996) reanalysed existing data and concluded that fertility is in fact reduced after laying. This point of view has been criticised for being exclusively based on evidence from artificial insemination studies in non-passeriformes, whereas further studies in the wild do show increased mating rates in the early morning in some species (Drachmann et al. 1997). Also, important rates of early-morning mating have been found in the nightingale, a species in which females look for males and mate before dawn (Roth et al. 2009) and in the highly promiscuous superb fairy-wren (*Malurus cyaneus*) (Double and Cockburn 2000), species in which females perform extra-pair forays at night. A particularly strong case for the hypothesis is provided by the European starling Sturnus vulgaris, a species that does not lay eggs at dawn but in the late morning. In agreement with the fertility insurance hypothesis, this species shows a peak of singing and mating in the late mornings of the fertile period around the time of egg laying (Pinxten and Eens 1998).

At least two additional reasons for dawn singing in relation to female fertility can be proposed (Staicer et al. 1996). Firstly, irrespectively of the existence of a fertilisation window, if females are more likely to look for extra-pair copulations at dawn to escape their mate's guarding (Double and Cockburn 2000; Roth et al. 2009), it would make sense for males to show a good performance at this time in order to attract these females. This would explain why males keep on singing when their own partner is no longer fertile. In other words, the selective pressure would not just relate to ensuring the male's paternity with his own female (Mace 1987a; Møller 1991), but to increase the probability of fertilising other females. A prediction from this hypothesis is that the dawn chorus should be longer (seasonally speaking) with increasing levels of extra-pair paternity and breeding asynchrony in the population. An experimental study in which male great tit sleep was prolonged with melatonin led to a decrease in paternity (Greives et al. 2015), in agreement with data showing that early singing males are more successful at fathering extra-pair offspring in this species (Kempenaers et al. 2010).

A second reason for singing early around the female's fertile period may be to increase female egg allocation (Biard et al. 2009; Gil 2008). Females can modify their investment in eggs (in terms of egg size or of contents, such as carotenoids or hormones) as a function of male attractiveness (Cunningham and Russell 2000; Gil et al. 1999a, 2004). This process should lead males to show off their ornaments not just before pairing, but also during egg production, since females may adaptively regulate their investment as a function of male quality. If singing early represents a handicap (see hypothesis 5), it would pay males to show their quality by singing as early as possible at the time when the female is producing eggs (i.e. a few days before egg laying).

3.4.3 Social Interactions

Singing behaviour takes place in several types of social interactions, from mate attraction to male-male counter-singing. Important cues about the intended receivers of the dawn chorus could be given by the particular characteristics of the song at this time (e.g. singing style or song repertoire), to the extent that song directed to males may differ from that directed to females (Collins et al. 2009). However, since the two main functions of song may entail honest signalling (Berglund et al. 1996), differences in repertoire size or song rate between dawn and other times of the day may not tell us much about the likely target of the dawn chorus (Dalziell and Cockburn 2008). Instead, more subtle characteristics, such as singing patterns (song matching, overlapping, singing site within the territory, etc.), may be more informative.

Singing location. Studies on several species have shown that dawn song is often uttered at territory boundaries, suggesting that it is directed to neighbouring territories (Bolsinger 2000; Lein 2007; Liu and Kroodsma 2007). For instance, in the chestnut-sided warbler (*Setophaga pensylvanica*), the use of unaccented songs at dawn, predominantly used in male-male interactions, takes place at territory borders (Byers 1996). However, studies in other species offer a contrasting view. Thus, in great tits and collared flycatchers, males sing near the nest (Mace 1986; Pärt 1991), or in the centre of the territory in the case of the yellowhammer (Møller 1988). Conflicting demands of foraging (Atienza and Illera 1997) and avoiding predation (Møller 2011) are known to determine choice of song posts. Given that both foraging substrate and predation risk change with time of day, analysing changes in song post location seems a promising source of evidence on the function of song and constraints on singing.

Song characteristics. In addition to daily differences in singing sites, several studies have found that songs at dawn sound particularly loud in some species. For instance, in the banded wren (*Thryophilus pleurostictus*), dawn songs are louder and more complex than those sung later in the day, and they are reminiscent of those that birds use in agonistic interactions (Trillo and Vehrencamp 2005). In the chipping sparrow, dawn songs are shorter but sung at very high rates (Liu and Kroodsma 2007). The intensity of some dawn performances can also be assessed by how song is matched to neighbours. Thus, in the black-capped chickadee, the levels of song matching are particularly high at this time (Foote et al. 2008a). Several researchers suggest that these singing styles reflect a dynamic network of communication between territory holders (Brunner and Pasinelli 2010; Foote et al. 2008a; Naguib et al. 1999; Trillo and Vehrencamp 2005). It would be interesting to know by means of playback experiments whether dawn songs elicit stronger responses than songs used during the daytime.

Special dawn songs. Ever since the first observations on dawn singing, researchers have noted differences between dawn song and songs that are used later in the day (Fig. 3.5). In most species studied so far, dawn songs tend to be more complex and elaborate than daytime songs (Staicer et al. 1996). Early authors already remarked that the Eastern kingbird and the Eastern wood pewee (*Contopus*

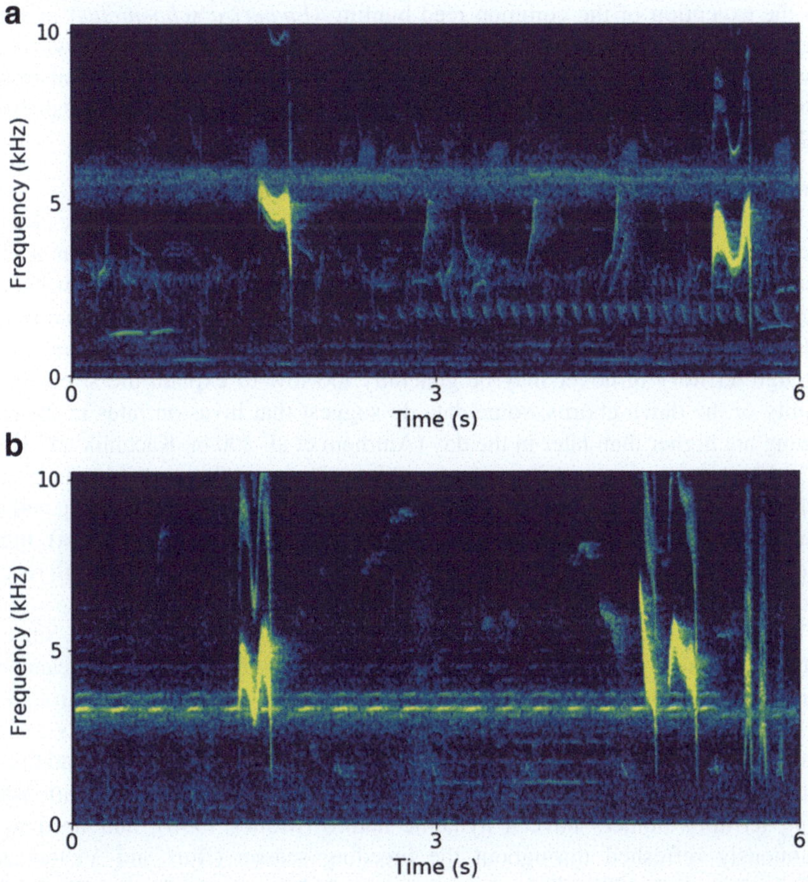

Fig. 3.5 Spectrograms of the songs of greenish elaenia (*Myiopagis viridicata*) at dawn (**a**) and dusk (**b**). Recordings by Andrew Spencer (Xenocanto codes: XC 116251 and XC116250). Figure created in Python v.3.7.4

virens) use a particular song before sunrise, which changes drastically after that (Allard 1930; Allen 1913). Similar patterns have been described in many species of the Americas (for a comprehensive list see: Staicer et al. 1996), mostly in the Tyrannidae and Parulidae (Kroodsma et al. 1989; Morse 1989; Spector 1992) and in some Emberizidae (Nelson and Croner 1991). Less is known about species that do not change repertoire in a dramatic way. Some examples show that differences may be subtle, with changes in general song composition. For instance, dawn repertoires are more complex and variable than daytime repertoires in Golden-cheeked warblers (*Setophaga chrysoparia*) (Bolsinger 2000), Canada warblers (*Cardellina canadensis*) (Demko et al. 2013) or field sparrows (*Spizella pusilla*) (Nelson and Croner 1991). Interestingly, we could not find examples among European species, with the exception of the common reed bunting (*Emberiza schoeniclus*) in which males sing at dawn a particular repertoire of long, highly versatile songs with short between-song intervals (Brunner and Pasinelli 2010), and the Eurasian blackbird, whose dawn song sounds louder than daytime song to human ears (Dabelsteen 1992).

Hypothesis 4: Territory Defence and Social Dynamics
One of the first explanations proposed for the dawn chorus was based on the idea that there should be a higher number of vacant territories in the morning than at other times of the day, because of overnight mortality, thus increasing territorial instability (Kacelnik and Krebs 1983). In that context, it would pay territorial birds to reinforce border defence at this time since new males would try to settle in these vacancies. Although territory turnover may be generally too low to explain the strength and ubiquity of the dawn chorus, some data do suggest that invasion rates in the early morning are higher than later in the day (Amrhein et al. 2004a; Kacelnik and Krebs 1983). Floaters in some species actively seek territories at this time (Amrhein et al. 2004a; Dalziell and Cockburn 2008), perhaps because all males are singing and it is easier to detect vacant areas than later on in the day (Amrhein et al. 2004a). In this sense, the hypothesis could generalise to the fact that dawn is a time of high risk for territory owners.

The hypothesis does not require a higher than average night mortality. Since no territories are taken up at night, there will always be a higher number of vacancies at dawn than in any other time of the day, even if mortality rate is constant through the day (Kacelnik and Krebs 1983). The realisation that the night represents a break from territory defence and signalling for most songbirds suggests to us an additional twist to this hypothesis. Even if no territorial vacancies appear, social relationships among territory holders have a dynamic nature (Møller 1990), and need to be continuously refreshed throughout the breeding season (Burt and Vehrencamp 2005; Staicer et al. 1996). This necessity may explain why birds sing through the day in most temperate songbird species, although singing rates are lower after dawn (Slagsvold 1977). This process is interrupted by the night and, thus, dawn would emerge as the first window of time to resettle hierarchies, territory borders and dominance relationships. Even if no territory vacancies have appeared during the night, the absence of communication during this period could select for a renewal on information exchanges that may transmit information on resource holding capacity, body condition, etc., since the probability of a change is higher given the extent of time that has passed.

Several sources of evidence support the importance of the dawn chorus for resettling territory and hierarchy structures (Foote et al. 2010). Firstly, mated males of many species show dawn song throughout the breeding season, well after female attraction and fertile periods, suggesting strong selection for signalling territory occupancy (Amrhein et al. 2004b; Liu and Kroodsma 2007). Secondly, we have seen before that in many species dawn songs are particularly loud, aggressive and directed to territory boundaries, often associated with a high degree of counter singing and song matching. For instance, a detailed study of singing

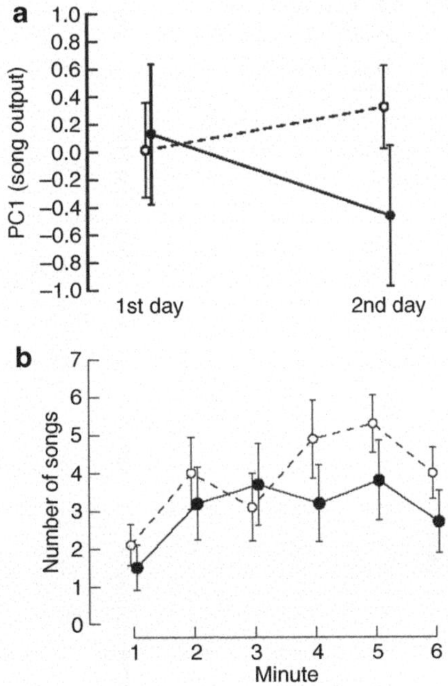

Fig. 3.6 Differences in the response of Eurasian wrens (*Troglodytes troglodytes*) depending on whether they received a stationary playback (black dots, solid lines) or a playback simulating an intruder changing song posts (white dots, dashed lines): (**a**) shows mean (± SE) dawn song output (PC1 scores loaded by song output and song earliness) shortly before (first day) and one day after (second day) the playback; and (**b**) shows number of songs per minute over 6 min in response to the two types of playbacks. Reprinted from Amrhein and Lerch (2010)

interactions in the banded wren has revealed a broad interactive network at this time, with complex matching patterns among many birds (Burt and Vehrencamp 2005). A third line of evidence comes from experimental manipulations of intruder presence that lead to changes in the dawn chorus in the morning following the manipulation (Amrhein and Erne 2006; Erne and Amrhein 2008; Foote et al. 2011; Liu 2004; Sexton et al. 2007; Xia et al. 2014). A clever experiment compared the next-morning song response of Eurasian wrens (*Troglodytes troglodytes*), depending on whether they had been exposed to a normal playback or a "scary" playback which involved several changes in loudspeaker position, simulating a particularly challenging intruder (Amrhein and Lerch 2010). The results showed that the wrens that had received the challenging playback started to sing earlier at dawn, and sang more and longer songs than those exposed to normal playback (Fig. 3.6).

3.4.4 Individual Differences

Following the suggestion that dawn singing could represent a handicap (Gil and Gahr 2002; Hutchinson et al. 1993; Montgomerie 1985), several studies have examined whether there are differences in dawn song timing, song rate or song length in relation to individual variation in fitness-related traits (e.g. quality, condition, social status, etc.). We should stress that, as far as the question of "why do birds sing at dawn" is concerned, the individual difference of interest is timing of singing,

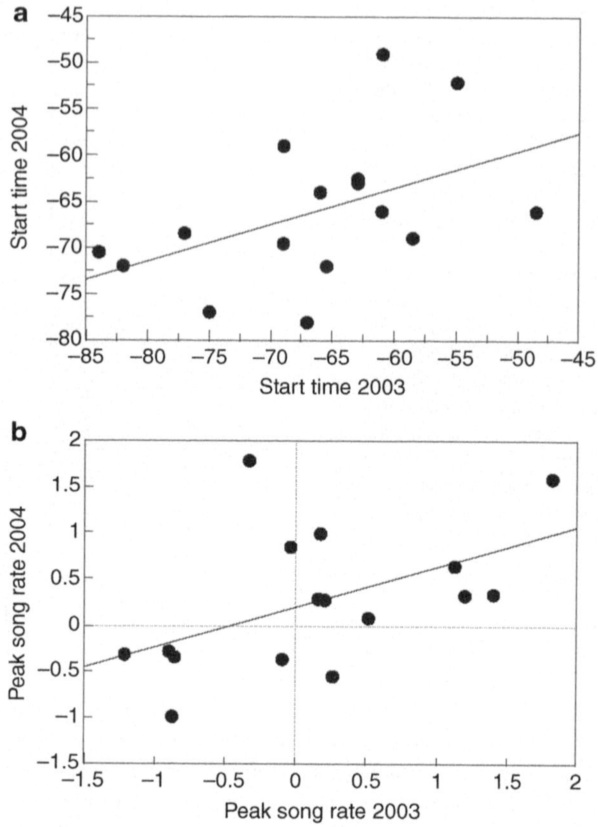

Fig. 3.7 Between-year repeatability for dawn song time (**a**) and rate (**b**) in Eastern kingbirds (*Tyrannus tyrannus*). Reprinted from Murphy et al. (2008), with permission from Elsevier

i.e. is there a relationship between male quality and time of first song at dawn? Timing considered as a sexually selected trait has been formalised by Hau et al. (2017), who propose that daily timing of trait expression is linked with circadian clock genes and can be subjected to sexual selection in birds and mammals.

Researchers have found repeatable differences between individuals in dawn song timing in several species (Amrhein and Erne 2006; Dunnett and Hinde 1953; Foote et al. 2011; Snijders et al. 2015). A remarkable study in the Eastern kingbird found timing to be a repeatable trait between different years (Fig. 3.7) (Murphy et al. 2008). This implies that males are consistent in how early they sing, a prediction from sexual selection models that propose a relationship between variation on this trait and individual condition or resource holding capacity (see hypothesis 5 below). Similarly, multiple parameters of sleep behaviour in the blue tit, such as sleep duration or awakening time, also show substantial individual repeatability across years (Steinmeyer et al. 2010).

Several studies have found that song timing can predict traits related to individual quality or condition. For instance, early singing males are larger, and possibly older in Eastern kingbirds (Murphy et al. 2008). Dominant black-capped chickadees sing earlier than subordinates (Otter et al. 1997). Also, older blue tits sing earlier than young birds (Poesel et al. 2006), although a non-significant tendency in the opposite direction was found for great tits (Snijders et al. 2015). Finally, early-singing willow tits (*Poecile montanus*) had stronger anti-predatory responses and were more likely to survive than late-singers (Welling et al. 1997b). Further evidence on the relationship between song timing and condition is provided by experiments that have found earlier song in food-supplemented birds (Barnett and Briskie 2007; Cuthill and Macdonald 1990; Grava et al. 2009; Montgomerie 1985; but see: Saggese et al. 2011; Thomas 1999; Thomas and Cuthill 2002).

An experiment on domestic roosters (*G. gallus domesticus*) has shown that dominant birds crow earlier, and that experimentally-induced changes in hierarchy lead to a correlated effect on timing; i.e. beta males sang earlier when the alpha male had been removed (Shimmura et al. 2015). This elegant experiment confirms observations on red jungle-fowl (*G. gallus murghi*) in the wild that link early crowing time to higher positions within the group hierarchy (Collias and Collias 1967). Although this example on a highly social and hierarchical species may not translate easily into the typically passerine-dominated dawn chorus, such a strong link between dominance and timing does suggest that males do signal something relevant by singing earlier than others.

A more definite test of the fitness advantage of early singing is provided by extra-pair paternity (EPP) studies. Several studies have linked the earliness of singing with higher success at siring extra-pair offspring. Eastern kingbirds that sang earlier had higher numbers of EPP chicks, and cuckolders sang earlier than cuckoldees (Dolan et al. 2007). Taking advantage of differences in dawn singing timing caused by artificial lighting of a part of the forest, Kempenaers et al. (2010) found that early blue tits, i.e. those exposed to artificial light, were more successful at obtaining EPP than late birds. This study replicated previous results in the same species, in which natural variation in dawn song was also identified as a good predictor of EPP success (Poesel et al. 2006). Similarly, male great tits treated with melatonin woke later and lost more paternity than control birds (Greives et al. 2015). It is not clear whether the mating advantages accrued by early birds was a consequence of males using the period before dawn for pursuing extra-pair mating, or was mediated by female choice of early singing birds. Evidence for the latter possibility is provided by a study showing that female Lincoln's sparrows (*Melospiza lincolnii*) preferentially chose songs that they had heard at very cold temperatures (Beaulieu and Sockman 2012), a situation associated with dawn song in this species.

Despite the previous examples, rather few studies have measured individual differences in dawn singing time, possibly because of the inherent difficulty of obtaining these data. The recent development of cheap automatic recording units (ARUs) and miniaturised sound-recording data loggers will improve the situation in the coming years.

Hypothesis 5: Early Singing as a Handicap

One of the leading hypotheses to account for the evolution and maintenance of male ornaments is Zahavi's handicap hypothesis and variations on it (Biernaskie et al.

2018; Grafen 1990; Kokko et al. 2006; Zahavi 1977). The hypothesis states that ornaments should be costly and that the cost should disproportionally increase with increasing expression of the signals. The suggestion that early singing in the dawn chorus could act as a handicap was proposed by Montgomerie (1985), on the assumption that singing at dawn is costlier than at other times of the day. It is by no means clear why this should be so, although some possibilities may be: temperature-dependent singing costs (Ward and Slater 2005) in the case of temperate-climate species, higher risk of predation by nocturnal birds of prey (Schmidt and Belinsky 2013), or costs of sleep deprivation (Cirelli and Tononi 2008).

However, as we saw above, several lines of evidence suggest that early singing, at least in some species, fulfils some of the requirements for handicaps: (1) there is individual repeatability in dawn singing time (Murphy et al. 2008); (2) some experiments and correlative studies show positive relationships between age, dominance, access to supplemental food and time of the dawn chorus (Cuthill and Macdonald 1990; Murphy et al. 2008; Otter et al. 1997; Poesel et al. 2006; Shimmura et al. 2015); (3) early singers may obtain a disproportional share of extra-pair fertilisations (Dolan et al. 2007; Kempenaers et al. 2010); and (4) predation by nocturnal predators may limit early singing (Lima 2009, but see Sect. 3.4.7).

Despite these suggestive findings, a true demonstration of a handicap requires some extra evidence both on the signal-production side (Cotton et al. 2004), and on the female response to it (e.g. Ballentine et al. 2004). A good test of this hypothesis would require experimental modifications of singing time, which are difficult to induce. Even if tampering with melatonin levels could modify singing time (Dominoni et al. 2013a), there are many correlated effects of this hormone that would render the results difficult to interpret (Greives et al. 2012).

3.4.5 Foraging

The role of food is central to several hypotheses about the dawn chorus (the inefficient foraging, unpredictable conditions and handicap hypotheses), and hence estimating the relationship between song and food availability is a crucial step to test these alternatives. Foraging also shows diel cycles, being constrained by both abiotic and biotic factors. Dawn does not appear to be a good time for a bird to forage, since low temperatures and dim light levels affect foraging success. It has been shown that the activity of arthropods is greatly reduced at low temperatures (Avery and Krebs 1984), decreasing the ability of birds to locate food. In addition, low light levels constrain vision, decreasing searching ability and foraging profitability (Kacelnik 1979), probably for both insectivorous and granivorous birds (Fig. 3.8).

Species differences in dawn chorus timing in relation to the type of diet were suggested early on by Armstrong (1963), who remarked that vermivorous and insectivorous species sing earlier than granivorous birds, and speculated that differences in vision and in the behaviour of prey may explain this pattern. However, we have not been able to find a proper test of this hypothesis. Most of the recent

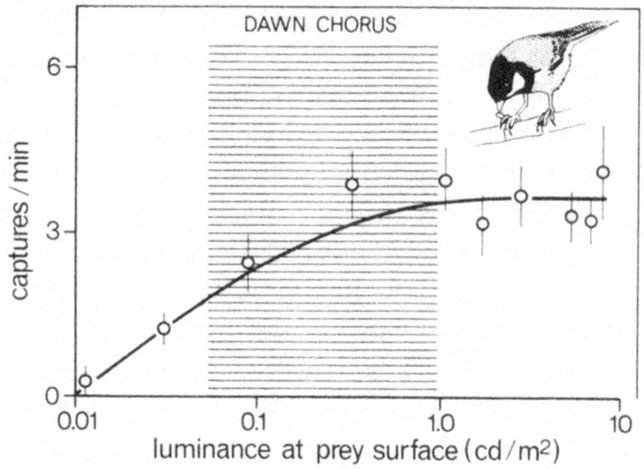

Fig. 3.8 Foraging profitability increases as a function of light levels in the great tit (*Parus major*). Shaded in grey is the time of dawn chorus, showing that light conditions are not ideal for foraging at this time. Reprinted from Kacelnik and Krebs (1983)

comparative analyses of dawn chorus timing dwell on differences in relative eye size with little mention of food guild (Berg et al. 2006; Gil et al. 2015; Thomas et al. 2002). Since eye size is related to food, i.e. birds that forage on mobile prey have relatively larger eyes than seed-eaters (Garamszegi et al. 2002), it seems that vision and feeding are so closely correlated that teasing apart these factors to explain song timing may be challenging (Chen et al. 2015).

Although some behavioural and ecological adaptations in relation to food guild have been studied (e.g. Falk et al. 1992; Wiens and Johnston 1977), we could find surprisingly little comparative information on how type of food affects behavioural patterns such as time-budgets or food search routines. In addition to differences in predictability between insects and seeds, important variation in metabolizable energy provided by arthropods, seeds or fruit (Karasov 1990) probably imposes differential constraints on time-budgets.

Hypothesis 6: Inefficient Foraging at Dawn

This hypothesis posits that dawn is a time when, because of low insect activity and poor visual conditions, it is not profitable to forage (Kacelnik 1979; Kacelnik and Krebs 1983). The lack of interference with feeding activities would favour birds that use this time for singing and communicating. Inefficient foraging is a clever hypothesis, but it has some difficulties. For instance, it has been found that birds breeding in continuous light in the Arctic also show dawn choruses (Staicer et al. 1996). This finding however would not necessarily refute the hypothesis since it is possible that foraging profitability in these birds may also change daily in response to variations in temperature or more subtle light differences.

An experimental test of this hypothesis in captivity failed to support one of its major predictions: birds did not modify their dawn singing activity in response to a dramatic change in food availability (Mace 1989). A possible line of research here could involve comparing male and female feeding time budgets in species in which females do not sing. Such a comparison could give us evidence as to whether males are using free time to sing, or else are trading off precious feeding time for singing (Greig-Smith 1983). The hypothesis has however received some theoretical support from a game-theory model, which predicts the occurrence of both dawn and dusk chorus by simply allowing feeding profitability to vary as originally suggested by Kacelnik (Hutchinson 2002). See section "Unpredictable overnight conditions" below for a combined discussion of these two hypotheses.

The inefficient foraging hypothesis also touches an important point as it suggests that dawn is "spare" time within the bird's time budget. For this to be true, it means that sleep in birds should require less time than the night duration. Several lines of evidence suggest that this is the case. Firstly, studies analysing differences in activity across latitudes show that birds are remarkably flexible in their sleep requirements, and that low light levels are a constraint in activity (Daan and Aschoff 1975). Secondly, it has been shown that late-singing species spend more time awake waiting to sing than early species (Scheer 1951). Although this latter evidence is largely anecdotal, it supports the idea that the duration of night in spring actually exceeds the requirements for sleeping time. Thirdly, studies of migratory restlessness show prolonged activity over night (Berthold 1973), or round-the-clock singing and mating patterns in nightingales and pectoral sandpipers (*Calidris melanotos*) (Lesku et al. 2012; Roth et al. 2012). Thus, if birds do not need to sleep for the whole night at the breeding period, it makes sense that dawn is used for an activity not requiring acute vision such as song (Siegel 2009). An alternatively view would propose that singing earlier may uses precious sleeping time (Cirelli and Tononi 2008), and that time of dawn song is a sexually-selected character (see handicap hypothesis above). This would suggest that early waking-up in birds may be sexually selected (Amrhein et al. 2002; Greives et al. 2015), although it is not clear what kind of costs are paid. It is possible that an increase in predation rather than costs of reduced sleep is the true currency of this mechanism. The study of comparative sleep in birds is in its infancy in comparison to mammals, and evidence so far shows that bird sleep has some important peculiarities (Roth et al. 2006). The ongoing development of small sleep dataloggers should allow researchers in the next decade to provide important insights on this topic (e.g. Lesku et al. 2012; Steinmeyer et al. 2010).

Hypothesis 7: Unpredictable Overnight Conditions

Stochastic-dynamic programming (SDP) models allow the prediction of optimal routines given quite realistic assumptions about foraging success, energy use and singing rewards (McNamara et al. 1987). Using this technique, a pioneering study by McNamara et al. (1987) showed that a dawn chorus could simply arise as a consequence of adaptive foraging decisions. The key aspect lies in the unpredictability of overnight costs. Briefly, since there is always some unpredictability of how demanding the night may be in terms of energy requirements, birds must go to sleep with an extra amount of energy. As dawn approaches, and the probability of overnight starvation decreases, birds can use that surplus of energy for singing at dawn. Interestingly, the model also predicts a smaller peak of singing at dusk, which is in agreement with many singing cycles in passerines (Cuthill and Macdonald 1990; Møller 1988; Morse 1989).

One of the most useful methods for testing the effect of energy managing on dawn chorus timing is by providing supplementary food. Experiments in four species have shown that individual males provided with supplementary food tend to sing earlier and at higher rates (Barnett and Briskie 2007; Cuthill and Macdonald 1990; Grava et al. 2009; Thomas 1999). In contrast to this finding, a study with great tits showed the opposite effect, with supplemented birds singing later than controls (Saggese et al. 2011). The main difference of the latter study is that it involved long-term instead of short-term modifications.

Although assumptions of the SDP models are somehow simplistic, they do make some interesting predictions on how variance in overnight requirements and foraging profitability may affect singing (McNamara et al. 1987). One key prediction is that the dawn chorus should be longer with increasing unpredictability of overnight requirements (Hutchinson et al. 1993; McNamara et al. 1987). In our opinion, the most relevant test of this model should focus on different populations of the same species facing contrasting environmental conditions, or on different species that differ in weather sensitivity. However, to our knowledge, no study so far has taken this approach.

At the individual level, an experiment in European robins (*Erithacus rubecula*) found that birds lost more weight when overnight temperature was lower, but they did not adjust their singing in relation to temperature (Thomas and Cuthill 2002). However, in this case it is possible that food supplementation may have altered the normal patterns of foraging and singing. Only one study that we know of has experimentally tested the prediction for a negative relationship between energy predictability and dawn song duration (Thomas 1999). In this experiment, Thomas provided supplemental feeding to territorial robins under two different treatments: predictable and unpredictable. His findings show that robins sang relatively less at dawn (and hence relatively more at dusk) after they have been fed in an unpredictable fashion, whereas the contrary was true when food supplementation was constant (Thomas 1999). The results supported the SDP model's prediction and imply that the high unpredictability of night time temperature together with food predictability combine to adjust singing times between dawn and dusk. It is regrettable that this hypothesis has not been tested further (Hutchinson and McNamara 2000), not just at the individual level, but also within a comparative perspective.

Further developments with this type of models have introduced more complex situations, such as differences in male quality and in singing costs across the time of day (Hutchinson et al. 1993). The latter effectively assumed that dawn song works as a strategic handicap (see hypothesis 5) within the framework of SDP predictions. By using SDP models, Hutchinson (2002) pitched Kacelnik's inefficient foraging hypothesis (see hypothesis 6) against hypotheses based on the stochasticity of overnight energy requirements, and found that both models were equally satisfactory in predicting dawn and dusk choruses.

Although these models are useful in generating ideas and predictions, in practice they are extraordinarily challenging to test (Hutchinson and McNamara 2000). For instance, they require a precise estimate of song costs, while the literature on this respect show ambiguous findings (Gaunt et al. 1996; Thomas 2002; Ward and Slater 2005; Ward et al. 2003). Generally speaking, data suggest that the greatest cost is that of lost foraging time, rather than a of singing per se (Gil and Gahr 2002).

Similarly, foraging profitability across the day depends on so many factors, and data are so scant on this, that some predictions are extremely difficult to test empirically.

3.4.6 Predation

Predation is a key factor that permeates every single aspect of bird behaviour and ecology, but its role in shaping dawn chorus is seldom considered (Lima 2009). This is regrettable because experiments have shown that predators do use acoustic signals to locate prey in a variety of taxa (Haskell 1994; Krams 2001; Zuk and Kolluru 1998), influencing the evolution of singing behaviour (e.g. Tuttle and Ryan 1981). It is not clear whether predation is higher at dawn or not: some researchers have predicted higher than average predation risk at dawn (Lima 2009), while others have proposed the opposite (Staicer et al. 1996).

Research shows that birds modify behaviour in response to predation risk, and that this can vary with changes in light or noise levels (Fernandez-Juricic and Tran 2007; Klett-Mingo et al. 2016). Since dawn is characterised by low light levels, foraging at this time could be risky if predator sight in these conditions is better than their prey's (Fernandez-Juricic and Tran 2007). It has been found that choice of perch during the day can be explained by predation risk (Møller 1991), and a study has shown that blue tits choose comparably safer perches from which to sing at dawn than later on in the day (Parker and Tillin 2006). However, we have not been able to find any experiment that has directly tested the influence of predation risk on singing at dawn. Nonetheless, a couple of studies on dusk singing show an influence of predation risk perception. Thus, a playback experiment with male veeries (*Catharus fuscescens*) shows reduced levels of dusk song in places where owl calls have been previously broadcast (Schmidt and Belinsky 2013). Similarly, a correlative study in tawny owls (*Strix aluco*) shows that dusk calling rates are reduced in areas with high calling rates of Eurasian eagle owls (*Bubo bubo*), a natural predator of the tawny owl (Lourenço et al. 2013).

Hypothesis 8: Low Predation at Dawn
It has been proposed that the dawn chorus is advantageous because of a presumably lower risk of predation at dawn (Staicer et al. 1996). However, as we have seen above, although there are some studies showing that the behaviour of singing birds is affected by predator risk, no study that we know of has analysed whether indeed dawn is less risky than other times of the day. Lower predation may also be a consequence of a dilution effect, since many birds singing simultaneously would reduce the chances of being located and predated (Hamilton 1971; Lima 2009). However, the dilution effect would not explain why dawn is the best time to sing, it would simply select birds to sing synchronously.

One of the predictions of this hypothesis, namely that birds should use more exposed perches at dawn than later on in the day (Staicer et al. 1996), was actually found to be the contrary in a study in blue tits (Parker and Tillin 2006). It is important to empirically establish the relationship between predation risk and time of day, because if predation is higher at dawn (Lima 2009), contrary to the predictions of this hypothesis, we would face a handicap situation (see hypothesis 5) with very different predictions. In agreement with the latter idea, Lima (2009) proposed that the relationship between relative eye size and time of singing at dawn could be explained by differences in predation risk, in addition to the more common explanations based on foraging profitability or unpredictable overnight conditions. Following this suggestion, the prediction would be that earliest-singing species, those with larger eyes, should excel at detecting predators at dawn.

3.4.7 Weather

Weather is also a pervasive factor influencing almost all aspects of bird ecology and behaviour. Most studies find that "bad" weather in general (rain, clouds, storms or low temperature) tends to delay the onset of the dawn chorus, although not all species respond similarly to these weather phenomena (Bruni et al. 2014; Keast 1994b). Several studies in northern temperate areas find a positive correlation between overnight temperature and duration of song, often implying earlier choruses (Bruni et al. 2014; Garson and Hunter 1979).

However, to the extent that bad weather limits singing in general (Curio 1959), this evidence does not shed much light on the reasons behind the dawn chorus. In other words, given that singing interferes with foraging, we expect a positive correlation between temperature (proxy of food availability, at least for insectivores) and singing activity irrespective of the time of day. Also, since bad weather typically implies a decrease in light levels, it is possible that the reasons for this relationship can also be explained by the effect of light (see Sect. 3.4.8).

Predictions about how bad weather should affect dawn singing are complex, and may depend on the particular aspect of weather (clouds vs. temperature) and the particular theory that is tested (Hutchinson 2002). A study in great tits found that the correlation between overnight temperature and dawn singing was lower (and non-significant) than that with overall morning singing (Garson and Hunter 1979). This pattern was explained as evidence of low foraging profitability at dawn: since profitability is low at dawn it would pay birds to sing irrespectively of temperature, whereas a finer adjustment would make sense later on in the morning, when feeding time is more rewarding and would trade-off with time devoted to sing.

Hypothesis 9: Better Sound Transmission Conditions at Dawn

A widely cited hypothesis for explaining the dawn chorus proposes that this particular time of the day is the best time for sound transmission (Henwood and Fabrick 1979). This proposal draws on the logic that sound attenuation and degradation are a function of the habitat and the atmospheric conditions in which the sound is produced and perceived (Morton 1975; Richards and Wiley 1980). Several studies have shown that atmospheric conditions before sunrise, characterised by low temperatures and absence of air turbulence, allow better sound transmission than those found later in the day, leading to a longer transmission distance and a larger active space (Brenowitz 1982; Henwood and Fabrick 1979; Larom et al. 1997). However, not all studies have been successful at finding better transmission efficiency at dawn, suggesting that this is by no means a general advantage (Brown and Handford 2003; Dabelsteen and Mathevon 2002). A refinement of this hypothesis proposed that dawn offers not better, but a more consistent sound transmission than other times of day (Brown and Handford 2003).

The sound transmission hypothesis has an important limitation (Mace 1987b): even if we accept that atmospheric conditions are on average best at dawn (Henwood and Fabrick 1979), or at least more constant (Brown and Handford 2003), the high number of birds singing at that time should impose huge perception costs (Langemann et al. 1998; Poesel et al. 2007; Pohl et al. 2009) that would most likely reduce this advantage (Mace 1987b). Indeed the dawn chorus can be so noisy that it has been suggested that switching singing to a dusk chorus could be a good strategy to reduce overlapping of song at dawn (Belinsky et al. 2012). In general, we believe that sound transmission does not offer a good explanation for the maintenance of the bird dawn chorus.

3.4.8 Light

From very early on, observers have found that higher light levels lead to earlier choruses and that, on cloudy mornings, birds sang later than usual (Allard 1930; Scheer 1952). From a comparative perspective, species with relatively large eyes for their size also start to sing earlier (Berg et al. 2006; Gil et al. 2015; Thomas et al. 2002), thus providing additional evidence for a role of light in song timing (see Sect. 3.2).

Studies that have tracked song timing with respect to the moon phase have shown that a full moon leads to earlier dawn choruses in several species (Bruni et al. 2014; York et al. 2014). In a study on the white-browed sparrow weaver (*Plocepasser mahali*), York et al. (2014) found that the effect of the moon phase was only detected when the moon was not hidden by clouds. In other words, it was the actual increase in light and not the lunar cycle that was responsible for the change in behaviour.

Recent interest on the effects of artificial lighting in bird behaviour has led to numerous studies that have shown earlier choruses in urban areas (Bergen and Abs 1997; Miller 2006). It is important to underline that correlative studies, simply comparing areas that differ in light levels, may be affected by other confounding factors, such as noise (see Sect. 3.4.9), and these may be difficult to disentangle statistically (Fuller et al. 2007; Nordt and Klenke 2013). In addition to effects on the dawn chorus, correlative and experimental data show that artificial light at night alters also the timing of activity (Dominoni et al. 2013a), so that earlier song could follow from earlier wake-up time. Furthermore, artificial light also advanced

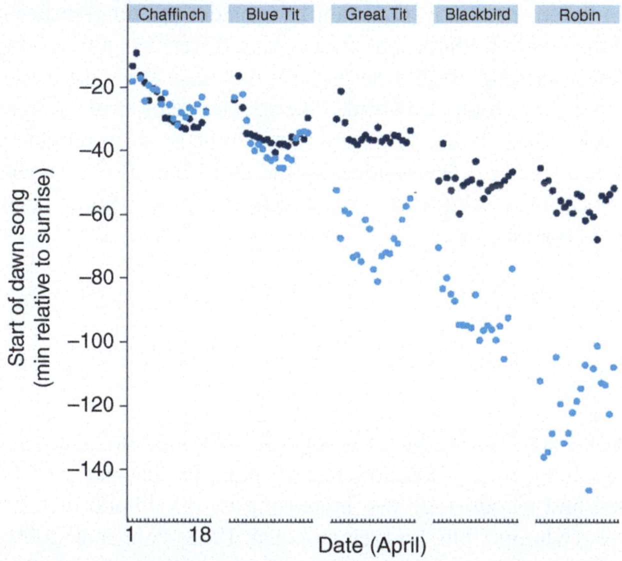

Fig. 3.9 Mean time of first songs in 5 European bird species in 6 control areas (dark blue dots) and 6 artificially illuminated sites (light blue dots) across 19 consecutive spring days. Reprinted from Kempenaers et al. (2010), with permission from Elsevier

reproductive cycles, leading to earlier gonad maturation, singing and breeding (Da Silva et al. 2015; Dominoni et al. 2013b; Gwinner and Brandstatter 2001).

A major study investigating the effects of night light pollution on avian behaviour found that four out of five species considered showed an advance in singing in suburban areas with artificial light (Kempenaers et al. 2010). This advance differed between species: earlier singers showed larger advances than late singers (Fig. 3.9); indeed, the last singer of the group, the common chaffinch (*Fringilla coelebs*), did not show an advance at all (Kempenaers et al. 2010). These patterns were partially replicated in an experiment in the wild in which light levels were manipulated, leading to immediate advances in singing for all studied species (Da Silva et al. 2016). Although the experimental results did not fully mirror the correlational data, the general pattern found in this and additional studies is that the earlier singing species are also those in which night light shows a stronger effect on song timing (Da Silva and Kempenaers 2017; Da Silva et al. 2014; Kempenaers et al. 2010). However, another similar experimental study testing different light colours found no effects of night lighting on singing, suggesting that fast timing plasticity is not present in all populations, and that previous selection may prime the response of birds to increased lighting (Da Silva et al. 2017).

The fact that birds sing earlier when nocturnal light levels are higher (due to moon lighting or artificial light pollution) does not explain why birds use dawn to sing rather other times of the day. However, to the extent that the size of the response varies between different species, this implies that the balance of costs and benefits also differs. In this respect is it important to consider the relationship of light with feeding. It is reasonable to think that effects may differ depending on type of food, since, for instance, arthropods relying on light may also show differences in behaviour. Some studies have found that birds take advantage of artificial light for feeding for a longer time (Russ et al. 2015), and that light is an important factor in the foraging/vigilance trade-off (Fernandez-Juricic and Tran 2007). Together, this evidence backs Kacelnik's suggestion (1979) that the dawn chorus occurs at a time when foraging is not profitable.

3.4.9 Noise

Background noise is a limiting factor in all types of communication and, in the case of acoustic signalling, it may lead to a reduction in the distance over which sounds can be detected and recognized, also increasing the likelihood of false alarms and missed positives (Klump 1996; Langemann et al. 1998; Pohl et al. 2009). As noise is predictable in some environments, many researchers have examined whether birds modify their singing regime with respect to the occurrence of noise around dawn.

Anthropogenic Noise Since noise in urban environments is reduced at night and in the first hours of the day, it makes sense for birds to try to reduce the overlap with urban noise by singing earlier than they would in the natural environment. This would lead to an earlier dawn chorus. The first study to detect this pattern in three common woodland species was not able to separate the contribution of light and that of noise pollution to the advance (Bergen and Abs 1997). However, research on three different species has shown that earlier dawn choruses in cities are due to increased noise levels after controlling for the effect of light (Dorado-Correa et al. 2016; Fuller et al. 2007; Nordt and Klenke 2013). This advance is considered advantageous, as it gives birds a window of communication at a time when noise levels are lower than later in the day (Fuller et al. 2007).

In these studies, it is important to demonstrate that earlier singing is not caused by noise disturbance, but rather that birds anticipate noise and sing earlier than that. An experiment in which noise was experimentally broadcast before dawn resulted in some species singing earlier, but it is unclear whether they were actually woken up by the sound rather than singing in anticipation of it (Arroyo-Solis et al. 2013). In contrast, true anticipation has been documented in studies that examined song timing near airports, in which birds were found to sing earlier than in control areas, thus anticipating the noisy start of airport operations (Dominoni et al. 2016; Gil et al. 2015; Sierro et al. 2017). Interestingly, in one of these studies, Gil et al. (2015) found that the effect size of the advance was inversely proportional to how early birds took part in the chorus. In other words, the advance was greater for the late birds than for the early birds. This made adaptive sense because the overlap with aircraft noise was greater for the late birds. On the same line, a study with blackbirds found that the advance was only detectable when airport timing overlapped with the dawn chorus

(Sierro et al. 2017). In striking contrast to these effects, advances in timing due to light pollution are found in early, and not in late singers (Kempenaers et al. 2010).

Biotic Noise In tropical environments, insects are a major source of noise for singing birds (Kirschel et al. 2009), and two different studies have examined whether birds avoid singing at the same time as cicadas. The first of these studies showed short-term avoidance, by which birds reduced the number of vocalizations or stopped singing when cicada song was present (Hart et al. 2015). The other study, conducted on Barro Colorado Island (Stanley et al. 2016), reports avoidance of insect-noise and also shows that birds whose songs overlap the frequency of the cicadas delayed their timing to avoid masking. However, the latter result may suffer from a basic methodological problem, as data collection was based on acoustic and not on visual observations, and singing birds could have lower detectability because of higher sound masking.

Although bird species start singing at different times, suggesting a division of the twilight-to-sunrise period, the truth is that, for a large part of the time, a respectable proportion of the bird community is singing simultaneously (although in some species there are strict slots seldom overlapped in time: Luther 2008). This realisation has led researchers to ask whether some form of time-sharing is at work during the dawn chorus, particularly in species-rich communities and among species that overlap extensively in acoustic frequency.

Avoidance of song overlap has been documented for several pairs of species coexisting in the same community (Brumm 2006; Cody and Brown 1969; Popp et al. 1985), but these findings only explain short-term adjustments that do not alter the general timing of singing activity. However, within a dawn chorus context, with many birds singing at the same time, it is relevant to ask if the avoidance of song overlap is related to the degree of song similarity between species. Similar songs lead to a greater degree of masking, and a higher probability of a potential receiver not detecting the signal (Klump 1996). The three studies that so far have tackled this question, all within Amazonian bird assemblages, have not found a common answer, possibly due to differences in methodology (Luther 2009; Planque and Slabbekoorn 2008; Tobias et al. 2014). A first study documented temporal avoidance between certain pairs of species, but not necessarily those with the largest overlap in frequency (Planqué and Slabbekoorn 2008). In contrast, working in a community with a high number of species, Luther (2009) found that birds that sang together within 30-min intervals at the same height in the forest showed lower overlap in frequency characteristics than birds picked at random (Luther 2009). This study suggested that species that sing at the same place and time reduce interference by shifting singing times (Luther 2009). In striking contrast, a more recent study by Tobias and co-authors (2014) found that species that sang within 10-min chorus sections had higher than expected song similarity and phylogenetic relatedness (Tobias et al. 2014). This finding could be explained by interspecific competition for resources leading to higher song similarity (Tobias et al. 2014), but the grounds for such claim are slightly speculative at this point.

This last study is one of the few examples pointing to interspecific communication in the dawn chorus. Earlier, Møller (1992) found increased singing activity of five different species after the playback of the song of the black wheatear (*Oenanthe leucura*), and suggested social facilitation as the underlying mechanism. Other attempts to test this hypothesis have failed to find support. For instance, a study in two species of thrushes (Turdidae) has shown that heterospecific playbacks do not lead to earlier choruses on the following days, whereas this is the case for species-specific playbacks (Hodgson et al. 2018).

In contrast to these minor temporal changes, the question of whether species-specific timing varies in response to competition between species for an acoustic space has received less attention. It is likely that the best answer to this question will come from comparative, rather than single-community studies. We are only aware of one study (as yet unpublished) that has compared whether differences in species richness affect the way that dawn chorus time is shared. Comparing dawn chorus patterns across the globe, Pollard (2009) found in her PhD thesis that, at sites with high species richness, the onset of dawn song was more variable than at sites of low richness. This pattern suggests that competition between species may indeed promote wider separation between the singing times of species singing in the same chorus. A limitation of this study is that most data on choruses came from the Northern Hemisphere, with a very small sample size for the Equator and the Southern Hemisphere (Pollard 2009).

3.5 Conclusions

Despite important efforts in the study of the dawn chorus in the last few decades, this review shows that there is still a good degree of confusion, and that much work is still needed. Four out of the nine hypotheses proposed to explain this phenomenon have not been empirically tested, while others present equivocal support as general theories. In Table 3.1, we have listed the key findings for each hypothesis, as well as our personal evaluation of the evidence supporting them and specific suggestions for further research.

In spite of this plethora of explanations, most hypotheses are not incompatible with each other. Their explanatory value changes depending on the species, and at any rate we should not expect a single explanation (Kacelnik and Krebs 1983; Mace 1987b). Although there might be common patterns in dawn song across bird species, the double function of male song (male vs. female receivers) is likely to be responsible for the contrasting patterns of dawn choruses of (for instance) great tits (Mace 1987a) and chipping sparrows (Liu and Kroodsma 2007) (Fig. 3.3). More importantly, the function of song may change across the season, and thus we can expect different functions at different times of the reproductive cycle for the same species.

Moreover, there is an interesting historical distinction among these hypotheses. Many of them arose from an optimality standpoint, and postulated that early singing is selected for reasons that should benefit all signalling males (low predation, good sound transmission, use of spare time and energy, etc.). In contrast, the handicap hypothesis proposed a distinct view, suggesting that singing early is costly and reflects male quality. This implies a shift from optimality to costliness. These two views are not incompatible, since sexual selection can permeate processes of signal production, modifying optimal patterns and shaping individual variation. In the case of bird song, we expect that whatever general selective advantage is accrued by singing early, between-individual variation in performance based on condition can provide a way to assess male quality for individuals of both sexes (Gil and Gahr 2002). In other words, while some hypotheses explain early singing, the handicap hypothesis aims to explain earlier-than-average singing.

Our personal view is that the best-supported hypotheses for early singing can be gathered around three complementary lines of explanation. Firstly, energetic constraints (inefficient foraging at dawn and unpredictable overnight conditions—hypotheses 6 and 7) allow a window of time in which singing has a low cost. Secondly, and depending on the species, it pays males to sing early to manipulate female mating (hypothesis 3) or settle territory boundaries and assert ownership (hypothesis 4). And, thirdly, costs of early singing (hypothesis 5) should push song timing earlier than optimal through a handicap mechanism that prevents dishonest signalling (but see Hutchinson et al. 1993).

One interesting possibility is that a main selective pressure on dawn singing is synchronicity in signalling. This would not explain why it pays to sing at dawn, but rather that there is strong selection for signalling in synchrony. Synchronous signalling results in a dilution of predation risk and a decrease in invasion rate (Kacelnik and Krebs 1983), allows ritualization and comparison among signallers (West-Eberhard 1979; Zahavi 1980), as well as favours the development of useful communication networks (Burt and Vehrencamp 2005; McGregor 2005; Mennill et al. 2003). We believe that this possibility has not been sufficiently explored.

In the near future, the development of new techniques for experimentally manipulating hormone levels, as well as remote-sensing devices for assessing singing performance (Sugai et al. 2019) or studies integrating singing with spatial movements of conspecifics (Amy et al. 2010), may offer suitable methods to test hypotheses that are still poorly explored (e.g., circadian cycles and handicap hypotheses). Further studies will also require detailed data at the individual scale and comparative approaches, which can provide more precision and a more relevant baseline to disentangle factors determining species- and population-specific patterns.

Ernst Mayr (1997) concluded that most, if not all, biological processes are probabilistic, the result of several factors with a high degree of randomness. This pluralism naturally leads to multiple hypotheses being correct at the same time, often including the traditional bimodal perspective in biology (i.e. proximal vs. evolutionary explanations). We believe that research on the bird dawn chorus represents a good example of this pluralism of factors and illustrates how several hypotheses can coexist simultaneously. We hope that this review may help future research to refine our knowledge of this puzzling phenomenon.

Acknowledgments We wish to thank Peter Slater, Barbara Helm, Marc Naguib, John Hutchinson, Valentin Amrhein, Constantino Macías, Thierry Aubin and Nicolas Mathevon for their insightful comments on previous versions of this review. Thiago Bicudo kindly provided the Brazilian data for Fig. 3.1. DL acknowledges post-doctoral study grants by the Fondation Fyssen (France) and the Comunidad de Madrid (Atracción de Talento Investigador, 2016-T2/AMB-1722, CAM, Spain). DG research was supported during the writing of this review by research grant CGL2017-83843-C2-1-P from the Spanish Ministry for Science, and DL research by grant CGL2017-88764-R from the Spanish Ministry for Science.

References

Allard HA (1930) The first morning song of some birds of Washington, DC: its relation to light. Am Nat 64:436–469

Allen FH (1913) More notes on the morning awakening. Auk 30:229–235

Alward BA, Cornil CA, Balthazart J, Ball GF (2018) The regulation of birdsong by testosterone: multiple time-scales and multiple sites of action. Horm Behav 104:32–40

Amrhein V, Erne N (2006) Dawn singing reflects past territorial challenges in the winter wren. Anim Behav 71:1075–1080

Amrhein V, Lerch S (2010) Differential effects of moving versus stationary territorial intruders on territory defence in a songbird. J Anim Ecol 79:82–87

Amrhein V, Korner P, Naguib M (2002) Nocturnal and diurnal singing activity in the nightingale: correlations with mating status and breeding cycle. Anim Behav 64:939–944

Amrhein V, Kunc HP, Naguib M (2004a) Non-territorial nightingales prospect territories during the dawn chorus. Proc R Soc B 271:S167–S169

Amrhein V, Kunc HP, Naguib M (2004b) Seasonal patterns of singing activity vary with time of day in the nightingale (*Luscinia megarhynchos*). Auk 121:110–117

Amrhein V, Johannessen LE, Kristiansen L, Slagsvold T (2008) Reproductive strategy and singing activity: blue tit and great tit compared. Behav Ecol Sociobiol 62:1633–1641

Amy M, Sprau P, de Goede P, Naguib M (2010) Effects of personality on territory defence in communication networks: a playback experiment with radio-tagged great tits. Proc R Soc B 277:3685–3692

Armstrong EA (1963) A study of bird song. Oxford University Press, London

Arroyo-Solis A, Castillo JM, Figueroa E, Lopez-Sanchez JL, Slabbekoorn H (2013) Experimental evidence for an impact of anthropogenic noise on dawn chorus timing in urban birds. J Avian Biol 44:288–296

Aschoff J (1966) Circadian activity pattern with two peaks. Ecology 47:657–662

Aschoff J, Meyer-Lohmann J (1955) Die Aktivität gekäfigter Grünfinken im 24-Stunden-Tag bei unterschiedlich langer Lichtzeit mit und ohne Dämmerung. Z Tierpsychol 12:254–265

Atienza JC, Illera JC (1997) Tree species selection to perform singing and foraging behaviour by Great and Blue Tits: a trade-off between food gathering and territorial behaviour? Bird Study 44:117–119

Au WW, Mobley J, Burgess WC, Lammers MO, Nachtigall PE (2000) Seasonal and diurnal trends of chorusing humpback whales wintering in waters off western Maui. Mar Mamm Sci 16:530–544

Avery MI, Krebs JR (1984) Temperature and foraging success of great tits (*Parus major*) hunting for spiders. Ibis 126:33–38

Ballentine B, Hyman J, Nowicki S (2004) Vocal performance influences female response to male bird song: an experimental test. Behav Ecol 15:163–168

Barnett CA, Briskie JV (2007) Energetic state and the performance of dawn chorus in silvereyes (*Zosterops lateralis*). Behav Ecol Sociobiol 61:579–587

Beaulieu M, Sockman KW (2012) Song in the cold is 'hot': memory of and preference for sexual signals perceived under thermal challenge. Biol Lett 8:751–753

Belinsky KL, Hogle JL, Schmidt KA (2012) Veeries experience more varied acoustic competition at dawn than at dusk. Wilson J Ornithol 124:265–269

Bentley GE (2001) Unraveling the enigma: the role of melatonin in seasonal processes in birds. Microsc Res Tech 53:63–71

Berg KS, Brumfield RT, Apanius V (2006) Phylogenetic and ecological determinants of the neotropical dawn chorus. Proc R Soc B 273:999–1005

Bergen F, Abs M (1997) Verhaltensökologie Studie zur Gesangsaktivität von Blaumeise (*Parus caeruleus*), Kohlmeise (*Parus major*) und Buchfink (*Fringilla coelebs*) in einer Grosstadt. J Ornithol 138:451–467

Berglund A, Bisazza A, Pilastro A (1996) Armaments and ornaments: an evolutionary explanation of traits of dual utility. Biol J Linn Soc 58:385–399

Berthold P (1973) Relationships between migratory restlessness and migration distance in six Sylvia species. Ibis 115:594–599

Biard C, Gil D, Karadas F, Saino N, Spottiswoode CN, Surai PF, Moller AP (2009) Maternal effects mediated by antioxidants and the evolution of carotenoid-based signals in birds. Am Nat 174:696–708

Biernaskie JM, Perry JC, Grafen A (2018) A general model of biological signals, from cues to handicaps. Evol Lett 2:201–209

Birkhead TR, Møller AP (1992) Sperm competition in birds: evolutionary causes and consequences. Academic, London

Birkhead TR, Møller AP (1998) Sperm competition and sexual selection. Academic, London

Birkhead TR, Cunningham EJA, Cheng KM (1996) The insemination window provides a distorted view of sperm competition in birds. Proc R Soc B 263:1187–1192

Bolsinger JS (2000) Use of two song categories by golden-cheeked Warblers. Condor 102:539–552

Brandt MJ, Cresswell W (2009) Diurnal foraging routines in a tropical bird, the rock finch Lagonosticta sanguinodorsalis: how important is predation risk? J Avian Biol 40:90–94

Brenowitz EA (1982) The active space of red-winged blackbird song. J Comp Physiol 147:511–522

Brockway BF (1967) The influence of vocal behavior on the performer's testicular activity in budgerigars (*Melopsittacus undulatus*). Wilson Bull 79:328–334

Brown TJ, Handford P (2003) Why birds sing at dawn: the role of consistent song transmission. Ibis 145:120–129

Brumm H (2006) Signalling through acoustic windows: nightingales avoid interspecific competition by short-term adjustment of song timing. J Comp Physiol A 192:1279–1285

Bruni A, Foote JR (2014) Dawn singing of eastern phoebes varies with breeding stage and brood number. Wilson J Ornithol 126:500–507

Bruni A, Mennill DJ, Foote JR (2014) Dawn chorus start time variation in a temperate bird community: relationships with seasonality, weather, and ambient light. J Ornithol 155:877–890

Brunner P, Pasinelli G (2010) Variation in singing style use in the reed bunting *Emberiza schoeniclus*: influencing factors and possible functions. J Avian Biol 41:388–397

Burt JM, Vehrencamp SL (2005) Dawn chorus as an interactive communication network. In: McGregor PK (ed) Animal communication networks. Cambridge University Press, Cambridge, pp 320–343

Byers BE (1996) Messages encoded in the songs of chestnut-sided warblers. Anim Behav 52:691–705

Cassone VM, Westneat DF (2012) The bird of time: cognition and the avian biological clock. Front Mol Neurosci 5:32

Cassone VM, Bartell PA, Earnest BJ, Kumar V (2008) Duration of melatonin regulates seasonal changes in song control nuclei of the House Sparrow, *Passer domesticus*: independence from gonads and circadian entrainment. J Biol Rhythms 23:49–58

Catchpole CK (1973) The functions of advertising song in the sedge warbler (*Acrocephalus schoenobaenus*) and the reed warbler (*A. scirpaceus*). Behaviour 46:300–319

Catchpole CK (1982) The evolution of bird sounds in relation to mating and spacing behaviour. In: Kroodsma DE, Miller EH (eds) Acoustic communication in birds. Vol 1. Production, perception and design features of sounds. Academic, New York, pp 253–295
Catchpole CK, Slater PJB (2008) Bird song: biological themes and variations, 2nd edn. Cambridge University Press, Cambridge
Chen W-M, Lee Y-F, Tsai C-F, Yao C-T, Chen Y-H, Li S-H, Kuo Y-M (2015) Dawn chorus variation in East-Asian tropical montane forest birds and its ecological and morphological correlates. Contrib Zool 84:255–265
Cheng M-F (1992) For whom does the female dove coo? A case for the role of vocal self-stimulation. Anim Behav 43:1035–1044
Cirelli C, Tononi G (2008) Is sleep essential? PLoS Biol 6:e216
Cody ML, Brown JH (1969) Song asynchrony in neighbouring bird species. Nature 222:778
Collias NE, Collias EC (1967) A field study of the red jungle fowl in north-central India. Condor 69:360–386
Collins SA, de Kort SR, Perez-Tris J, Telleria JL (2009) Migration strategy and divergent sexual selection on bird song. Proc R Soc B 276:585–590
Cotton S, Fowler K, Pomiankowski A (2004) Do sexual ornaments demonstrate heightened condition-dependent expression as predicted by the handicap hypothesis? Proc R Soc B 271:771–783
Cresswell W (1994) Song as a pursuit deterrent signal, and its occurrence relative to other anti-predation behaviors of skylark (*Alauda arvensis*) on attack by merlins (*Falco columbarius*). Behav Ecol Sociobiol 34:217–223
Cunningham EJA, Russell AF (2000) Egg investment is influenced by male attractiveness in the mallard. Nature 404:74–77
Curio E (1959) Verhaltensstudien am Trauerschnäpper: Beiträge zur Ethologie und Ökologie von Musicapa h. hypoleuca Pallas. Z Tierpsychol (Beiheft) 3:1–118
Cuthill IC, Macdonald WA (1990) Experimental manipulation of the dawn and dusk chorus in the blackbird *Turdus merula*. Behav Ecol Sociobiol 26:209–216
D'Spain G, Batchelor H (2006) Observations of biological choruses in the Southern California Bight: a chorus at midfrequencies. J Acoust Soc Am 120:1942–1955
Da Silva A, Kempenaers B (2017) Singing from North to South: latitudinal variation in timing of dawn singing under natural and artificial light conditions. J Anim Ecol 86:1286–1297
Da Silva A, Samplonius JM, Schlicht E, Valcu M, Kempenaers B (2014) Artificial night lighting rather than traffic noise affects the daily timing of dawn and dusk singing in common European songbirds. Behav Ecol 25:1037–1047
Da Silva A, Valcu M, Kempenaers B (2015) Light pollution alters the phenology of dawn and dusk singing in common European songbirds. Philos Trans R Soc Lond B Biol Sci 20140126:370
Da Silva A, Valcu M, Kempenaers B (2016) Behavioural plasticity in the onset of dawn song under intermittent experimental night lighting. Anim Behav 117:155–165
Da Silva A, de Jong M, van Grunsven RHA, Visser ME, Kempenaers B, Spoelstra K (2017) Experimental illumination of a forest: no effects of lights of different colours on the onset of the dawn chorus in songbirds. R Soc Open Sci 4:16063
Daan S, Aschoff J (1975) Circadian rhythms of locomotor activity in captive birds and mammals: their variations with season and latitude. Oecologia 18:269–316
Dabelsteen T (1992) Interactive playback: a finely tuned response. In: McGregor PK (ed) Playback and studies of animal communication. Springer, New York, pp 97–110
Dabelsteen T, Mathevon N (2002) Why do songbirds sing intensively at dawn? Acta Ethol 4:65–72
Dalziell AH, Cockburn A (2008) Dawn song in superb fairy-wrens: a bird that seeks extrapair copulations during the dawn chorus. Anim Behav 75:489–500
Demko AD, Reitsma LR, Staicer CA (2013) Two song categories in the canada warbler (*Cardellina canadensis*). Auk 130:609–616
Doggett F (1974) Romanticism's singing bird. Stud Engl Lit 1500–1900(14):547–561

Dolan AC, Murphy MT, Redmond LJ, Sexton K, Duffield D (2007) Extrapair paternity and the opportunity for sexual selection in a socially monogamous passerine. Behav Ecol 18:985–993

Dominoni DM, Goymann W, Helm B, Partecke J (2013a) Urban-like night illumination reduces melatonin release in European blackbirds (*Turdus merula*): implications of city life for biological time-keeping of songbirds. Front Zool 10:1–10

Dominoni DM, Helm B, Lehmann M, Dowse HB, Partecke J (2013b) Clocks for the city: circadian differences between forest and city songbirds. Proc R Soc B 280:20130593

Dominoni DM, Greif S, Nemeth E, Brumm H (2016) Airport noise predicts song timing of European birds. Ecol Evol 6:6151–6159

Dorado-Correa AM, Rodriguez-Rocha M, Brumm H (2016) Anthropogenic noise, but not artificial light levels predicts song behaviour in an equatorial bird. R Soc Open Sci 3:160231–160231

Double M, Cockburn A (2000) Pre-dawn infidelity: females control extra-pair mating in superb fairy-wrens. Proc R Soc B 267:465–470

Drachmann J, Komdeur J, Boomsma JJ (1997) Copulation behaviour in the linnet *Carduelis cannabina* and the insemination window hypothesis. J Avian Biol 28:191–196

Dunnett GE, Hinde RA (1953) The winter roosting and awakening behaviour of captive Great Tits. Br J Anim Behav 1:91–95

Eriksson D, Wallin L (1986) Male bird song attracts females: a field experiment. Behav Ecol Sociobiol 19:297–299

Erne N, Amrhein V (2008) Long-term influence of simulated territorial intrusions on dawn and dusk singing in the Winter Wren: spring versus autumn. J Ornithol 149:479–486

Falk H, Biebach H, Krebs JR (1992) Learning a time-place pattern of food availability: a comparison between an insectivorous and a granivorous weaver species (*Ploceus bicolor* and *Euplectes hordeaceus*). Behav Ecol Sociobiol 31:9–15

Falls JB (1988) Does song deter territorial intrusion in White-throated Sparrows (*Zonotrichia albicollis*)? Can J Zool 66:206–211

Farina A, Ceraulo M, Bobryk C, Pieretti N, Quinci E, Lattanzi E (2015) Spatial and temporal variation of bird dawn chorus and successive acoustic morning activity in a Mediterranean landscape. Bioacoustics 24:269–288

Fernandez-Juricic E, Tran E (2007) Changes in vigilance and foraging behaviour with light intensity and their effects on food intake and predator detection in house finches. Anim Behav 74:1381–1390

Fisler GF (1962) Variation in the morning awakening time of some birds in south-central Michigan. Condor 64:184–198

Foerster K, Poesel A, Kunc H, Kempenaers B (2002) The natural plasma testosterone profile of male blue tits during the breeding season and its relation to song output. J Avian Biol 33:269–275

Foote JR, Fitzsimmons LP, Mennill DJ, Ratcliffe LM (2008a) Male chickadees match neighbors interactively at dawn: support for the social dynamics hypothesis. Behav Ecol 19:1192–1199

Foote JR, Fitzsimmons LP, Mennill DJ, Ratcliffe LM (2008b) Tied to the nest: male black-capped chickadees decrease dawn chorus movement behaviour when their mate is fertile. Anim Behav 76:1227–1233

Foote JR, Fitzsimmons LP, Mennill DJ, Ratcliffe LM (2010) Black-capped chickadee dawn choruses are interactive communication networks. Behaviour 147:1219–1248

Foote JR, Fitzsimmons LP, Mennill DJ, Ratcliffe LM (2011) Male black-capped chickadees begin dawn chorusing earlier in response to simulated territorial insertions. Anim Behav 81:871–877

Forstmeier W, Balsby TJ (2002) Why mated dusky warblers sing so much: territory guarding and male quality announcement. Behaviour 139:89–111

Fuller RA, Warren PH, Gaston KJ (2007) Daytime noise predicts nocturnal singing in urban robins. Biol Lett 3:368–370

Garamszegi LZ, Moller AP, Erritzoe J (2002) Coevolving avian eye size and brain size in relation to prey capture and nocturnality. Proc R Soc B 269:961–967

Garson PJ, Hunter ML (1979) Effects of temperature and time of year on the singing behaviour of wrens *Troglodytes troglodytes* and great tits *Parus major*. Ibis 121:481–487
Gaunt AS, Bucher TL, Gaunt SLL, Baptista LF (1996) Is singing costly? Auk 113:718–721
Gil D (2008) Hormones in avian eggs: physiology, ecology and behavior. Adv Stud Behav 38:337–398
Gil D, Gahr M (2002) The honesty of bird song: multiple constraints for multiple traits. Trends Ecol Evol 17:133–141
Gil D, Graves JA, Hazon N, Wells A (1999a) Male attractiveness and differential testosterone investment in zebra finch eggs. Science 286:126–128
Gil D, Graves JA, Slater PJB (1999b) Seasonal patterns of singing in the willow warbler: evidence against the fertility announcement hypothesis. Anim Behav 58:995–1000
Gil D, Leboucher G, Lacroix A, Cue R, Kreutzer M (2004) Female canaries produce eggs with greater amounts of testosterone when exposed to preferred male song. Horm Behav 45:64–70
Gil D, Honarmand M, Pascual J, Pérez-Mena E, Macías-García C (2015) Birds living near airports advance their dawn-chorus and reduce overlap with aircraft noise. Behav Ecol 26:435–443
Grafen A (1990) Biological signals as handicaps. J Theor Biol 144:517–546
Grava T, Grava A, Otter KA (2009) Supplemental feeding and dawn singing in black-capped chickadees. Condor 111:560–564
Green AJ (1990) Determinants of chorus participation and the effects of size, weight and competition on advertisement calling in the tungara frog, *Physalaemus pustulosus* (Leptodactylidae). Anim Behav 39:620–638
Greenfield MD (2015) Signal interactions and interference in insect choruses: singing and listening in the social environment. J Comp Physiol A 201:143–154
Greig-Smith P (1983) Use of perches as vantage points during foraging by male and female stonechats *Saxicola torquata*. Behaviour 86:215–236
Greives TJ, Kingma SA, Beltrami G, Hau M (2012) Melatonin delays clutch initiation in a wild songbird. Biol Lett 8:330–332
Greives TJ et al (2015) Costs of sleeping in: circadian rhythms influence cuckoldry risk in a songbird. Funct Ecol 29:1300–1307
Gwinner E, Brandstatter R (2001) Complex bird clocks. Philos Trans R Soc Lond B Biol Sci 356:1801–1810
Haldar C, Singh SS (2001) Seasonal changes in melatonin and immunological adaptations in birds. J Endocrinol Reprod 5:13–24
Halfwerk W, Bot S, Buikx J, van der Velde M, Komdeur J, ten Cate C, Slabbekoorn H (2011) Low-frequency songs lose their potency in noisy urban conditions. Proc Natl Acad Sci U S A 108:14549–14554
Hamilton WD (1971) Geometry for the selfish herd. J Theor Biol 31:295–311
Hart PJ, Hall R, Ray W, Beck A, Zook J (2015) Cicadas impact bird communication in a noisy tropical rainforest. Behav Ecol 26:839–842
Hasan NM (2011) Effect of seasonal variations, altitude and geographical location on the onset of dawn chorus in three bird species in the Middle East. Open Ornithol J 4:30–34
Haskell D (1994) Experimental evidence that nestling begging behaviour incurs a cost due to nest predation. Proc R Soc B 257:161–164
Hau M (2007) Regulation of male traits by testosterone: implications for the evolution of vertebrate life histories. Bioessays 29:133–144
Hau M et al (2017) Timing as a sexually selected trait: the right mate at the right moment. Philos Trans R Soc Lond B Biol Sci 372:20160249
Helm B, Visser ME (2010) Heritable circadian period length in a wild bird population. Proc R Soc B 277:3335–3342
Henwood K, Fabrick A (1979) Quantitative-analysis of the dawn chorus: temporal selection for communicatory optimization. Am Nat 114:260–274
Hodgson L, Waas JR, Foote JR (2018) Early singers attend to conspecific but not heterospecific behavioural cues at dawn. J Avian Biol 49:e01749

Hutchinson JMC (2002) Two explanations of the dawn chorus compared: how monotonically changing light levels favour a short break from singing. Anim Behav 64:527–539
Hutchinson JMC, McNamara JM (2000) Ways to test stochastic dynamic programming models empirically. Anim Behav 59:665–676
Hutchinson JMC, McNamara JM, Cuthill IC (1993) Song, sexual selection, starvation and strategic handicaps. Anim Behav 45:1153–1177
Kacelnik A (1979) Foraging efficiency of great tits (*Parus major L.*) in relation to light-intensity. Anim Behav 27:237–241
Kacelnik A, Krebs JR (1983) The dawn chorus in the great tit (*Parus major*): proximate and ultimate causes. Behaviour 83:287–309
Karasov WH (1990) Digestion in birds: chemical and physiological determinants and ecological implications. In: Morrison ML, Ralph CJ, Verner J, Jehl JR (eds) Avian foraging: theory, methodology, and applications. Studies in avian biology, vol 13. Cooper Ornithological Society, Lawrence, pp 391–415
Keast A (1994a) The dawn chorus in a eucalypt forest bird community, seasonal shifts in timing and contribution of individual species. Corella 18:133–140
Keast A (1994b) Temporal vocalization patterns in members of a eucalypt forest bird community: the effects of weather on song production. Emu 94:172–180
Kempenaers B, Borgstroem P, Loes P, Schlicht E, Valcu M (2010) Artificial night lighting affects dawn song, etra-pair siring success, and lay date in songbirds. Curr Biol 20:1735–1739
Kirschel ANG, Blumstein DT, Cohen RE, Buermann W, Smith TB, Slabbekoorn H (2009) Birdsong tuned to the environment: green hylia song varies with elevation, tree cover, and noise. Behav Ecol 20:1089–1095
Klett-Mingo JI, Pavón I, Gil D (2016) Great tits (*Parus major*) increase vigilance time and reduce feeding effort during peaks of aircraft noise. Anim Behav 115:29–34
Klump GM (1996) Bird communication in the noisy world. In: Kroodsma DE, Miller EH (eds) Ecology and evolution of acoustic communication in birds. Cornell University, Ithaca, pp 321–338
Kokko H, Jennions MD, Brooks R (2006) Unifying and testing models of sexual selection. Annu Rev Ecol Evol Syst 37:43–66
Krams I (2001) Communication in crested tits and the risk of predation. Anim Behav 61:1065–1068
Krebs JR, Ashcroft R, Webber M (1978) Song repertoires and territory defence in the Great Tit. Nature 271:539–542
Kroodsma DE, Byers BE (1991) The function(s) of bird song. Am Zool 31:318–328
Kroodsma DE, Bereson RC, Byers BE, Minear E (1989) Use of song types by the Chestnut-sided Warbler: evidence for both intra- and inter-sexual functions. Can J Zool 67:447–456
Kunc HP, Amrhein V, Naguib M (2005) Seasonal variation in dawn song characteristics in the common nightingale. Anim Behav 70:1265–1271
Kunc HP, Foerster K, Vermeirssen ELM, Kempenaers B (2006) Experimentally elevated plasma testosterone levels do not influence singing behaviour of male blue tits (*Parus caeruleus*) during the early breeding season. Ethology 112:984–992
Lambrechts M, Dhondt AA (1986) Male quality, reproduction, and survival in the great tit (*Parus major*). Behav Ecol Sociobiol 19:57–63
Langemann U, Gauger B, Klump GM (1998) Auditory sensitivity in the great tit: perception of signals in the presence and absence of noise. Anim Behav 56:763–769
Langmore NE (1998) Functions of duet and solo songs of female birds. Trends Ecol Evol 13:136–140
Larom D, Garstang M, Payne K, Raspet R, Lindeque M (1997) The influence of surface atmospheric conditions on the range and area reached by animal vocalizations. J Exp Biol 200:421–431
Laucht S, Dale J, Mutzel A, Kempenaers B (2011) Individual variation in plasma testosterone levels and its relation to badge size in house sparrows *Passer domesticus*: it's a night-and-day difference. Gen Comp Endocrinol 170:501–508

Lein MR (2007) Patterns of dawn singing by Buff-breasted Flycatchers. J Field Ornithol 78:343–351
Leopold A, Eynon AE (1961) Avian daybreak and evening song in relation to light intensity. Condor 63:269–293
Lesku JA et al (2012) Adaptive sleep loss in polygynous pectoral sandpipers. Science 337:1654–1658
Lima SL (2009) Predators and the breeding bird: behavioral and reproductive flexibility under the risk of predation. Biol Rev 84:485–513
Liu W-C (2004) The effect of neighbours and females on dawn and daytime singing behaviours by male chipping sparrows. Anim Behav 68:39–44
Liu W-C, Kroodsma DE (2007) Dawn and daytime singing behavior of chipping sparrows (*Spizella passerina*). Auk 124:44–52
Lourenço R, Goytre F, del Mar Delgado M, Thornton M, Rabaça JE, Penteriani V (2013) Tawny owl vocal activity is constrained by predation risk. J Avian Biol 44:461–468
Luther DA (2008) Signaller: receiver coordination and the timing of communication in Amazonian birds. Biol Lett 4:651–654
Luther D (2009) The influence of the acoustic community on songs of birds in a neotropical rain forest. Behav Ecol 20:864–871
Mace R (1986) Importance of female behavior in the dawn chorus. Anim Behav 34:621–622
Mace R (1987a) The dawn chorus in the great tit *Parus major* is directly related to female fertility. Nature 330:745–746
Mace R (1987b) Why do birds sing at dawn? Ardea 75:123–132
Mace R (1989) The relationship between daily routines of singing and foraging: an experiment on captive great tits *Parus major*. Ibis 131:415–420
Martin GR (1993) Producing the image. In: Zeiger HP, Bischof HJ (eds) Vision, brain and behaviour in birds. MIT Press, Cambridge, pp 5–24
Martínez-Ortega C, Santos ESA, Gil D (2014) Species-specific differences in relative eye size are related to patterns of edge avoidance in an Amazonian rainforest bird community. Ecol Evol 19:3736–3745
Mayr E (1997) This is biology: the science of the living world. Cambridge University Press, Cambridge
McCabe RA (1951) The song and song-flight of the Alder Flycatcher. Wilson Bull 63:89–98
McDonald MV (1989) Function of song in Scott's seaside sparrow *Ammodramus maritimus peninsulae*. Anim Behav 38:468–485
McGregor PK (2005) Animal communication networks. Cambridge University Press, Cambridge
McNamara JM, Mace RH, Houston AI (1987) Optimal daily routines of singing and foraging in a bird singing to attract a mate. Behav Ecol Sociobiol 20:399–405
Mennill DJ, Boag PT, Ratcliffe LM (2003) The reproductive choices of eavesdropping female black-capped chickadees, *Poecile atricapillus*. Naturwissenschaften 90:577–582
Miller MW (2006) Apparent effects of light pollution on singing behavior of American robins. Condor 108:130–139
Møller AP (1988) Spatial and temporal distribution of song in the yellowhammer *Emberiza citrinella*. Ethology 78:321–331
Møller AP (1990) Changes in the size of avian breeding territories in relation to the nesting cycle. Anim Behav 40:1070–1079
Møller AP (1991) Why mated songbirds sing so much: mate guarding and male announcement of mate fertility status. Am Nat 138:994–1014
Møller AP (1992) Interspecific response to playback of bird song. Ethology 90:315–320
Møller AP (2011) Song post height in relation to predator diversity and urbanization. Ethology 117:529–538
Møller AP, Erritzøe J (2010) Flight distance and eye size in birds. Ethology 116:458–465
Montgomerie RD (1985) Why do birds sing at dawn? In: Proceedings of the 19th International Ethol Conference, p 242

Morse DH (1989) Song patterns of warblers at dawn and dusk. Wilson Bull 101:26–35
Morton ES (1975) Ecological sources of selection on avian sounds. Am Nat 109:17–34
Mountjoy DJ, Lemon RE (1991) Song as an attractant for male and female European starlings, and the influence of song complexity on their response. Behav Ecol Sociobiol 28:97–100
Murphy MT, Sexton K, Dolan AC, Redmond LJ (2008) Dawn song of the eastern kingbird: an honest signal of male quality? Anim Behav 75:1075–1084
Naguib M, Fichtel C, Todt D (1999) Nightingales respond more strongly to vocal leaders of simulated dyadic interactions. Proc Biol Sci 266:537–543
Naguib M, van Rooij EP, Snijders L, Van Oers K (2016) To sing or not to sing: seasonal changes in singing vary with personality in wild great tits. Behav Ecol 27:932–938
Needham KB, Dochtermann NA, Greives TJ (2017) Consistent individual variation in day, night, and GnRH-induced testosterone concentrations in house sparrows (*Passer domesticus*). Gen Comp Endocrinol 246:211–217
Nelson DA, Croner LJ (1991) Song categories and their functions in the field sparrow (*Spizella pusilla*). Auk 108:42–52
Nordt A, Klenke R (2013) Sleepless in town: drivers of the temporal shift in dawn song in urban European blackbirds. PLoS One 8:e71476
Odom KJ, Hall ML, Riebel K, Omland KE, Langmore NE (2014) Female song is widespread and ancestral in songbirds. Nat Commun 5:3379
Ord TJ (2008) Dawn and dusk "chorus" in visually communicating Jamaican anole lizards. Am Nat 172:585–592
Otter K, Ratcliffe L (1993) Changes in singing behavior of male black-capped chickadees (*Parus atricapillus)* following mate removal. Behav Ecol Sociobiol 33:409–414
Otter K, Chruszcz B, Ratcliffe L (1997) Honest advertisement and song output during the dawn chorus of black-capped chickadees. Behav Ecol 8:167–173
Ottinger MA, Follett BK (1987) Circulating levels of luteinizing hormone and androgen in mature male Japanese Quail over 24 hours. Poult Sci 66:1411–1413
Parker TH, Tillin H (2006) Male Blue Tits *Cyanistes caeruleus* choose early-leafing tree species during spring dawn chorus. Bird Study 53:253–257
Pärt T (1991) Is dawn singing related to paternity insurance: the case of the collared flycatcher. Anim Behav 41:451–456
Peters J, Lebrasseur O, Deng H, Larson G (2016) Holocene cultural history of Red jungle fowl (*Gallus gallus*) and its domestic descendant in East Asia. Quat Sci Rev 142:102–119
Phelps SM, Rand AS, Ryan MJ (2006) The mixed-species chorus as public information: túngara frogs eavesdrop on a heterospecific. Behav Ecol 18:108–114
Pinxten R, Eens M (1998) Male starlings sing most in the late morning, following egg-laying: a strategy to protect their paternity? Behaviour 135:1197–1211
Planqué R, Slabbekoorn H (2008) Spectral overlap in songs and temporal avoidance in a Peruvian bird assemblage. Ethology 114:262–271
Poesel A, Foerster K, Kempenaers B (2001) The dawn song of the blue tit *Parus caeruleus* and its role in sexual selection. Ethology 107:521–531
Poesel A, Kunc HP, Foerster K, Johnsen A, Kempenaers B (2006) Early birds are sexy: male age, dawn song and extrapair paternity in blue tits, *Cyanistes (formerly Parus) caeruleus*. Anim Behav 72:531–538
Poesel A, Dabelsteen T, Pedersen SB (2007) Implications of conspecific background noise for features of blue tit, *Cyanistes caeruleus*, communication networks at dawn. J Ornithol 148:123–128
Pohl NU, Slabbekoorn H, Klump GM, Langemann U (2009) Effects of signal features and environmental noise on signal detection in the great tit, *Parus major*. Anim Behav 78:1293–1300
Pollard AL (2009) Visual constraints upon avian behaviour. PhD thesis. School of Biosciences, Cardiff University
Popp JW, Ficken RW, Reinartz JA (1985) Short-term temporal avoidance of interspecific acoustic interference among forest birds. Auk 102:744–748

Quispe R, Sèbe F, Da Silva ML, Gahr M (2016) Dawn-song onset coincides with increased HVC androgen receptor expression but is decoupled from high circulating testosterone in an equatorial songbird. Physiol Behav 156:1–7

Rattner BA, Sileo L, Scanes CG (1982) Oviposition and the plasma concentrations of LH, progesterone and corticosterone in bobwhite quail (*Colinus virginianus*) fed parathion. J Reprod Fertil 66:147–155

Refinetti R (2016) Circadian physiology. CRC, Boca Raton

Richards DG, Wiley RH (1980) Reverberations and amplitude fluctuations in the propagation of sound in a forest: implications for animal communication. Am Nat 115:381–399

Riebel K, Hall ML, Langmore NE (2005) Female songbirds still struggling to be heard. Trends Ecol Evol 20:419–420

Rodrigues M (1996) Song activity in the chiffchaff: territorial defence or mate guarding? Anim Behav 51:709–716

Rodríguez AB, Terrón MP, Durán J, Ortega E, Barriga C (2001) Physiological concentrations of melatonin and corticosterone affect phagocytosis and oxidative metabolism of ring dove heterophils. J Pineal Res 31:31–38

Roth TC, Lesku JA, Amlaner CJ, Lima SL (2006) A phylogenetic analysis of the correlates of sleep in birds. J Sleep Res 15:395–402

Roth T, Sprau P, Schmidt R, Naguib M, Amrhein V (2009) Sex-specific timing of mate searching and territory prospecting in the nightingale: nocturnal life of females. Proc R Soc B 276:2045–2050

Roth T, Sprau P, Naguib M, Amrhein V (2012) Sexually selected signaling in birds: a case for Bayesian change-point analysis of behavioral routines. Auk 129:660–669

Rusak B (2013) Vertebrate behavioral rhythms. In: Aschoff J (ed) Biological rhythms. Plenum, New York

Russ A, Rueger A, Klenke R (2015) Seize the night: European Blackbirds (*Turdus merula*) extend their foraging activity under artificial illumination. J Ornithol 156:123–131

Saggese K, Korner-Nievergelt F, Slagsvold T, Amrhein V (2011) Wild bird feeding delays start of dawn singing in the great tit. Anim Behav 81:361–365

Scheer G (1951) Uber die zeitliche Differenz zwischen Erwachen und Gesangsbeginn. Die Vogelwarte 16:13–15

Scheer G (1952) Beobachtungen und Untersuchungen über die Abhangigkeit des Frühgesangs der Vogel von inneren und äußeren Faktoren. Biol Abhandlungen 3(4):1–68

Schmidt KA, Belinsky KL (2013) Voices in the dark: predation risk by owls influences dusk singing in a diurnal passerine. Behav Ecol Sociobiol 67:1837–1843

Seltmann S, Trost L, Ter Maat A, Gahr M (2016) Natural melatonin fluctuation and its minimally invasive simulation in the zebra finch. Peer J 4:e1939

Sexton K, Murphy MT, Redmond LJ, Dolan AC (2007) Dawn song of eastern kingbirds: intra-population variability and sociobiological correlates. Behaviour 144:1273–1295

Shakespeare W (1597) An excellent conceited tragedie of Romeo and Juliet. John Danter, London

Sheldon BC, Burke T (1994) Copulation behavior and paternity in the chaffinch. Behav Ecol Sociobiol 1994:149–156

Shimmura T, Yoshimura T (2013) Circadian clock determines the timing of rooster crowing. Curr Biol 23:R231–R233

Shimmura T, Ohashi S, Yoshimura T (2015) The highest-ranking rooster has priority to announce the break of dawn. Sci Rep 5:11683

Siegel JM (2009) Sleep viewed as a state of adaptive inactivity. Nat Rev Neurosci 10:747

Sierro J, Schloesing E, Pavón I, Gil D (2017) European blackbirds exposed to aircraft noise advance their chorus, modify their song and spend more time singing. Front Ecol Evol 5:68

Slagsvold T (1977) Bird song activity in relation to breeding cycle, spring weather, and environmental phenology. Ornis Scand 8:197–222

Slagsvold T (1996) Dawn and dusk singing of male American Robins in relation to female behavior. Wilson Bull 108:507–515

Snijders L, van Rooij EP, Henskens MFA, van Oers K, Naguib M (2015) Dawn song predicts behaviour during territory conflicts in personality-typed great tits. Anim Behav 109:45–52

Spector DA (1992) Wood-warbler song systems. In: Power DM (ed) Current Ornithology, vol 9. Springer, Boston, MA, pp 199–238

Staicer CE, Spector DA, Horn AG (1996) The dawn chorus and other diel patterns in acoustic signalling. In: Kroodsma DE, Miller EH (eds) Ecology and evolution of acoustic communication in birds. Cornell University Press, Ithaca, pp 426–453

Stanley CQ, Walter MH, Venkatraman MX, Wilkinson GS (2016) Insect noise avoidance in the dawn chorus of Neotropical birds. Anim Behav 112:255–265

Steinmeyer C, Schielzeth H, Mueller JC, Kempenaers B (2010) Variation in sleep behaviour in free-living blue tits, *Cyanistes caeruleus*: effects of sex, age and environment. Anim Behav 80:853–864

Sugai LSM, Silva TSF, Ribeiro JW Jr, Llusia D (2019) Terrestrial passive acoustic monitoring: review and perspectives. Bioscience 69:15–25

Thomas RJ (1999) The effect of variability in the food supply on the daily singing routines of European robins: a test of a stochastic dynamic programming model. Anim Behav 57:365–369

Thomas RJ (2002) The costs of singing in nightingales. Anim Behav 63:959–966

Thomas RJ, Cuthill IC (2002) Body mass regulation and the daily singing routines of European robins. Anim Behav 63:285–292

Thomas RJ, Szekely T, Cuthill IC, Harper DGC, Newson SE, Frayling TD, Wallis PD (2002) Eye size in birds and the timing of song at dawn. Proc R Soc B 269:831–837

Tobias JA, Planque R, Cram DL, Seddon N (2014) Species interactions and the structure of complex communication networks. Proc Natl Acad Sci U S A 111:1020–1025

Trillo PA, Vehrencamp SL (2005) Song types and their structural features are associated with specific contexts in the banded wren. Anim Behav 70:921–935

Tuttle MD, Ryan MJ (1981) Bat predation and the evolution of frog vocalizations in the neotropics. Science 214:677–678

Vabishchevich AP (2011) Dawn singing in pied flycatchers: mated males sing highly versatile songs in the early morning. Ann Zool Fenn 48:376–380

Van Duyse E, Pinxten R, Eens M (2003) Seasonal fluctuations in plasma testosterone levels and diurnal song activity in free-living male great tits. General Comp Endocrino 134:1–9

Van Duyse E, Pinxten R, Snoeijs T, Eens M (2005) Simultaneous treatment with an aromatase inhibitor and an anti-androgen decreases the likelihood of dawn song in free-living male great tits, *Parus major*. Horm Behav 48:243–251

Wada M (1986) Circadian rhythms of testosterone-dependent behaviors, crowing and locomotor activity, in male Japanese quail. J Comp Physiol A 158:17–25

Wang G, Harpole CE, Trivedi AK, Cassone VM (2012) Circadian regulation of bird song, call, and locomotor behavior by pineal melatonin in the zebra finch. J Biol Rhythm 27:145–155

Ward S, Slater PJ (2005) Raised thermoregulatory costs at exposed song posts increase the energetic cost of singing for willow warblers *Phylloscopus trochilus*. J Avian Biol 36:280–286

Ward S, Speakman JR, Slater PJ (2003) The energy cost of song in the canary, *Serinus canaria*. Anim Behav 66:893–902

Welling P, Koivula K, Orell M (1997a) Dawn chorus and female behaviour in the Willow Tit *Parus montanus*. Ibis 139:1–3

Welling PP, Rytkonen SO, Koivula KT, Orell MI (1997b) Song rate correlates with paternal care and survival in willow tits: advertisement of male quality? Behaviour 134:891–904

West-Eberhard MJ (1979) Sexual selection, social competition, and evolution. Proc Am Philos Soc 123:222–234

Wiens J, Johnston R (1977) Adaptive correlates of granivory in birds. In: Pinowski J, Kendeigh SC (eds) Granivorous birds in ecosystems: their evolution, populations, energetics, adaptations, impact and control. Cambridge University Press, Cambridge, pp 301–340

Wright HW (1912) Morning awakening and even-song. Auk 29:307–327

Xia C, Wei C, Lloyd H, Liu J, Wu Q, Zhang Y (2014) Dawn singing intensity of the male brownish-flanked bush warbler: effects of territorial insertions and number of neighbors. Ethology 120:324–330

York JE, Young AJ, Radford AN (2014) Singing in the moonlight: dawn song performance of a diurnal bird varies with lunar phase. Biol Lett 10:20130970

Zahavi A (1977) The cost of honesty: further remarks on the handicap principle. J Theor Biol 67:603–615

Zahavi A (1980) Ritualization and the evolution of movement signals. Behaviour 72:77–80

Zuk M, Kolluru GR (1998) Exploitation of sexual signals by predators and parasitoids. Q Rev Biol 73:415–438

Chapter 4
Coding Human Languages for Long-Range Communication in Natural Ecological Environments: Shouting, Whistling, and Drumming

Julien Meyer

Abstract Human languages represent very complex coding systems that can be decoded by the human brain after a long acquisition phase. In their acoustic form, human languages can be expressed through different natural speech types. Modal speech is the most common one but several other registers have evolved around the world to enable interlocutors to speak from far. This chapter first provides a large overview of the limits of modal speech for distance communication and lists the major acoustic constraints that interfere with spoken communications in rural outdoor settings. Next, it describes how speech has been naturally adapted to these constraints in different populations by transforming the sounds of spoken languages in shouted speech, whistled speech, or drummed speech. These three registers represent different ways of coding the same linguistic targets as modal speech. Their comparison in a wide variety of languages of the world highlights the great productive and perceptual flexibility of humans to transmit messages of linguistic attitude for telecommunication purposes in natural surroundings.

4.1 Introduction

Human language, in its acoustic form, has the flexibility to be emitted in very different ways according to the communication context. For example, while a sentence is simply spoken in modal speech if the interlocutors are close enough—which is the most comfortable way to maintain a dialog by using the human voice, it would rather be shouted if the interlocutors are located at distance or in a noisy environment. On the other hand, if the conversation has to remain discreet, it would be whispered.

J. Meyer (✉)
Université Grenoble Alpes, CNRS, GIPSA-Lab, Grenoble, France

Research Association 'The World Whistles' (Le Monde Siffle), Paris, France
e-mail: julien.meyer@gipsa-lab.fr

T. Aubin, N. Mathevon (eds.), *Coding Strategies in Vertebrate Acoustic Communication*, Animal Signals and Communication 7,
https://doi.org/10.1007/978-3-030-39200-0_4

Around the world, several languages are also expressed through other special and complementary natural speech registers which are particularly adapted to long distance transmission of human language by transforming speech into either modulated whistles or drummed series of beats. These whistled and drummed forms of languages have developed in rural contexts in response to the frequent need of individuals to communicate from far during their daily activities in rural settings favoring the isolation of individuals and imposing several acoustic degradations to the signal. Their acoustic forms are adapted to the pressure exerted by ecological environments on sounds during propagation and therefore constitute true ancient telecommunication systems. Here, we first explain why modal speech is limited in distance and which of its constituent acoustic cues are the more resistant to the variety of natural background noise. Next, we describe precisely how speech adapts into three different registers—shouted speech, whistled speech, and drummed speech—showing that they exploit differently some of the acoustic cues characterizing the polyvalent phonetic strategy of modal speech to encode human language. We also see how they transform its phonetics and phrasing, reducing the signal but remaining understandable at several kilometers when the conditions are favorable. These registers, due to their drastic reductions of spoken phonetics, sometimes require additional training in both production and perception, but their coding and decoding is possible because of the productive and perceptive flexibility of humans.

4.2 Environmental Constraints and Their Impact on Modal Speech

In the real world, human speech recognition nearly always involves listening in background noise. Speech-in-noise research has revealed that voiced speech signals incorporate several acoustic properties that contribute to compensating for signal distortions and noisy interferences (see Assmann and Summerfield 2004 for a review). In parallel, our cognitive system was found to be adapted to overcome speech degradations and is able, to some extent, to overcome distortions and fragmentations of the signal. This means that our perceptual and cognitive systems perform highly sophisticated mechanisms of informational shielding (Palmer and Shamma 2004; Meyer et al. 2013).

However, recent research has shown that modal speech loses rapidly its efficiency with distance in situations approaching ecological valid contexts. Indeed the impact of noise on modal speech signals and on intelligibility performance increases significantly with the distance emitter–receiver (or speaker–listener), even when the perturbation is limited to spherical spreading with an interfering noise that is the most simple and frequently encountered outdoors, such as "natural quiet" background noise (Meyer et al. 2013, 2015). These studies tested, at variable distances (between 11 and 33 m), the impact on speech recognition of the noise that remains when no geophony, no biophony and no anthropophony interfere in the

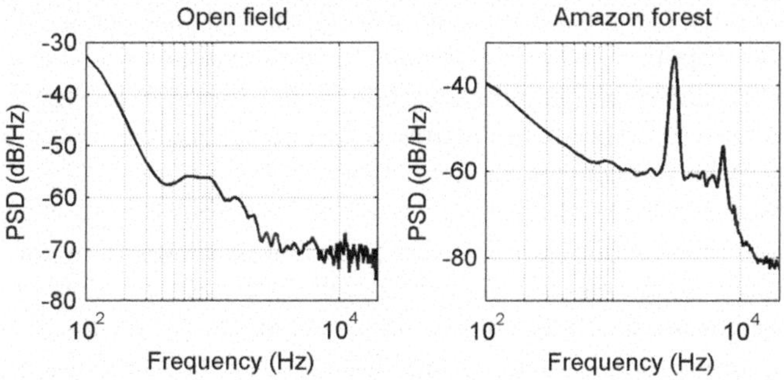

Fig. 4.1 Long-term spectrum of a typical natural quiet background noise (left), and long-term spectrum of the background noise in the Amazonian forest (Park of Gunma near Belém, Brazil) (right) [reproduced from Meyer (2015)]

foreground with the quiet murmur of the background.[1] This kind of noise was tested because it represents the most common acoustic basic properties underlying any kind of natural background noise. Such a typical "natural quiet" background noise is firstly characterized by a nonuniform distribution of noisy frequencies that emphasize low-frequency content (Fig. 4.1). This is a general trend in every ecological milieu because low frequencies travel farther and are less attenuated by obstacles, so they are more present everywhere. A second important aspect is that power levels decrease rapidly as a function of increasing frequencies, more rapidly than the ones of a pink noise, for instance. Moreover, at high frequencies the general relative shape of power level distribution in natural quiet rather resembles speech-shaped noises. In the high frequency range of human voice, reasonably at above 1 kHz, a natural quiet noise is weak, so the signal-to-noise ratio of frequencies situated in this frequency range remains generally high enough to be clearly heard at distance if they are strong enough. This is why, in Meyer et al. (2013, 2015), while word recognition decreased in the average proportion of correct answers from 77.8% at 11 m to 35.9% at 33 m, some phonetic segments (some consonants and all vowels) were found to resist well and to be very well recognized at the longest distances tested. The best recognized phonemes were first the ones characterized by strong sibilant frications (such as in [s] or [ʃ][2]) for which recognition scores remained above 90% at 33 m, or by formant like resonance patterns, such as in vowels for which recognition scores remained above 80% at 33 m because they are by far the most energetic phonemes due to their sonorant sounds characterized by well-formed formant resonances. The fact that [ʃ]

[1]An optimal natural quiet is generally found at night, when the power of sounds is low (often approximately 30–35 dB). In dense tropical forests, natural quiet is rare because of the activity of birds, anurans, and insects but it is pervading in the background. In temperate climates its occurrence depends greatly on the season, besides its underlying presence all the year (Fig. 4.1).

[2][ʃ] is the phonetic symbol used in the International Phonetic Alphabet to represent a consonantal sound used in many languages, including English and French. In English it is usually spelled "sh," as in "*ship*." It is called a voiceless palato-alveolar fricative.

and [s] were better recognized than all vowels showed that energy alone cannot explain a good response. A human sibilant frication has other advantages such as remaining easily above the natural background noise (because its signal is encoded at frequencies above 3 kHz), being narrow band and also falling within the range of the best perceived frequencies in human hearing (Calliope 1989). However, a sibilant frication is really efficient if it is associated with a high energy level to provide a good transmission performance, as shown for example by the case of the sibilant [z] which ranked much lower in recognition performance than [ʃ, s] because its fricative sounds are much less intense in amplitude. Other results of Meyer et al. (2013, 2015) are instructive to understand how human speech coding resists to the most pervading type of noise in natural outdoor surroundings. For example, they show the important role of bursts in strong plosive consonants such as [k] or [t], which ranked just after phonemes with formant-like resonances. They also suggest that listeners had access to other acoustic cues associated with formant peaks, such as formant transitions and/or rapid spectral changes, particularly for consonants [l, j, m, n]. Finally, they show that modal speech reserves different functions to different categories of phonemes: vowels playing a central role in the word-detection step that precedes the word recognition step, whereas consonants in general are mainly used by listeners for lexical identification in isolated words. As a consequence, just by taking place in the conditions of a simple ecological setting, these studies underline that normally spoken words—produced at around 65 dB(A) of intensity—do not project very far, but also highlight that modal spoken speech is very diverse and polyvalent in the acoustic forms it uses and in the functions it gives them to encode linguistic information, with sibilant, formants like sounds, and burst properties being some of the most important parameters helping phoneme recognition at distance.

Modal speech is not targeted at distance communication, and it is only through the modern invention of telephony that it could gain this advantage. Without technology, however, listening to distant speakers or speaking at distance is rather common tasks in daily life, both in urban or rural contexts. When humans need to communicate farther than just a few twenty meters or so, they generally emphasize some of the parameters that the studies cited previously highlighted, with a priority given to increasing energy in the signal to optimize the Signal-to-Noise Ratio. Humans have developed several different strategies to adapt to such situations and to cope with the multiple outdoor constraints they may encounter when trying to reach a distant interlocutor. Before describing these different adaptions of speech in the next section, we will draw a short list of the very large and variable span of disturbing contexts they have to overcome. These are the result of the combined effects of background noise and sound propagation outdoors.

Background noise is ubiquitous in natural environments and rural background noise is known to be rather variable, even when it does not include mechanical sources of noise. It depends on the geographical situation, the terrain, the vegetation, the meteorological circumstances, the bio-noises such as animal calls (biophony) and the hydronoises such as rivers or sea rumble (geophony). Background noise that is loud, near the receiver, and similar to the signal is often called "foreground noise" because it creates greater problems than noise that is quiet, distant, and dissimilar (Bradbury and Vehrencamp 1998). The types of natural foreground noises that may interfere most with human communications are animal acoustic communications and the acoustic effects of wind. Most noise from abiotic sources has energy below 1 kHz. But rivers, torrents, waterfalls, and sea rumble are important exceptions because they show strong amplitudes in the entire frequency spectrum (much as does white noise) and therefore are very good complete maskers of human communications. Moreover, some species of birds, anurans, insects, and mammals produce songs that use the same bands of frequencies as human speech. Arthropods tend to produce sounds in the 3–10 kHz range. They generally produce signals with regular patterns that are often quasi-stationary and sometimes in chorus. They can be very annoying for human speech communication. For example, in the Amazon forest, some very current species are nearly omnipresent and produce long-lasting frequency bands in the 2.5–3.5 kHz range during the day, masking important acoustic cues of the human voice and even of some of the highest frequency bands of whistling (see Fig. 4.1). Anuran calls may have a wide frequency range and be very disturbing to human communication also, but they are much more limited in time and space. Most birds and mammals have vocalizations with frequencies between 1 and 4 kHz (Luther 2008), bands of frequencies that are so important for human speech that they are the ones that have been privileged for telephony. Calls are generally much shorter than for arthropods and are much less stationary. Therefore, birdsongs and mammal calls as well as anuran calls may greatly interfere with human voiced speech but in a different way than that found for arthropods. Most of them consist of the repetition of stereotyped modulated patterns, which allow humans to listen between the gaps and to grasp a part of the masked portion of the signal to cognitively reconstruct the rest, much as they do in speech in noise and in speech listening tasks (e.g., Hoen et al. 2007; Varnet et al. 2012). Humans also benefit from their ability to focus attention on the dynamics of a specific signal, on the direction of its source and on its distance, which helps the receiver follow a speech signal in such intermittent and frequency-limited foregrounds. All of these biotic and abiotic sources define a wide and diverse span of situations, ranging from very noisy conditions to optimal ones.

Some effects of propagation are also important to consider as they alter significantly human spoken speech. We already spoke about the main of them, spherical spreading, which is theoretically predicted by the inverse square law of amplitude attenuation. It corresponds to a decrease in intensity of approximately 6 dB for each doubling of distance. However, other factors change the characteristics of the signal during the transmission between two interlocutors, especially through forests (Wiley and Richards 1978, 1982; Blumenrath and Dabelsteen 2004). Absorption associated with multiple scattering by vegetation attenuates sounds and thus reduces the signal-to-noise ratio (Michelsen 1983). Reflection and absorption on the ground may also significantly alter the signal, depending on the topography and the terrain. The duration of silences between sound elements is modified because of sound reverberation that elongates sounds with tails of echoes (Holland et al. 2001). Furthermore, selective frequency filtering, reverberation, and atmospheric turbulences distort amplitude and frequency patterns over time, giving recorded sounds a blurred aspect (Wiley and Richards 1982; Dabelsteen et al. 1993). Moreover, atmospheric temperature gradients, which influence the changing celerity of sound in the air, depend on wind and humidity and are difficult to quantify. It is only at night and in rainy weather that these gradients are equalized because in such conditions, there are no air slides. All of these effects profoundly alter the voice. However, some effects of propagation are also interesting for signal detection and separation from natural background noise. For example, sounds acquire slow (<50 Hz), large envelope fluctuations during atmospheric propagation because of microturbulence (Richards and Wiley 1980). Such backgrounds can elicit comodulation masking release (CMR) in humans, which is an adaptation of the auditory system for detecting signals over naturally occurring separable backgrounds (Nelken et al. 1999).

4.3 Different Strategies of Adaptation for Distant Dialogs

In several situations, mainly associated with rural natural environments which represent the dominant setting for the vast majority of human evolution, it can be very useful to communicate at long distance with all the communicative richness of the local spoken language. For example, to signal a danger or give important news to someone located on the other side of a river or of a valley, or to call an interlocutor difficult to locate because he/she is hidden by the vegetation in a forest, or even by the houses in a village. In response to this combination of environmental and social pressures, human speech has evolved into several different complementary speech registers adapted to distance speech transmission. These are of three main types, based on (1) shouting, (2) whistling, and (3) drumming. Shouting is generally well known but whistled speech and drummed speech are much less reported and studied in the literature. However, they are very striking as they define real ancient telecommunication systems that consist in emulating some key salient phonetic traits of spoken words and sentences with sounds which are very different from the human voice. Strikingly, all three registers represent different bioacoustics strategies to resist to the acoustic degradation imposed by environmental acoustic constraints. They are also adapted to different ranges of communication: (1) the human voice can transmit information normally only up to a few hundred meters in shouted speech. (2) Whistled speech enables the populations which have adopted it to go much further: it is not rare to find records of whistled dialogs at 1 or 2 km when the environment is favorable (for instance in Busnel and Classe 1976; Meyer 2015; Meyer and Díaz Reyes 2018). (3) Drum communication systems can even extend this range by a factor of up to 20—reaching up to 20 km in some cases—through emulating speech in sequences of drumbeats (Thiesen 1969; Meyer et al. 2012; Seifart et al. 2018). Here we describe in detail how these three registers manage to encode human language and we present also their overall distribution worldwide. Next, we compare some of their characteristics.

4.3.1 Shouted Speech

We have seen in the preceding section that human modal speech adopts a polyvalent acoustic strategy to encode meaning in human languages. The most common response to the need to communicate at distance while speaking is to try to keep this polyvalent strategy by simply increasing energy power in the signal: that is, to shout. In shouting, to increase the range of ordinary speech or to overcome noise, individuals adjust their voices by raising tacitly amplitude levels. During this vocal effort, called the "Lombard effect" (Lombard 1911), the spoken voice becomes "strong" and progressively passes into the register of the shouted voice. Effort is also intensified with the tendency to prolong syllables, to reduce the flow of speech and to increase the fundamental frequency. There is a large body of literature on this phenomenon for speech under noisy conditions at close range (e.g., Dreher and O'Neill 1957; Garnier and Henrich 2014). However, there are far fewer studies on variations in talker-to-listener distance in natural outdoor conditions (Meyer 2008; Cheyne et al. 2009; Fux 2012; Meyer et al. 2018). The studies that address distance shouting all find that the fundamental frequency (F0) increases regularly with distance, parallel to an intensified vocal effort. For example, Cheyne and his colleagues find that F0 is increased by approximately 40 Hz when distance is doubled at distances between 1 and 32 m. The data of Fux, measured for distances between 2 and 100 m, provides approximately the same results. Fux also finds that most of the increase in duration is carried by vowels and that the consonants, depending of their position, may suffer strong degradations during this process. The analysis of Meyer (2008) is performed from the viewpoint of the receiver of the message. Therefore, he analyzes the second harmonics (2∗F0) because they stand out the best from natural background noise. His results show a quasilinear increase of frequency between 50 and 300 m (in an open valley and in natural quiet conditions). In this study, tiring of the vocal folds occurs at approximately 90–100 dB for a distance of 200 m as reanalyzed by Meyer (2015). Beyond that limit, the speakers attempt to compensate with a drastic lengthening of the vowels and with a type of screamed voice Due to the biological limits of the vocal folds, most shouted dialogs are brief. The limits of shouted speech in both distance and time of dialog are mainly due to the facts that the vocal cords get tired and hurt with effort.

4.3.2 Whistled Speech

Whistled speech is a natural speech register used mainly to project messages beyond shouting distances, by transposing spoken languages into whistles (for a review, see Meyer 2015). In whistled speech, people articulate words while whistling, and thereby transform spoken utterances by simplifying them, syllable by syllable, into whistled melodies. Whistled speech is therefore always based on a spoken language

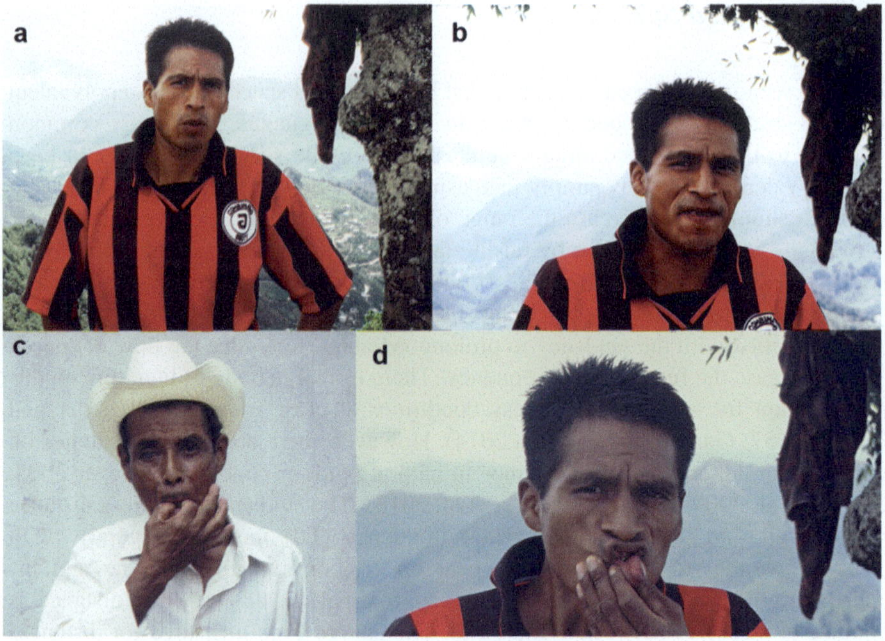

Fig. 4.2 Different whistling techniques presented by two Mazatec whistlers of Mexico. (**a**) Bilabial protrusion; (**b**) Lingo-dental technique with the retroflexed tongue on the lower jaw; (**c**) Insertion of fingers in the mouth; (**d**) Pulled lower lip (this is the only technique of this image produced with ingressive air flow) [Photos **a**, **b**, **c**, **d** courtesy of Julien Meyer/Laure Dentel (© Julien Meyer/Laure Dentel. All Rights Reserved)]

and it is not a separate language or dialect from a native tongue but rather an extension of it. This is why the term "whistled language" that is sometimes used is somewhat of a misnomer and we prefer to use "whistled speech." Indeed, language-based whistles do not rely on whistled codes that mean by themselves concepts like "goat" or "boat," but they require the combination of articulated whistled linguistic elements to form whistled words which in turn are combined into sentences, as is the case in the spoken form. This allows practitioners to issue any type of message that has not been previously agreed upon.

In contrast to shouting, the production of whistles, rather exploit sounds that resemble fricatives of modal speech at a stronger energy,[3] a strategy that is extended to the whole phonemes and which does not cause the vocal cords to suffer because they are only used to control the force of the airflow and not in vibrant mode. The sounds are generated by a stream of compressed air in the mouth that is molded by the tongue, the jaw, the lips, and eventually the fingers (Fig. 4.2). Sometimes the

[3]Whistles are among the most powerful acoustic productions that can be produced by the human vocal tract, as they can easily reach 120 dB at 1 m of the emitter with the strongest techniques that imply fingers or retroflexed tongue.

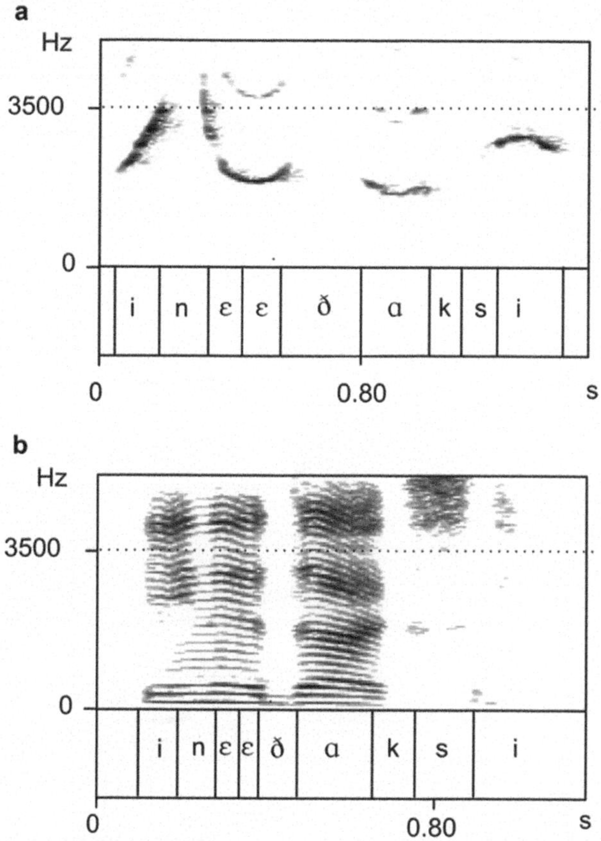

Fig. 4.3 Spectrogram of the Greek sentence "ειναι ενταξει" [inɛ ɛðaksi], or "ine entaksi" in Roman transliteration (meaning "all right") in whistled (**a**) and spoken (**b**) speech. Here, the whistled form consists of the emulation of the quality of spoken vowels and consonants (formant-based whistling) [reproduced from Meyer (2015)]

sound source is produced with other elements such as a leaf that vibrates between the lips or with the resonance of the airflow inside a cavity formed by the hands in front of the mouth (Meyer 2015). Whistles have the advantage to be produced by the vocal tract and therefore to enable the simultaneous articulation of speech (Figs. 4.3 and 4.4).

Many populations in the world living in natural environments that favor the isolation of individuals during traditional subsistence activities (such as shepherding, hunting, hill agriculture) have developed a whistled form of their language to communicate at a distance. Whistling is one of the multiple modes of expression for these languages of the world, as well as whispering, shouting, or singing. Distant interlocutors need powerful and clear signals, little disturbed by natural ambient noise and physical obstacles that affect the propagation of sound. As we just told earlier, the physical characteristics of the whistles are well adapted to these environmental limitations because they allow a lot of power in a narrow band of frequencies (1000–4000 Hertz) corresponding to the best audibility and sound discrimination of the human auditory system. Human whistles also fall higher than the frequencies were natural background noise is powerful. Hence, a strong whistle goes much

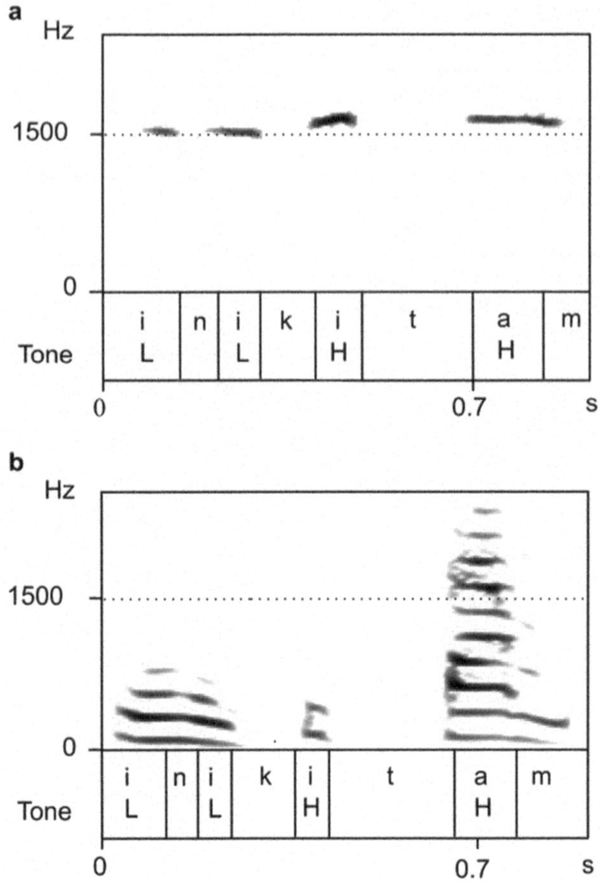

Fig. 4.4 Spectrogram of the Gavião words [ini kitam], meaning "hammock rope," in whistled (**a**) and spoken (**b**) speech. Here, the whistled form consists in the emulation of the pitch of the spoken form (pitch-based whistling) [reproduced from Meyer (2015)]

further than any shouted sound, however stentorian, and thus prevents the interlocutors from traveling long distances to talk to each other (Fig. 4.5). When a whistler wants to reach a faraway interlocutor, his whistles can easily reach maximum levels of amplitude of approximately 120 dB (at 1 m) with finger techniques (Fig. 4.2). Maximum amplitudes are more commonly approximately 100 dB with the linguo-dental technique. For middle-distance communications (approximately 200 m), levels of production are approximately 80–90 dB. The different peaks of energy of the amplitude envelope of a whistled sentence represent the different syllables, just such as in spoken speech (Fig. 4.6). Other fine aspects of the amplitude envelope help whistlers separate consonants into various classes as described shortly later in this chapter (continuous, near-continuous, or clearly interrupted signals) (see Figs. 4.5 and 4.6). Moreover, the sentences are produced at a general flow from 10 to 50% slower than standard speech (Moles 1970; Meyer 2015). Finally, whistled speech increases further intensity, frequency, and duration of speech, in comparison to shouted speech. The three fundamental parameters of speech can be more

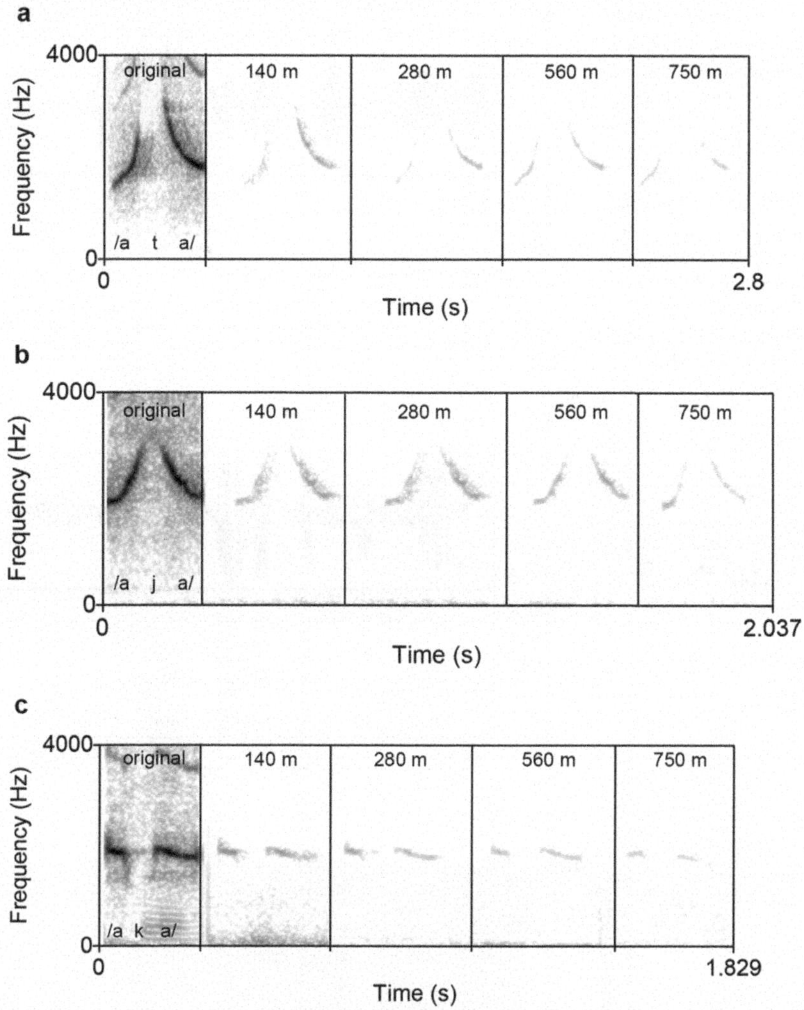

Fig. 4.5 (**a**) Spectrogram of a /ata/ whistled utterance in its original form in whistled Turkish and recorded back after transmission by loudspeaker at 140, 280, 560, and 750 m, (**b**) Spectrogram of a /aja/ whistled utterance in its original form and recorded back at the same distances, (**c**) Spectrogram of a /aka/ whistled utterance in its original form and recorded at the same distances [reproduced from Meyer (2015), where the full experiment of transmission is described]

comfortably adapted to the distance that standard speech. Whistled speech is even efficient in the most adverse situations, such as in natural broadband noises resembling white noise. For example, in the region of Turkey situated near the village of Kusköy, several villages are crossed by torrents that represent non-negligible sources of broadband noise that mask spoken and shouted communications nearby. Some activities, such as trout breeding, require proximity with the torrent (Fig. 4.7).

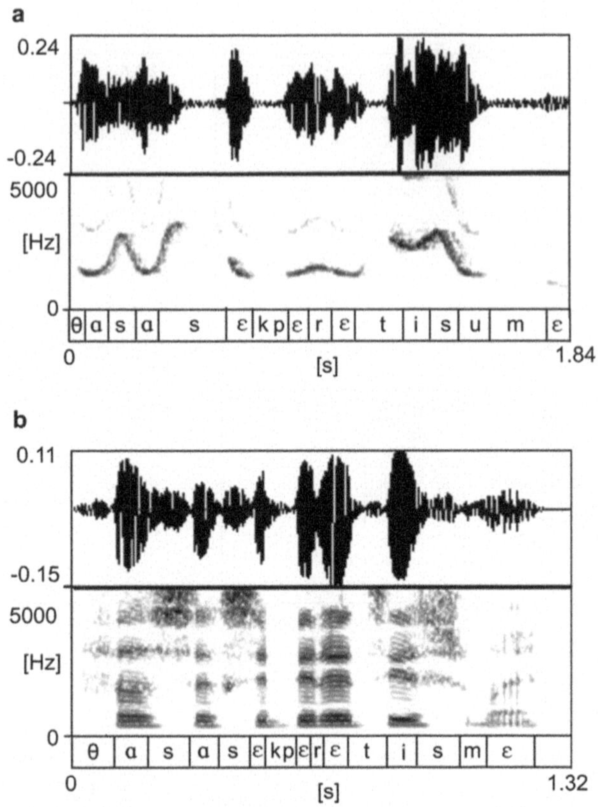

Fig. 4.6 Spectrogram of the whistled (**a**) and spoken (**b**) Greek sentence "θα σας εξυπηρετήσουμε" (meaning "at your service"). The rapid rate of speech of these examples explains some elisions (the [u] in spoken speech; the [s] in the [ksp] cluster in whistled and spoken speech) and some assimilations, such as the one of [irε] to [ε] in [spirε]]. [reproduced from Meyer (2015)]

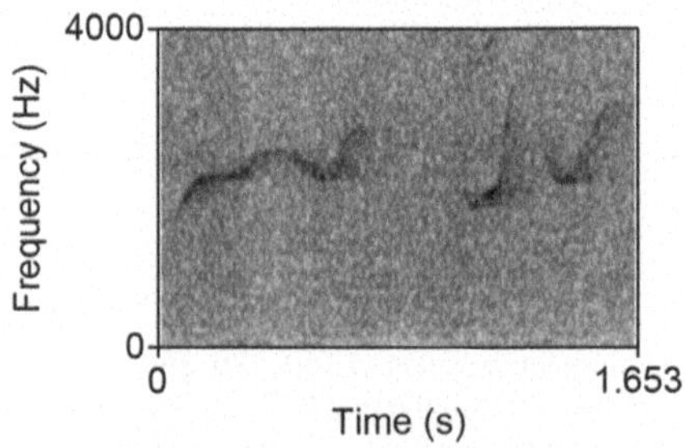

Fig. 4.7 Whistled Turkish signal in the broadband noise of a river. The recording was made at 30 m from the whistler with the microphone near the listener on the other side of the river. It was recorded in May 2004 when the water flow was particularly strong. Whistling is well adapted to communication in such conditions because the characteristic modulations of whistled speech still clearly emerge from the noise [reproduced from Meyer (2015)]

Whistled forms of languages represent a widespread language practice on all inhabited continents. However, it is a little-known practice because it was not always identified as a linguistic phenomenon by researchers and because it quickly loses vitality with the modernization of rural life, although it occurs in a wide range of languages and language families in all continents populated of the planet (Meyer 2015). According to scientific references, more than 70 languages have been found in the world with whistled language. About 40 of them have been studied or recorded in their particular whistled form. The languages that present whistling modality exist in linguistic families as different as Afro-Asian and Niger-Congo in Africa; Algonquian, Esquimo-Aleut, and Oto-Manguean in Central America and North America; Arawak, Jê, and Tupí in South America; Altaic, Indo-European, and Sino-Tibetan in Eurasia; and Torricelli, Trans-New Guinea, and Sepik in Oceania. This linguistic diversity shows that this practice is not limited to a type of linguistic structure or a type of linguistic family. Quite the contrary, it suggests that all human languages can develop a whistled form if ecological and social conditions require it. In addition, it was developed for both tonal languages (such as Mixtec and Chinantec in Mexico, or Akha and Hmong in Southeast Asia) and for nontonal languages (such as Greek, Spanish, Turkish, Béarnese, Wayãpi, or Tamazight) and even for a rare case of language with nascent tones (Chepang language, from southern Nepal).

Phonetics and Typology of Whistled Languages Whistled speech relies on the whistlers' selection of salient features of a given language. The strong frequency reduction and articulatory constraints operating during this procedure explain why this register requires additional long training in both production and perception. From previous research it is known that whistled speech profoundly modifies the phonetics of modal speech, speech with voice, applying a reduction in the frequency domain: transforming the complex frequency spectrum of the voice into an imitation based on a simple variation of the tone of the whistle (Figs. 4.3, 4.4, 4.5, and 4.6). During this procedure, certain phonetic details present in the modal speech are lost: by using only the anterior part of the mouth, the precision with respect to the spoken language is reduced, which causes frequent cases of ambiguity in the reception of the whistled messages. The interlocutors resolve this lack thanks to the context of the situation or through clarifying questions. However, high levels of intelligibility can be observed in particular in nontonal languages or in tonal languages with a high number of distinctive phonological tones (Busnel 1970; Rialland 2005; Meyer 2015; Sicoli 2016). When comparing the whistled versions of several languages of the world, it is observed that each one adapts to several aspects of the structure of the language that is being replaced. Among the languages that are whistled, different strategies of transposition from spoken speech to whistled speech have been observed depending on a major typological distinction: tonal (Fig. 4.4) versus nontonal languages (Fig. 4.3). What differentiates tonal languages from nontonal languages is that in tonal languages, the pronunciation of tones influences the meaning of words, something that does not occur in nontonal languages. While tone and vowel quality can be conveyed simultaneously by the timbre of spoken voice, the use of whistles forces the whistlers to choose between the two. The whistlers of tonal languages will emulate in priority the tone carried by the pitch of a vowel because it changes the meaning or the grammar of words[4] (Fig. 4.4); whereas whistlers of languages without tones will rather render the vowel quality, which characterizes its identity and is carried by the timbre of spoken voice. In nontonal languages the whistlers approximate as precisely as possible the articulation of the vocal apparatus used in the spoken form; this approach causes a whistled adaptation of the quality of the vowels and consonants made by the timbre of the voice. For instance, the whistled /i/ is the most acute because its spoken timbre in modal speech is originally compact in the upper frequencies (with close higher formants forming a salient perceptual frequency band), while the /o/ is low and the /e/ intermediate (Meyer 2005, 2008; Díaz Reyes 2008) as illustrated in Figs. 4.3 and 4.6. It was recently shown that this type of distribution of vowels whistled on a scale of simple frequencies corresponds to a perceptual reality that allows discriminating the vowels not only for fluent whistlers in a whistled language but also for listeners without any knowledge about this phenomenon (Meyer et al. 2016, 2017). It was also shown that individuals adapt both in production (Meyer 2015) and in perception (Meyer et al. 2017) to the specificities of pronunciation of each vowel in each language.

When considering all languages, we observe that the whistlers adapt their whistling either to the number of tones, or to the number of vowels in the language and to the way these tones and vowels are articulated with the consonants. The consonants are represented by continuous, near-continuous, or interrupted modulations of these tonal and vowel frequencies, when they are between two vowels (see Figs. 4.3, 4.4, 4.5, and 4.6; and Meyer 2015 for more details on all languages). They are made imitating their articulation and sound in the spoken language as well as possible. Consonants are whistled by performing frequency and amplitude modulations of the tonal vocalic frequencies that surround them. For example, when the amplitude modulation shuts off the whistle to emulate rapid amplitude modulations of standard speech, whistled consonants are also characterized by silent gaps (such as in /t/ or /k/ in Fig. 4.5).

All together, these descriptions show that the whistled version adapts to each language and to great typological distinctions of the world's languages.

4.3.3 Drummed Speech

Drummed speech is a traditional practice found in several languages around the world that consists in adapting speech to enable very long distance communication in natural surroundings. It transmits information by emulating some key salient

[4]Therefore, in the whistled form of a tonal language, the vocal quality is completely excluded. This exclusion occurs even when the functional load of information transported by tones is less than that corresponding to the vocal quality, as if there were a functional or perceptual precedence of lexical tone that guides the emulation of spoken speech.

Fig. 4.8 Two examples of drums traditionally used to speak. Left: Bendré used by the Mossi People speaking Mooré (see also Junzo 1998). Right. Manguaré drum pair used by the Bora people of West Amazon, Peru. The drums are hung in parallel and the drummer stands in the middle to beat them with two wooden mallets covered with natural rubber. In order to speak with the drums, the drummer uses two pitches, one on each drum on the inside lip [Photos: courtesy of Julien Meyer/Laure Dentel (© Julien Meyer/Laure Dentel. All Rights Reserved)]

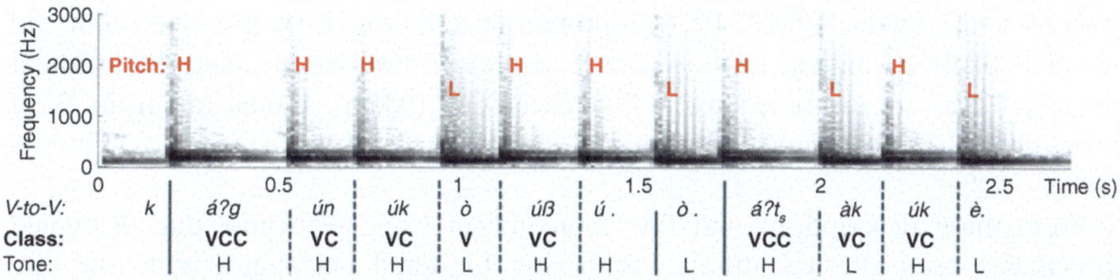

Fig. 4.10 Spectrogram of the drummed speech form of the Bora phrase */káʔgúnúkòúβú ò áʔtsàkúnè/* (meaning "I am finishing the cahuana (manioc starch drink)"). Red annotations explain the pitch of the drum. Bottom rows provide transcription and segmentation into vowel-to-vowel intervals as well as spoken tone levels (in Bora language, just as in Akan, a High is written with an acute accent and Low tone is written with a grave accent) [Figure adapted from Seifart et al. (2018)]

linguistics cues of spoken languages through rhythms and melodies. The source of the signal is a drum (see for example Figs. 4.8 and 4.10) and speech is encoded in sequences of drumbeats which remain comprehensible to the receivers (Stern 1957; Sebeok and Umiker-Sebeok 1976; Niangoran-Bouah 1980; Agawu 1995). Among natural long distance speech encoding systems, drummed speech can reach the longest distances but also employs the most radical reduction in acoustic complexity in all three main dimensions of acoustic signals: frequency, amplitude, and time. By comparison, whistled speech consists of a much less radical acoustic reduction, primarily in the frequency domain (see Sect. 4.3.2. in this chapter). Acoustically, drumming is most similar to the bursts of spoken plosives that we identified earlier as quite resistant to natural quiet background noise. It generalizes the strategy of using beat-like sounds to encode language to every acoustic cue of the message. Drummed forms of languages also generally employ low frequency sounds in which a pitch can be perceived. Drummed signals exploit the natural bioacoustic properties of percussions for optimal sound propagation in natural environments (Rossing 2000). Their low pitch frequencies are not blocked by large vegetation (Bass et al. 1990; Padgham 2004) and their high amplitudes and narrow frequency band reduce further noise-masking effects (Zwicker 1982; Green 1985).

Drummed speech messages are generally addressed to a general audience to give instructions, recite texts, call someone. Another general characteristic of drummed sentences is that, to ease the decoding of the message, the syntax of a drummed sentence is simplified by surrounding the variable content of the message with stereotyped general formulas that are also drummed word by word and that have the function to contextualize the new information by: opening the call, naming an addressee, informing about the general topic of the call (invitation to an event, request to do an action), and finally closing the call (Carrington 1949; Cloarec-Heiss 1999; Seifart et al. 2018). A wide variety of messages are communicated through this special speech register which has been sometimes also called “drummed surrogate,” “drummed substitute,” or “drummed abridgement” of language. This kind of practice is attested in Africa, South America, Asia, and Oceania and it should not be confounded with other complex drummed systems that do not represent elements spoken languages. The most comprehensive review on drummed speech (Sebeok and Umiker-Sebeok 1976) reprinted descriptions from the nineteenth and first half of the twentieth century and it provides information about 18 different languages that are drummed, including Ewe, Twi (Akan), Banen (all from West Africa), Chin (from Burma), and an early, short report on Bora (Amazonia) (Thiesen 1969).

From these descriptions and few more recent ones, we know that drummed speech has been attested almost exclusively for tonal languages, with one rare exception recently described (Tang 2007; Winter 2014), but for which precise phonetic analysis of drumming is still needed. In the large diversity of tonal languages that have been described so far for this practice, drummed speech systems encode tonal and rhythmic patterns of spoken sentences in sequences of drumbeats. Generally, each drumbeat represents a syllable and emulates its tonal level with the pitch of the drum. For example, in a two-tone language such as Bora, a High tone (H) is played on the high pitch slit log drum and a Low tone (L) is played on the low pitch slit log drum (see Figs. 4.8 (right) and 4.10). But there are some exceptions: when a rising or falling contour tone occurs on a single syllable, this particularity is encoded by a succession of two beats of different pitches for this syllable [H-L for falling and L-H for rising such as in Banda Linda of Central Africa (Cloarec-Heiss 1999)].

The relation between speech rhythm and the timing of the sequences of beats has been scarcely studied. However, the case studies of the few languages for which the drummed speech rhythmic encoding has been described already provide an original insight into speech rhythm. So far, only Akan of Africa (Nketia 1976), Bora of Amazonia (Meyer et al. 2012; Seifart et al. 2018), and, to a lesser degree, Banda

(a) **bèrèmpɔ̀ń** **' nobleman '** 

(b) **ɔ̀déáfɔ̀ɔ́** **' witch '** 

(c) **àsààsé** **' earth '** 

Fig. 4.9 Written transcriptions—in API—of three different Akan words and their corresponding drummed versions explained with musical notes of different rhythmic durations and different pitch heights. The written contrasting tone levels are annotated above each syllable nucleus: here, High (acute accent) and Low (grave accent). Note that in this language, a single sonorant consonant can function as asyllabic nucleus (as in the last syllable of (**a**)). These examples provide an illustration of the three rhythmic categories found by Nketia (1976) [adapted with information taken from Nketia (1976)]

Linda of Africa (Cloarec-Heiss 1999) have been investigated for this aspect. Nketia provided the first extensive study with a large sample of examples even if no statistics were published. He found that three relative lengths of interbeat duration are encoded in drummed Akan, in accordance with some phonological and prosodic structures of the spoken language: a long interbeat duration represents closed syllables[5] (which always have a nasal in coda in Akan, see example (a) Fig. 4.9). An extra-short interbeat duration represents open syllables followed by a voiced consonant or a vowel (see Fig. 4.9 (a) for a syllable followed by a voiced consonant, and Fig. 4.9 (b or c) for syllables followed by just a vowel). A simple short interbeat duration represents all other syllables. In Banda Linda, Cloarec Heiss found that interbeat duration may give indications about consonant qualities: liquid consonants (/l, r, v/) correspond to shorter interbeat durations than other consonants. Not much examples are provided in this latter publication so the results of this study are only impressionistic, but we mention it as it is one of the rare papers dealing with this topic. Finally, for Bora, we provided strong statistical evidence that the rhythmic units encoded in interbeat durations of drummed Bora match vowel-to-vowel (V-to-V) intervals more closely than syllables, i.e., that drumbeats align with vowel nuclei irrespective of onset consonants within the same syllable, which are associated with preceding intervals (Meyer et al. 2012; Seifart et al. 2018). We found that four rhythmic units are encoded in the length of pauses between beats, word-internally. These units correspond to different numbers of consonants and vowel lengths, and therefore, in linguistic terms, to different "weights"[6] (Fig. 4.10). In the order of increasing duration of interbeat pauses these rhythmic units of drummed

[5]In linguistics, a syllable is considered "closed" if the nucleus of the syllable (the vowel most of the time) is followed by a consonant which pertains to this same syllable. The syllable is considered "open" if the nucleus is not closed by a consonant.

[6]The notion of linguistic "weight" generally correlates with duration. It may be applied to syllables but also to V-to-V intervals. In any case, a "heavy" unit is composed of long vowels and/or is ending with consonant(s) (Lunden 2017).

Bora are: V < VC < VVC < VCC (where V is a short vowel, VV a long vowel, C is a consonant, and CC represents two different consonants).

These different studies on speech rhythm in drumming show that interbeat durations correlate with linguistic weights. Moreover, the latest findings on Bora drummed speech suggest that weight distinctions are better represented—in duration—by vowel-to-vowel intervals than syllables. These results must be confirmed on other languages but we can already say that they are coherent with recent advances in language studies which support the vowel-to-vowel (V-to-V) intervals as relevant units for phonetic duration (Lunden 2017). Research in psycholinguistics on the notion of "perceptual center" (p-center)—the perceived moment of the occurrence of an acoustic event—also go in this direction as the p-center is commonly noncongruent with the physical signal onset (Morton et al. 1976) and is more closely aligned to the beginning of the vowel than the beginning of the syllable (Otterbein et al. 2012; Ryan 2014; see also Seifart et al. 2018 for further discussions on these aspects).

Finally, drummed speech, because of its clearly rhythmic nature to encode linguistic messages—with beaten acoustic cues that are simple to detect—provides an original and relevant way to analyze speech rhythm, which is an otherwise complex notion to deal with due to the great number of acoustic cues that may encode rhythm in modal speech.

4.3.4 Some Important Elements of Comparison Between These Three Registers

The comparison of shouted, spoken, and drummed speech forms reveals important continuities and breaks in acoustic strategies and/or production means. The following paragraphs describe the ones that we identified as particularly important.

Use of the Vocal Tract for Production Shouting still relies on the vocal fold vibration as a sound source while whistling and drumming change completely the sound production. Whistled speech still uses the vocal tract as a sound source and resonator (Fig. 4.2), whereas drummed speech completely externalizes the sound source (and its eventual resonances) to different kinds of tools (drums with different properties depending of the cultures).

Control of Strong Energy All three strategies have in common to increase significantly the levels of maximum energy encoded in the signal in comparison to modal speech. In each of these three speech registers, an efficient emission relies on a homogeneous, powerful, relaxed, and precise control of the physiological constraints imposed by word pronunciation. Compared to modal speech, shouting and whistling increase muscle tension in the vocal tract. These tensions reinforce the concentrations of energy in the signal. The shouted voice uses the vocal cords and a vocal tract often modified by a low and large pharynx. Whistled speech production is much more focused on the front oral part of the vocal tract. Drummed speech is rather characterized by the use of a hand beaten tool, far from the mouth, and targeted to produce precise concentration of strong energy levels in short beats.

Universal Implicit Learning Versus Additional Training Shouting was universally adopted for distance talk. It is generally developed from birth without any particular distinct explicit extra learning. By contrast, whistled or drummed speech need special training in both production and perception, in addition to the implicit acquisition developed by growing up in linguistic communities or families that extensively use them. An additional effort is necessary to master the emulation of speech with sounds that are very different from the ones of the human voice, even if they imitate some of their salient informative properties. Moreover, the activities that involve them are often outdoor ones that are accessible to children at a certain age only. One consequence is that, worldwide, whistled and drummed forms of speech appear to be less frequent than shouting, which is also nowadays partly due to the general loss of vitality of these very traditional practices.

Different Bioacoustic Strategies to Reach Long Range Most importantly for the present chapter, each speech register represents a different strategy to resist to signal scattering and to consecutive information degradation with distance. (1) Shouting, by reinforcing vocal fold vibration and opening the resonance chamber develops principally the "formant-like" strategy. The resulting signals bear large band complex frequency spectra characteristic of the human voice. (2) Whistling rather concentrates energy in a narrow band of frequencies corresponding to the best auditory performance of humans and to rather quiet background noises (pretty much similar to the "strong sibilant" sounds of modal speech, and extending this strategy to the whole language). (3) Drumming exploits short and narrow band signals (with some similarities to what happens in plosive sounds of modal speech) characterized low frequencies which travel far due to their long wavelength.

4.4 Discussion and Conclusions

The present chapter was the occasion to revisit previous findings on distance communication in human language and to explore the different coding strategies corresponding to language-based telecommunication systems that have evolved in human rural communities all over the world: shouted speech, whistled speech, and drummed speech. We first explained that modal speech does not enable long-distance spoken communication but that its polyvalent acoustic strategy to encode spoken phonemes is a good starting point to explore the speech acoustic cues that are the most resistant to degradation in outdoor natural rural surroundings. The main point of this chapter is to reveal a correlation between these acoustic cues—belonging to the categories of "sonorants (formant-like)," "sibilants," and "plosives"—and the three different types of speech adaptations represented by respectively shouting, whistling, and drumming.

We show here that the longest the distance to reach, the more the acoustic encoding cues of the speech signal are reduced in order to concentrate energy and information in sounds which are very resistant to natural background noise and outdoor propagation. This process begins with shouting, the most universal response to such ecological constraints. It enables humans to prolong the polyvalent strategy of modal speech by emphasizing more particularly the formant-like speech cues which bear most of the energy. But this strategy is limited due to the rapid tiring of vocal folds. Next, the process continues with whistled speech which preserves the use of the vocal tract as a sound source but reduces drastically the phonetics of speech in a simple frequency band modulated in amplitude and frequency. This constraint leads to a typological distinction in encoding strategies, depending mainly of the tonal/nontonal structures of the languages. Whistled speech users rely more on the context of the conversation but are still able to exchange complex—not stereotyped—sentences in real dialogs. This practice is less widely found than shouting but still represents a large diversity of languages and language structures. Finally, to reach the longest distances, drummed speech reduces languages to series of beats produced outside the vocal tract with a drum. It is not used for real dialogs but rather for short two-way exchanges or for one-way public announces. It requires an adaptation of the spoken messages through procedures consisting in enphrasing the original spoken sentences. This type of enphrasing relies on formulaic sentences which contextualize the core of the message and somehow simplify the syntax of speech while making sentences longer to ease decoding (thus following one of the main principles of Information Theory). This kind of drummed register is rather common in Africa (especially West Africa) but is also found in Asia, Oceania, and Amazonia. Yet it is the rarest of the registers described here and the great majority of the languages in which it was found are tonal.

Whistling and drumming operate drastic changes in comparison to modal and shouted speech in order to create real telecommunication systems. However, the messages they encode are perfect imitations of selected phonetic cues of their spoken equivalent. For this reason, these different registers are very informative for advancing research on speech. They provide an alternative point of view on human languages. The three different adaptations in speech production that are presented here demonstrate the amazing productive flexibility of humans. Such a diversity of encoding strategies is only possible because humans have at the same time a great capacity in terms of perceptual flexibility, i.e., the ability to recognize a stimulus as belonging to a category even if partly deviant from the acoustic form to which they are used. Strikingly, all these registers exploit the capacity of humans to still recognize speech when part of it is degraded, blurred, or decayed (typically shown previously in the large amount of studies investigating speech-in-noise conditions or artificially degraded speech). In the speech domain, perceptual flexibility also corresponds to the ability to recognize words or other language units with novel pronunciations, such as those encountered in unfamiliar dialects and accents or in many different register of speech (Bent 2015). This is an essential skill for children's receptive language development for example, but also to learn a second language or understand someone who has a particular pronunciation. Studying the adaptation of speech encoding for long-range communication corresponds to studying acoustic perceptual flexibility in its most complex and extreme, yet natural, manifestations. As a consequence, the perspectives opened by the present chapter are numerous, with several possibilities of experimental testing to reinforce and discuss our observations.

References

Agawu VK (1995) African rhythm: a northern ewe perspective. Cambridge University Press, Cambridge

Assmann PF, Summerfield Q (2004) Perception of speech under adverse conditions. In: Greenberg S et al (eds) Speech processing in the auditory system. Springer, New York, pp 231–308

Bass HE, Sutherland LC, Zuckerwar AJ (1990) Atmospheric absorption of sound: update. J Acoust Soc Am 88:2019–2021

Bent T (2015) Development of perceptual flexibility. In: Proceedings of the 18th International Congress of Phonetics Sciences, Glasgow

Blumenrath SH, Dabelsteen T (2004) Degradation of great tit *Parus major* song before and after foliation: implications for vocal communication in deciduous forests. Behavior 8:935–958

Bradbury JB, Vehrencamp SL (1998) Principles of animal communication. Sinauer Associates, Sunderland, MA

Busnel RG (1970) Recherches expérimentales sur la langue sifflée de Kusköy. Revue de phonétique appliquée 14–15:41–57

Busnel RG, Classe A (1976) Whistled languages. Springer, Berlin

Calliope (1989) La parole et son traitement automatique. Masson, Paris

Carrington JF (1949) Talking drums of Africa. Carey Kingsgate Press, London

Cheyne HA, Kalgaonkar K, Clements MA, Zurek P (2009) Talker-to-listener distance effects on speech production and perception. J Acoust Soc Am 126:2052–2060

Cloarec-Heiss F (1999) From natural language to drum language: an economical encoding procedure in Banda-Linda, Central African Republic. In: Fuchs C, Robert S (eds) Language diversity and cognitive representations. John Benjamins, Amsterdam, Philadelphia, pp 145–157

Dabelsteen T, Pedersen SB, Larsen ON (1993) Habitat-induced degradation of sound signals: quantifying the effects of communication sounds and bird location on blur ratio, excess attenuation and signal-to-noise ratio. J Acoust Soc Am 93:2206–2220

Díaz Reyes D (2008) El lenguaje silbado en la Isla de El Hierro. Excmo, Cabildo Insular de El Hierro, Santa Cruz de Tenerife

Dreher JJ, O'Neill J (1957) Effects of ambient noise on speaker intelligibility for words and phrases. J Acoust Soc Am 29:1320–1323

Fux T (2012) Vers un système indiquant la distance d'un locuteur par transformation de sa voix. Université de Grenoble, France, PhD Dissertation

Garnier M, Henrich N (2014) Speaking in noise: how does the Lombard effect improve acoustic contrasts between speech and ambient noise? Computer Speech and Language 28:580–597

Green DM (1985) Temporal factors in psychoacoustics. In: Michelsen A (ed) Time resolution in auditory systems. Springer, Berlin, pp 122–140

Hoen M, Meunier F, Grataloup C, Pellegrino F, Grimault N et al (2007) Phonetic and lexical interferences in informational masking during speech-in-speech comprehension. Speech Comm 49:905–916

Holland J, Dabelsteen T, Bjørn CP, Pedersen SB (2001) The location of ranging cues in wren song: evidence from calibrated interactive playback experiments. Behaviour 138:189–206

Junzo K (1998) La voix: étude d'ethno-linguistique comparative. Éd. de l'École des hautes études en sciences sociales, Paris

Lombard E (1911) Le signe de l'élévation de la voix. Annales des maladies de l'oreille, du larynx, du nez et du pharynx 37:101–119
Lunden A (2017) Syllable weight and duration: a rhyme/intervals comparison. Proc Linguistic Soc Am 2(33):1–12. https://doi.org/10.3765/plsa.v2i0.4084
Luther DA (2008) The evolution of communication in a complex acoustic environment. PhD dissertation, Chapel Hill University
Meyer J (2005) Description typologique et intelligibilité des langues sifflées: approche linguistique et bioacoustique. Ph.D. dissertation Université Lyon 2, France
Meyer J (2008) Acoustic strategy and typology of whistled languages; phonetic comparison and perceptual cues of whistled vowels. J Int Phon Assoc 38:69–94
Meyer J (2015) Whistled languages – a worldwide inquiry on human whistled speech. Springer, Berlin
Meyer J, Díaz Reyes D (2018) Geolinguistica de los lenguajes silbados del mundo, con un enfoque en el español silbado. Andean Geol 17:99–124
Meyer J, Dentel L, Seifart F (2012) A methodology for the study of rhythm in drummed forms of languages: application to Bora Manguaré of Amazon. In: Proceedings of Interspeech 2012, Portland, USA, pp 686–690
Meyer J, Dentel L, Meunier F (2013) Speech recognition in natural background noise. PLoS One 9 (1). https://doi.org/10.1371/journal.pone.0079279
Meyer J, Meunier F, Dentel L (2015) Speech recognition experiment in 'natural quiet' background noise. In: Proceedings of International Congress of Phonetic Sciences, 2015, Glasgow, UK
Meyer J, Dentel L, Meunier F (2016) Categorization of natural Spanish whistled vowels by naïve Spanish listeners. Proceedings of Interspeech 2016. San Francisco, USA
Meyer J, Dentel L, Meunier F (2017) Categorization of natural whistled vowels by naïve listeners of different language background. Front Psychol 8:25. https://doi.org/10.3389/fpsyg.2017.00025
Meyer J, Meunier F, Dentel L, Do Carmo Blanco N, Sèbe F (2018) Loud and shouted speech perception at variable distances in a forest. In: Proceedings of the 19th International Congress of Phonetic Sciences 2018, Hyderabad, India
Michelsen A (1983) Biophysical basis of sound communication. In: Lewis B (ed) Bioacoustics, a comparative approach. Academic Press, New York, pp 3–38
Moles A (1970) Etude sociolinguistique de la langue sifflée de Kusköy. Revue de Phonétique Appliquée 14(15):78–118
Morton J, Marcus S, Frankish C (1976) Perceptual centers (P-centers). Psychol Rev 83:405–408. https://doi.org/10.1037/0033-295X.83.5.405
Nelken I, Rotman Y, Bar-Yosef O (1999) Response of auditory cortex neurons to structural features of natural sounds. Nature 397:154–157
Niangoran-Bouah G (1980) Introduction à la drummologie. G. N. B, Abidjan
Nketia JHK (1976) Surrogate languages of Africa. In: Sebeok TA, Umiker-Sebeok DJ (eds) Speech surrogates: drum and whistle systems. Mouton, The Hague, pp 825–864
Otterbein S, Abel C, Heinemann LV, Kaiser J, Schmidt-Kassow M (2012) P3b reflects periodicity in linguistic sequences. PLoS One 7:e51419. https://doi.org/10.1371/journal.pone.0051419
Padgham M (2004) Reverberation and frequency attenuation in forests – implications for acoustic communication in animals. J Acoust Soc Am 115:402–410. https://doi.org/10.1121/1.1629304
Palmer AR, Shamma SA (2004) Physiological representation of speech. In: Greenberg S et al (eds) Speech processing in the auditory system. Springer, New York, pp 163–230
Rialland A (2005) Phonological and phonetic aspects of whistled languages. Phonology 22(2): 237–271
Richards DG, Wiley RH (1980) Reverberations and amplitude fluctuations in the propagation of sound in a forest: implications for animal communication. Am Nat 115:381–399
Rossing TD (2000) Science of percussion instruments. World Scientific, Singapore; River Edge, NJ
Ryan KM (2014) Onsets contribute to syllable weight: statistical evidence from stress and meter. Language 90:309–341. https://doi.org/10.1353/lan.2014.0029

Sebeok TA, Umiker-Sebeok DJ (eds) (1976) Speech surrogates: drum and whistle systems. Mouton, The Hague

Seifart F, Meyer J, Grawunder S, Dentel L (2018) Reducing language to rhythm: Amazonian bora drummed language exploits speech rhythm for long-distance communication. R Soc Open Sci 5:170354. https://doi.org/10.1098/rsos.170354

Sicoli MA (2016) Repair organization in Chinantec whistled speech. Language 92(2):411–432

Stern T (1957) Drum and whistle 'languages': an analysis of speech surrogates. Am Anthropol 59:487–506. https://doi.org/10.1525/aa.1957.59.3.02a00070

Tang P (2007) Masters of the sabar: wolof griot percussionists of Senegal. Temple University Press, Philadelphia, PA

Thiesen W (1969) The bora signal drums. Lore 19:101–103

Varnet L, Meyer J, Hoen M, Meunier F (2012) Phoneme resistance during speech-in-speech comprehension. In: Proceedings of Interspeech 2012, Portland, USA, pp 598–602

Wiley RH, Richards DG (1978) Physical constraints on acoustic communication and the atmosphere: implications for the evolution of animal vocalizations. Behav Ecol Sociobiol 3:69–94

Wiley RH, Richards DG (1982) Adaptations for acoustic communication in birds: sound transmission and signal detection. In: Kroodsma EH, Miller DE (eds) Acoustic communication in birds, vol 1. Academic Press, New York, pp 131–181

Winter Y (2014) On the grammar of a Senegalese drum language. Language 90:644–668. https://doi.org/10.1353/lan.2014.0069

Zwicker E (1982) Psychoakustik. Springer, Berlin

Chapter 5
Coding of Static Information in Terrestrial Mammal Vocal Signals

Benjamin D. Charlton, Katarzyna Pisanski, Jordan Raine, and David Reby

Abstract Understanding how information is encoded in vocal signals, and what function it serves during social and sexual interactions, is a key objective of animal communication research. In this chapter, we focus on the origins and functions of vocal cues to stable, long-term attributes of callers (such as identity, sex, and body size) in terrestrial mammals, including humans. We show that relatively static vocal cues largely originate from biomechanical constraints linked to the animal's anatomy or physiology that affect the possible ranges of key frequency components (fundamental frequency and formant spacing) in their calls. We illustrate how the source–filter theory of voice production provides a useful framework for determining the biomechanical origins of the information content of mammalian vocal signals, and how dedicated tools enable researchers to test the functions and social outcomes of relevant acoustic variation in controlled playback experiments. The body of research reviewed in this chapter illustrates how combining observational (anatomy, acoustics), experimental (playback studies), and comparative approaches enables researchers to draw general conclusions about the selection pressures driving the evolution of vocal production and perception in mammals.

All authors equally contributed to this chapter.

B. D. Charlton
San Diego Zoo's Institute for Conservation Research, Escondido, CA, USA

K. Pisanski · D. Reby (✉)
Equipe de Neuro-Ethologie Sensorielle, ENES/CRNL CNRS UMR5292, INSERM UMRS1028, University of Lyon/Saint-Etienne, Saint-Etienne, France
e-mail: david.reby@univ-st-etienne.fr

J. Raine
Mammal Vocal Communication and Cognition Research Group, School of Psychology, University of Sussex, Brighton, UK

T. Aubin, N. Mathevon (eds.), *Coding Strategies in Vertebrate Acoustic Communication*, Animal Signals and Communication 7,
https://doi.org/10.1007/978-3-030-39200-0_5

Evidence that mammal vocal signals mediate social and sexual behaviour was initially slow to accumulate when compared to other animals groups, such as birds, anurans and insects (Bradbury and Vehrencamp 2011). Notwithstanding this, over the last 30 years a growing number of studies have shown that the acoustic structure of mammalian vocalisations encodes a wide range of information about the physical (or static) attributes of callers, like their identity, sex, and body size, as well as 'dynamic' information about short-term fluctuations in internal state and/or physiology that are of potential importance in a range of social contexts (Taylor et al. 2016). The aim of this chapter is to provide an overview of the origins and functions of vocal characteristics that have the potential to communicate socially relevant information about relatively static attributes of the caller (such as sex and size) in terrestrial mammals, including humans.

5.1 The Source–Filter Model of Vocal Production

Knowledge of vocal production is an essential prerequisite for understanding how the acoustic characteristics of mammal vocalisations encode information. The generalisation of the source–filter theory of speech production (Chiba and Kajiyama 1958; Fant 1960) to the production of nonhuman vertebrate vocal signals has provided a useful framework for linking vocal production to signal properties in terrestrial mammals (Taylor et al. 2010). This theory states that vocal production follows a two-stage process. In the first stage, air expelled from the lungs passes through the glottis (the space between the vocal folds) and forces the vocal folds inside the larynx (*the source*) to open. Air pressure differences across the glottis (termed Bernoulli forces), in combination with the elastic recoil of the vocal fold tissue, cause the folds to then rapidly close after they have been forced open (Titze 1994). The rate at which the vocal folds open and close determines the fundamental frequency (F0) of the vocal signal. Because the opening and closing phases of the vocal folds typically differ in duration, the sound spectrum of most mammal vocal signals also contains harmonics (or overtones) that occur at integer multiples of F0 (see Fig. 5.1). It is possible to predict F0 as a function of basic characteristics of the vocal apparatus producing the signal using the following formula, where L is vocal fold length, σ is the stress applied to the vocal folds, and ρ is the tissue density of the vocal folds:

$$F_0 = \frac{1}{2L}\sqrt{\frac{\sigma}{\rho}} \qquad (5.1)$$

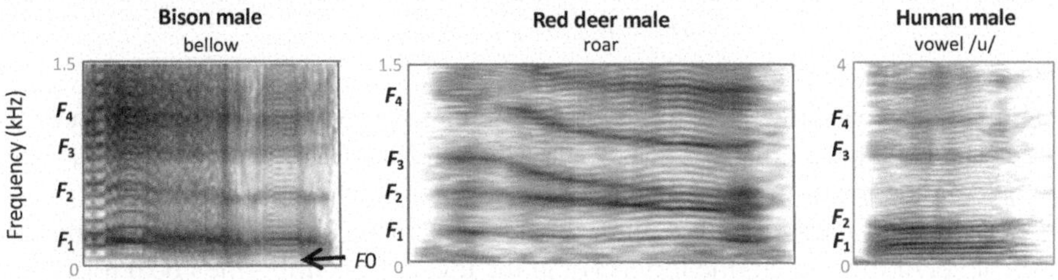

Fig. 5.1 Spectrograms illustrating vocalisations produced by a male bison, red deer and human. Frequency is represented on the *y*-axis (kHz), wherein fundamental frequency (F0) is indicated with an arrow, and harmonics are shown as integer multiples of the F0. The first four formant frequencies (F1–F4) are indicated by adjacent labels, and according to source–filter theory, these vocal tract resonances are largely independent of F0 and harmonics. The central spectrogram illustrates formant modulation for size exaggeration in a red deer stag, wherein formants progressively lower (and formant spacing, ΔF, progressively decreases) over the course of the roar as the male lowers his larynx in the vocal tract. Despite this dynamic vocal modulation, minimum formant frequencies are anatomically constrained by vocal tract length, thus ensuring that some degree of reliable information regarding body size is retained ("honesty in an exaggerated signal")

In the second stage of mammal vocal signal production, the glottal waveform produced by vocal fold vibration is filtered by the supra-laryngeal vocal tract (*the filter*), which consists of the pharyngeal, oral and/or nasal cavities. The air passing through these cavities vibrates preferentially at certain resonance frequencies that are determined by the size and shape of the vocal tract (Titze 1994). These vocal tract resonances shape the frequency spectrum of the sound wave created by the source by boosting the energy of frequencies corresponding to them and damping frequencies that do not. This process produces frequency peaks in the spectrum, termed formants (from *formare*, meaning 'to shape' in Latin) (Fant 1960; Titze 1994) (see Fig. 5.1). In calls radiated through the mouth, and under the assumption that the nasal tract is not involved, the vocal tract can be modelled as a tube with a uniform cross-sectional area across its length that is closed at one end (representing the glottis) and open at the other (the mouth). Such a tube acts as a quarter wavelength resonator. The relationship between formants and vocal tract length is given by the following formula, where c is the speed of sound in air (approximated as 350 m/s in a mammal's vocal tract), F_i is the ith Formant, and VTL represents vocal tract length measured along a curve from the glottis to the lips:

$$F_i = \frac{(2_i - 1)c}{4\text{VTL}} \tag{5.2}$$

The average formant spacing (ΔF), representing the spacing between any two consecutive formants, can thus be derived as:

$$\Delta F = F_{i+1} - F_i = \frac{c}{2\mathrm{VTL}} \quad (5.3)$$

Equation (5.3) shows that ΔF is inversely related to vocal tract length. Thus, if larger animals have longer vocal tracts, then formant frequencies and ΔF are expected to be lower in the calls of relatively larger individuals.

A key assumption of source–filter theory is that acoustic components produced at the level of the source (the larynx) or at the level of the filter (the vocal tract) are separate and additive, and can vary independently of one another (Fant 1960; Fitch and Hauser 1995). As a consequence, both F0 and formants have the potential to independently encode information about physical, and typically socially relevant characteristics of the caller. Understanding how these acoustic parameters co-vary with the dimensions and tissue properties of the vocal apparatus can therefore lead to informed predictions about what information vocalisations contain, and playback experiments can be used to determine the function (e.g. social, reproductive) of acoustic variation in different behavioural contexts. In addition to using natural (unmanipulated) vocalisations in playback experiments, dedicated re-synthesis techniques, such as Pitch Synchronous Overlap and Add (PSOLA) (Moulines and Charpentier 1990), allow researchers to independently manipulate acoustic components and create 're-synthesised' calls that mimic acoustic variation in potentially important biosocial information that can be presented to conspecific receivers using playback experiments (Reby et al. 2005, 2010; Charlton et al. 2007b, 2010, 2013c; Charrier et al. 2011; Pitcher et al. 2015). The use of re-synthesis techniques has proved to be an especially valuable tool for understanding the functional relevance, and thus, the evolutionary origins of acoustic variation in mammal vocalisations.

5.2 The Source

5.2.1 Information Encoded by F0

Mean F0 is highly individualised in several terrestrial mammal species (Charrier et al. 2003; McComb et al. 2003; Searby and Jouventin 2003; Torriani et al. 2006; Charlton et al. 2009a). This is not surprising because the range of possible F0s an individual can produce is ultimately constrained by factors such as the length and stiffness of the vocal folds (Titze 1994, 2011) and these factors are likely to vary among individuals. Other aspects of the F0 contour are also individualised in terrestrial mammals. For instance, maximum F0 is individually distinctive in Iberian wolves (Palacios et al. 2007) and infant chimpanzees (Levrero and Mathevon 2013) whereas F0 modulation characteristics have been found to encode caller identity in other terrestrial mammal species, including giant pandas (*Ailuropoda melanoleuca*) (Charlton et al. 2009a) and female goats (*Capra hircus*) (Briefer and McElligott 2011). In some cases, highly distinctive F0 contours are also thought to constitute a uniquely identifiable 'vocal signature'. Although this has been most widely studied in non-terrestrial mammals, such as the bottlenose dolphin (*Tursiops truncate*) (Janik et al. 2006), there is good empirical evidence that some bats use similar acoustic cues to identify specific individuals (Carter et al. 2008; Melendez and Feng 2010).

In addition, because larger animals can be expected to have larger larynges with longer vocal folds, they should produce calls with lower F0s (see Eq. 5.1). This expectation is verified across many terrestrial mammals (Charlton and Reby 2016; Bowling et al. 2017; Fig. 5.2a). For example, African elephants (*Loxodonta africana*) can produce calls with mean F0s as low as 16.8 Hz (McComb et al. 2003), whereas house mice (*Mus musculus*) can produce F0s of up 97,000 Hz (Gourbal et al. 2004). In a number of mammalian species, however, F0 does not predict body size within age and sex classes (e.g. when comparing adult males of a single species) (Reby and McComb 2003; Charlton et al. 2011b), with only a few notable exceptions (Pfefferle and Fischer 2006; Charlton et al. 2009b). This lack of a relationship between F0 and body size in same-sex adults is most likely due to the fact that laryngeal growth is largely unconstrained by surrounding skeletal structures (Fitch and Hauser 2002; Garcia et al. 2017). Vocal fold tension and length can also be adjusted to vary F0 within a producible range that is ultimately dictated by the length of the vocal folds (Titze 2011). Humans, for example, have a particularly high degree of neural control over their vocal folds and can therefore effortlessly volitionally raise or lower F0 (e.g. in speech production or singing: Pisanski et al. 2016a). Such voice modulation is likely to contribute to the weak relationship between F0 and height in humans, where F0 explains less than 2% of the variance in men's and 0.5% of the variance in women's heights (Pisanski et al. 2014b).

Although, as we have seen, F0 tends to be a relatively poor predictor of body size within age and sex classes, it can be a reliable cue to the caller's age or maturity because the larynx typically grows larger as an animal develops (Pfefferle and Fischer 2006; Briefer and McElligott 2011; Stoeger and Baotic 2016). In some mammals, F0 is also a reliable cue to a caller's sex (Rendall et al. 2004; Charlton 2015). This is particularly true in humans (Puts et al. 2016), among which vocal fold length is affected by differential androgen exposure between the sexes during puberty, leading to a male speaking voice with a 50–80% lower F0 (mean 120 Hz) than that of females (mean 210 Hz) (Hollien et al. 1994; Pisanski et al. 2014b). The mean F0 of human nonverbal vocalisations, including laughter (Bachorowski et al. 2001) and tennis grunts (Raine et al. 2017), retains a similar degree of sexual dimorphism as that observed in modal speech.

The F0 characteristics of human speech are also individually distinctive. Indeed, while the average difference in mean F0 between sexes is around 90 Hz, differences within adults of each sex can span an even greater range, with individual differences in mean F0 ranging from 80 to 180 Hz in men, and 160 to 275 Hz in women (Pisanski et al. 2014b; Kreiman and Sidtis 2011). Such individual differences in the F0 of men's voices remain stable during adulthood and may even be determined

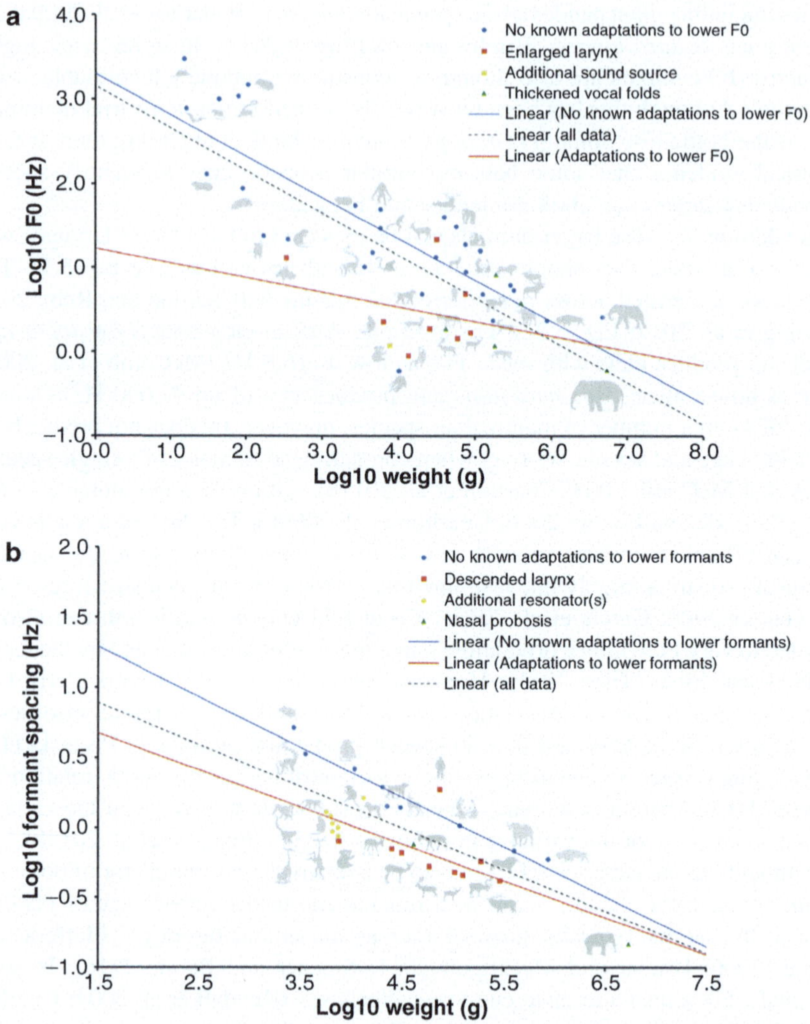

Fig. 5.2 (**a**) The relationship between male body mass and F0 across 67 species of terrestrial mammal, and (**b**) male body mass and formant frequency spacing (ΔF) across 35 terrestrial mammal species. The data were generated using phylogenetic generalised least-squares regressions that control for the shared ancestry of each species and the species-typical habitat. In both plots, the red line shows the relationship for species with adaptations that allow them to lower F0 or ΔF, and the blue line shows the relationship for species that are not known to possess adaptations (adapted from Charlton and Reby 2016)

before puberty (Fouquet et al. 2016). Furthermore, inter-individual differences in the F0 of 4-month-old babies' cries predict inter-individual differences in their speech F0 at age 5, suggesting a possible role of in-utero androgen exposure in shaping F0 (Levrero et al. 2018).

There is also growing evidence that F0 may provide reliable information about circulating sex hormone levels in humans. For example, men with higher levels of salivary testosterone have lower F0 (Dabbs and Mallinger 1999; Cartei et al. 2014). This relationship appears strongest in men with low cortisol levels (Puts et al. 2016), where concurrently high testosterone and low cortisol levels also predict higher ratings of attractiveness (Rantala et al. 2012) and dominance (Mehta and Josephs 2010) in playback experiments. Men with relatively lower F0 also score higher on various indices of dominance (see e.g. Fig. 5.3b). While the information communicated by F0 in women's voices is understudied relative to men, female voice F0 appears linked to dynamic changes in oestrogen and androgen levels, and may communicate reproductive status. For example, women's voice F0 decreases permanently following menopause (Abitbol et al. 1999), decreases temporarily following pregnancy (Pisanski et al. 2018), and also varies dynamically across the menstrual cycle (Bryant and Haselton 2009; Fischer et al. 2011).

In contrast to research on humans, there is relatively little evidence that F0 is a reliable cue to hormonal quality in nonhuman mammal species. Indirect evidence that increased androgen exposure during development may lead to lower adult F0 exists in sheep (*Ovis aries*) (Beckford et al. 1985). In their study, Beckford and colleagues showed that castrated lambs given higher doses of testosterone and dihydrotestosterone went on to develop larger larynges (Beckford et al. 1985), which may in turn be expected to increase vocal fold length and thus lower F0 in these individuals. In addition, F0 is inversely correlated with male dominance rank in fallow deer (*Dama dama*) (Vannoni and McElligott 2008) (Fig. 5.3), Père David's deer (*Elaphurus davidianus*) (Liu et al. 2016) and gelada baboons (*Theropithecus gelada*) (Benitez et al. 2016), raising the possibility that these relationships are mediated by androgen levels. In contrast, F0 is positively correlated with dominance status in male chacma baboons (*Papio ursinus*) (Fischer et al. 2004). One plausible explanation for this positive relationship is that dominant male chacma baboons produce louder calls with higher sub-glottal pressure, leading to higher F0s.

To our knowledge, definitive evidence that circulating androgen levels influence F0 in nonhuman mammals only exists in male chimpanzees (*Pan troglodytes*) (Fedurek et al. 2016), white-handed gibbons (*Hylobates lar*) (Barelli et al. 2013) and giant pandas (*A. melanoleuca*) (Charlton et al. 2011c). Both chimpanzees and white-handed gibbons with higher androgen levels produce long-distance calls with a higher dominant frequency (the frequency with the highest amplitude) or a higher F0 (Barelli et al. 2013; Fedurek et al. 2016) compared to conspecifics with lower androgen levels, and male giant pandas with higher androgen levels produce bleats with relatively faster rates of F0 modulation (Charlton et al. 2011c). In future studies, researchers should now examine the effect of androgens on laryngeal development in other mammal species to determine whether low F0 is indicative of high androgen levels in nonhuman mammals, as it is in humans.

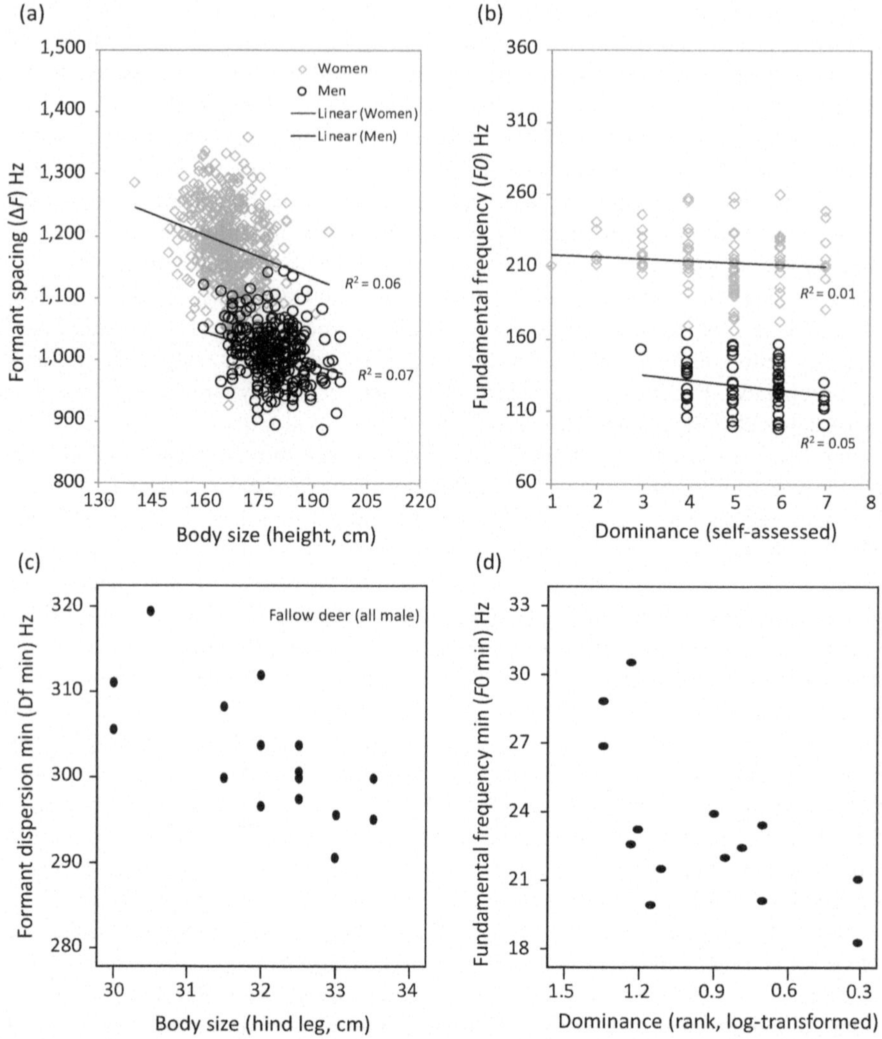

Fig. 5.3 Scatterplots illustrating negative relationships between formants and body size and between fundamental frequency and dominance in two species of mammals. Top panels represent these relationships in human adult males and females [(**a**) $N = 700$ cross-cultural sample; (**b**) $N = 200$, Cuban sample; males represented as circles], whereas bottom panels show the relationships in male fallow deer [(**c**) $N = 17$; (**d**) $N = 14$]. These graphs were created or adapted with permission from the following sources: (**a**) Pisanski et al. (2014b); (**b**) unpublished data [Pisanski K.]; (**c** and **d**) Vannoni and McElligott (2008)

5.2.2 Function of F0

Although F0 is a highly variable and salient feature of nonhuman mammal calls, the function of F0 variation within calls remains poorly understood. Nevertheless, F0 modulation appears to be important for a mother's identification of her offspring in white-winged vampire bats (*Diaemus youngi*) (Carter et al. 2008) and for Australian fur seal (*Arctocephalus tropicalis*) pups to recognise their mothers' calls (Charrier et al. 2009), wherein F0 and the first two harmonics are sufficient for mothers to recognise the vocalisations of their own pups in this species (Charrier et al. 2002). In addition, a recent study on horses (*Equus caballus*) revealed that more fertile stallions produce whinnies with relatively lower F0, and that oestrous females preferentially orient towards and spend more time in close proximity to playback speakers broadcasting low F0 whinnies (Lemasson et al. 2015). It must be noted, however, that this study did not use re-synthesis techniques to systematically vary F0 whilst leaving other acoustic features unchanged. As a consequence, the reported female preference could have been based on other features of the call that co-varied with F0. To date, the only playback studies to have used re-synthesis techniques to test the functional relevance of F0 variation in mate choice contexts among nonhuman mammals were conducted on red deer (*Cervus elaphus*). Although initial investigations failed to reveal a differential response from females to male roars with low versus high F0 (McComb 1991; Charlton et al. 2008), further work on oestrous hinds revealed a preference for male roars with higher F0 (Reby et al. 2010), consistent with observations that males who produce higher F0 roars also have higher reproductive success (Reby and McComb 2003). While the adaptive significance of a preference for high F0 is not clear, these findings emphasise that experimental investigation of female choice in mammals should be conducted during the female's peak fertility.

F0 could also play an important role during male–male assessment in nonhuman mammals. For example, male fallow deer (*D. dama*) that produce relatively low F0 groans have higher reproductive success and dominance ranks (Vannoni and McElligott 2008) (Fig. 5.3d), and playback studies have shown that groans with experimentally lowered F0 are perceived as more threatening by males during the breeding season (Pitcher et al. 2015). In the closely related red deer (*C. elaphus*), however, males do not show differential responses to different F0 variants, indicating that F0 may not be an important component of competitive interactions in this species (Garcia et al. 2013). Finally, many species of mammal produce bleat-like calls in which the F0 is rapidly, and periodically modulated. The rate of F0 modulation in male giant panda (*A. melanoleuca*) bleats is a reliable cue to inter-individual variation in androgen levels (Charlton et al. 2011c) and both male and female giant pandas have been shown to attend to this cue during the breeding season (Charlton et al. 2012b).

Although the functional relevance of F0 variation in nonhuman terrestrial mammals is not clear, the increase in F0 dimorphism observed in polygamous relative to monogamous anthropoids (Puts et al. 2016) strongly indicates that F0 is a sexually selected component in this phyletic group. The fact that F0 is disproportionately lower in terrestrial mammal species with relatively smaller testes (Charlton and Reby 2016) also confirms that F0 is likely to be an important, sexually selected component of pre-copulatory signalling in many mammals. Additional playback studies that use re-synthesis techniques to systematically vary F0 are now required to further elucidate the role of F0 characteristics in terrestrial mammal sexual communication.

In humans, playback studies have consistently shown a strong influence of F0 on listeners' assessments of a range of evolutionarily relevant traits. Largely due to strong sexual dimorphism in F0, human listeners can judge the sex of adult speakers with almost perfect accuracy from natural speech (Kreiman and Sidtis 2011), and are more than 90% accurate when judging sex from nonverbal vocalisations, such as tennis grunts (Raine et al. 2017). Within sexes, men with lower and less variable F0 are judged as more dominant (both by themselves, Fig. 5.3b, and by others) and as less cooperative than are men with relatively higher and more variable F0 (Puts et al. 2007; Hodges-Simeon et al. 2010; Knowles and Little 2016). This is consistent with the aforementioned negative correlation between speech F0 and androgen levels in human males. Lower mean F0 is also judged as more attractive (Feinberg et al. 2005; Apicella and Feinberg 2009; Šebesta et al. 2017) more masculine (Cartei et al. 2014), and more strongly indicative of resource acquisition capability in men (Apicella and Feinberg 2009). At the same time, men with lower F0 are assigned a higher perceived likelihood of infidelity (O'Connor et al. 2011; Hughes and Harrison 2017) and a lower likelihood of investing resources into a romantic relationship and potential offspring (O'Connor et al. 2012). Thus, while men with relatively lower F0 voices may benefit from a higher level of social dominance and reproductive success (i.e., a higher total number of offspring), and are often preferred by women as short-term mates, such men are less likely to be perceived as suitable long-term mates (Puts et al. 2012b).

Like men, women with lower F0 are judged as more dominant and as better at acquiring resources relative to women with higher pitched voices (Apicella and Feinberg 2009; Borkowska and Pawlowski 2011; Knowles and Little 2016). Women with relatively lower F0 also rate themselves as more dominant (Leongómez et al. 2017). In contrast, women with comparatively higher F0 voices are judged as more feminine, younger, and, in some contexts, as more attractive, though studies investigating the relationship between F0 and perceived attractiveness in women have produced mixed results (Pisanski and Feinberg 2018). As previously described, women's F0 is linked to dynamic changes in oestrogen and androgen levels, and as such, appears to communicate information about women's fertility status and reproductive age. While this suggests that women's F0 may serve as a signal of fertility, few playback studies have tested the extent to which listeners are able to accurately gauge fertility-related information from women's F0, and this constitutes an important avenue for future work. Importantly, perceptual biases linked to low F0 in humans have also been shown to influence social behaviour beyond a mate choice context, including voting preferences (Pavela Banai et al. 2017), access to emergency medical care (Boidron et al. 2016), men's aggressive intent (Zhang and Reid 2017), and even women's trust (Montano et al. 2017), thus highlighting the broad implications of F0 perception on human behaviour.

5.2.3 Departures from F0/Size Allometry

We have described how larger terrestrial mammal species tend to produce calls with relatively lower F0s (Charlton and Reby 2016; Bowling et al. 2017) (Fig. 5.2a). Yet, in the absence of strong external constraints on vocal fold morphology, many species have developed adaptations that further decouple F0 and body size at both interspecific and intraspecific levels. For example, some bats (Suthers and James 1973) and primates (Schön-Ybarra 1995) produce higher than expected vocal frequencies due to thin vocal membranes lining the edge of their vocal folds. Other species possess hypertrophied larynges (howler monkeys, *Alouatta palliata*, Kelemen and Sade 1960; hammer-headed bats, *Hypsignathus monstrosus*, Bradbury 1977) or fleshy padded vocal folds (Mongolian gazelles, *Procapra gutturosa*, Frey and Gebler 2003; lions and tigers, *Panthera* sp., Klemuk et al. 2011) enabling them to produce calls with lower than expected F0s. The koala (*Phascolarctos cinereus*) has even evolved an additional sound production organ, termed the 'velar vocal folds' located outside of the larynx (Charlton et al. 2013a), that allows this species to produce extremely low F0s (circa 30 Hz) for its relatively small size. The North American elk (*Cervus canadensis*) also appears to use a non-laryngeal mechanism during the production of bugle vocalisations. In this species, air flow from the larynx is directed through a narrow supra-glottic constriction to produce a very high-frequency aerodynamic whistle (Reby et al. 2016). Together these adaptations clearly indicate that F0 can be under strong, often divergent sexual selection in terrestrial mammals.

5.3 The Filter

5.3.1 Information Encoded by Formant Frequencies

Formants are likely to be individually distinctive in terrestrial mammal vocal signals due to intra-individual differences in vocal tract morphology. This prediction has now been verified in a wide range of species, including humans (*H. sapiens*) (Bachorowski and Owren 1999), rhesus macaques (*Macaca mulatta*) (Rendall et al. 1998), African elephants (*L. africana*) (McComb et al. 2003; Stoeger and Baotic 2016), male koalas (*P. cinereus*) (Charlton et al. 2011a), male fallow deer (*D. dama*) (Vannoni and McElligott 2007), Corsican deer (*Cervus elaphus corsicanus*) (Kidjo et al. 2008), red deer (*C. elaphus*) (Reby et al. 2006), giant pandas (*A. melanoleuca*) (Charlton et al. 2009a), meerkats (*Suricata suricatta*) (Townsend et al. 2014), red-bellied lemurs (*Eulemur rubriventer*) (Gamba et al. 2012), southern hairy-nosed wombats (*Lasiorhinus latifrons*) (Charlton 2014) and chacma baboons (*P. ursinus*) (Rendall 2003). Vocal signals with a broadband and noisy source or rapid F0 modulation are especially well suited for highlighting a given caller's distinctive formant pattern (Owren and Rendall 2001; Charlton et al. 2017).

According to Eq. (5.3), larger animals should also produce calls with lower formant frequency spacing (ΔF) because they have longer vocal tracts with lower resonant frequencies. This is clearly the case across terrestrial mammals (Charlton and Reby 2016; Bowling et al. 2017) (Fig. 5.2b). In addition, unlike F0, ΔF is a reliable cue to body size within sex and age classes in several terrestrial mammal species, including humans (Pisanski et al. 2014b), rhesus macaques (*M. mulatta*) (Fitch 1997), male red deer (*C. elaphus*) (Reby and McComb 2003), male fallow deer (*D. dama*) (Vannoni and McElligott 2008), male giant pandas (*A. melanoleuca*) (Charlton et al. 2009b), male and female koalas (*P. cinereus*) (Charlton et al. 2011b, 2015), male bison (*Bison bison*) (Wyman et al. 2012), and male and female domestic dogs (*Canis familiaris*) (Riede and Fitch 1999). In these species, formant frequencies have the potential to signal reliable size-related information to receivers.

Formants can also be reliable cues to a caller's sex in size dimorphic species (Rendall et al. 2004; Charlton 2015), and to a caller's maturity (Reby and McComb 2003; Charlton et al. 2009b; Briefer and McElligott 2011), as animals will inevitably grow larger as they approach adulthood. In humans, for example, testosterone causes the larynx to descend further in males than females during puberty, paralleling male-specific laryngeal descents in nonhuman mammal species (Fitch and Reby 2001; McElligott et al. 2006; Frey et al. 2011). This, in combination with males' increased growth in height (Gaulin and Boster 1985), elongates the male vocal tract and lowers formants by 15–20% more in adult men than women (Fant 1960; Fitch and Giedd 1999; Pisanski and Rendall 2011).

5.3.2 Function of Formants

Studies using re-synthesised playback stimuli, where formant spacing is rescaled to mimic individual differences in body size, have shown that nonhuman mammals perceive and respond to size-related formant variation in the species-specific calls of conspecifics. Habituation–discrimination playback experiments have shown that rhesus macaques (*M. mulatta*) (Fitch and Fritz 2006), female red deer (*C. elaphus*) (Charlton et al. 2007a), and male koalas (*P. cinereus*) (Charlton et al. 2012c) can all perceive shifts in formant frequency spacing in species-specific calls corresponding to natural variation in the body sizes of these species (Fitch and Fritz 2006). In addition, formants are used by red deer (*C. elaphus*) and koalas (*P. cinereus*) as acoustic cues to assess the body size of potential mates (Charlton et al. 2007b, 2012a) and rivals (Reby et al. 2005; Charlton et al. 2013c) during the breeding season. In both of these species, males and females respond to sexual calls with lower formants as more threatening and attractive, respectively.

Playback experiments on male fallow deer (*D. dama*) have shown that bucks attend to size-related formant variation in male groans, and treat groans with lower formants, mimicking larger males, as more threatening (Pitcher et al. 2015). Similar results have been found for male giant pandas (*A. melanoleuca*) (Charlton et al. 2010) and domestic dogs (*C. familiaris*) (Taylor et al. 2010). Conversely, male

Australian sea lions (*Neophoca cinerea*) respond more strongly to re-synthesised barks with higher formants, which are designed to represent smaller potential rivals (Charrier et al. 2011). It must be noted, however, that an inverse relationship between male formant spacing and body size has not yet been established in this species. Finally, domestic dogs (*C. familiaris*) have also been shown to spontaneously match size-related formant information with corresponding visual stimuli (Faragó et al. 2010; Taylor et al. 2011), and rhesus macaques (*M. mulatta*) associate lower formants with pictures of more mature conspecifics (Ghazanfar et al. 2007). Taken together, these studies demonstrate that terrestrial mammals are capable of perceiving and using size-related formant information in conspecific calls, which is particularly relevant in inter- and intra-sexual contexts. There is also good evidence that nonhuman mammals have the ability to match perceived acoustic size with visual size.

Human listeners can perceive shifts in formant frequencies of about 5% from baseline from strings of vowel sounds in human speech (Pisanski and Rendall 2011) and also pay attention to formants when assessing sex or body size from the voice (Rendall et al. 2007; Pisanski and Rendall 2011). While F0 is relatively more important than formants in influencing sex judgments (Markova et al. 2016), re-synthesis of both F0 and formants most effectively changes perceived sex (Hillenbrand and Clark 2009). Within sexes, listeners can assess relative differences in men's and women's heights, with accuracy increasing as the difference in height between two speakers increases (Pisanski et al. 2014a; Rendall et al. 2007). In fact, human listeners are able to assess size from the voice even in the absence of prior audio-visual experience (i.e. blind persons: Pisanski et al. 2016b, 2017), and from as early as 3 months of age (Pietraszewski et al. 2017). However, when the voices of speakers are randomly paired to represent a natural range of variation in the relative heights of speakers, listeners can correctly identify the taller of two men from their voices only 60% of the time, on average (Rendall et al. 2007; Pisanski et al. 2014b), with comparable or lower accuracy for assessments of women's relative heights (Gonzalez 2004; Pisanski et al. 2017), and exceptional performance only when relative height differences exceed about 15 cm (Pisanski et al. 2014a, 2016b; Rendall et al. 2007). This is consistent with work showing that just-noticeable differences in formant-based vocal tract length are roughly 4–7% in synthesised speech (Ives et al. 2005; Smith et al. 2005).

Despite F0 being a poor predictor of body size when controlling for sex and age (Pisanski et al. 2014b), accuracy of size estimation is nevertheless highly dependent on F0. Many studies report a consistent perceptual bias in human listeners to associate low F0 voices with larger body size within a given sex (Smith et al. 2005; Rendall et al. 2007; Pisanski and Rendall 2011), despite F0 being a poor predictor of body size when controlling for sex and age (Pisanski et al. 2014b). This bias may represent an overgeneralisation of between-sex and between-species sound–size relationships to the within-sex level, where F0–size relationships no longer hold (Rendall et al. 2007). Thus, while listeners can prioritise formant information over F0 cues in judging body size (Pisanski and Rendall 2011), listeners' assessments are more accurate when F0 is matched between exemplars in two-alternative forced-choice trials (Rendall et al. 2007).

Notably, F0 may also affect formant perception. Vocalisations with low F0s will have more closely spaced harmonics, which should increase the ability of receivers to resolve formant peaks (Ryalls and Lieberman 1982; Owren and Rendall 2001). Indeed, two studies on human listeners have shown that increased harmonic density yields relatively higher accuracy in body size estimation (Charlton et al. 2013b; Pisanski et al. 2014b). Although Charlton and colleagues found that low F0 did not significantly alter the behavioural responses of receivers to size-related formant information in red deer (*C. elaphus*) roars (Charlton et al. 2008) or giant panda (*A. melanoleuca*) bleats (Charlton et al. 2010), more recent work has shown that formant perception in non-speech sounds is improved by F0 modulation in nonhuman mammals (i.e., short-term vibrato-like fluctuations in F0 (Charlton et al. 2017)), and that mammals modulate F0 over greater frequency ranges when the number of harmonic overtones per formant is low. These findings suggest that F0 modulation may be a mechanism that is used to improve formant perception in calls with low spectral density, providing a plausible mechanism for the evolution of F0 modulation in nonhuman mammals (Charlton et al. 2017).

Given the strong influence of body size on social and reproductive outcomes, and the perceptibility of formant-based cues to size, it logically follows that formants in humans, as in nonhuman mammals, should influence listeners' fitness-related attributions. Indeed, men with lowered formant frequencies are perceived as more socially and physically dominant (Puts et al. 2007, 2012a), and as more masculine (Feinberg et al. 2005; Pisanski and Rendall 2011; Cartei et al. 2014), compared to men with higher formant frequencies. Yet the effect of men's formant frequencies on their perceived attractiveness is unclear, as attractiveness ratings may only be influenced by large formant differences (Feinberg et al. 2005; Pisanski and Rendall 2011) and relatively lower formants may only be attractive in certain cultures (Šebesta et al. 2017), and/or to fertile-phase women (Hodges-Simeon et al. 2010). The relationship may also be curvilinear in some populations, with formant values closer to the mean considered more attractive, thus potentially mapping onto preferences in both sexes for mates of average to slightly above-average height (Stulp and Barrett 2014).

5.3.3 *Adaptations for Producing Lower Formants*

Across mammals, larger species tend to produce calls with lower formants (Charlton and Reby 2016) (Fig. 5.2b). However, in mating systems with greater sexual size dimorphism (a proxy for sexual selection pressures), some male terrestrial mammals produce vocal signals with much lower ΔF (but not F0) than expected for their size (Charlton and Reby 2016) (Fig. 5.2b). These observations indicate that sexual selection pressures to exaggerate apparent body size have led to the evolution of anatomical adaptations in some terrestrial mammals that enable males to lower formants in calls, thus sounding larger than they actually are.

One way to sound larger is to lengthen the supra-laryngeal vocal tract by lowering the larynx. The most extreme form of this is seen in male red deer that retract the larynx to the sternum when they roar during the breeding season (Fitch and Reby 2001; Reby and McComb 2003). Male fallow deer (*D. dama*) (Fitch and Reby 2001; McElligott et al. 2006), Mongolian gazelles (*P. gutturosa*) (Frey et al. 2008) and goitered gazelles (*Gazella subgutturosa*) (Frey et al. 2011) also temporarily retract the larynx when they produce sexual calls. Big cats (*Panthera* sp.) (Weissengruber et al. 2002) and koalas (*P. cinereus*) (Charlton et al. 2011b) have a permanently descended larynx; however, it is not known whether these species actively lower the larynx further during call production. Interestingly, although humans (and particularly men) also have a descended larynx, humans produce higher than expected ΔF for their body size (Charlton and Reby 2016), possibly as a consequence of competing selection pressures to decrease facial size (thereby shortening the oral cavity), facilitating speech production, thermoregulation and locomotion (Charlton and Reby 2016). Other species can achieve disproportionately low formant spacing in their calls using other means. Male saiga (*Saiga tatarica*), proboscis monkeys (*Nasalis larvatus*) and elephant seals (*Mirounga leonina*) have enlarged noses that extend the nasal tract and lower formants (Frey and Hofmann 1996; Frey et al. 2007; Sanvito et al. 2007; Koda et al. 2018), whereas male black and white colobus monkeys (*Colobus guereza*) and howler monkeys (*Alouatta* sp.) have air sacs that appear to act as additional resonators, either adding vocal tract resonances or lowering vocal tract resonances in emitted calls (Harris et al. 2006; Dunn et al. 2015).

Crucially, while these specialisations exaggerate the acoustic impression of size at the species level (animals can achieve frequencies that are normally typical of larger species), the signals remain honest within a given species (Reby and McComb 2003; Charlton et al. 2011b), because at a given stage of evolution, all members of the species share the adaptation. This is especially the case when vocalisers utilise such adaptations to their anatomical limits (e.g. maximum laryngeal retraction in red deer (*C. elaphus*) male roars, see Fig. 5.1), ensuring that vocalisations contain honest information on body size, which is in turn attended to by receivers. At the same time, by increasing the range of producible frequencies, these anatomical adaptations effectively broaden the acoustic space available to the species. For example, in addition to modulating formants by extending the vocal tract for size exaggeration, lengthening the vocal tract may additionally indicate aggression and threat, whereas shortening the vocal tract may signal subordination. Such vocal modulation could provide an evolutionary pathway towards increased vocal tract control, and subserve the communication of motivational information by conventionalisation processes (Ohala 1984; Pisanski et al. 2016a).

5.4 Conclusion

In summary, we have shown how acoustic cues to static, long-term attributes of callers (e.g. identity, sex, body size) are present in the vocalisations of numerous terrestrial mammals, including humans, due to morphological and physiological constraints linked to the biomechanics of sound production. We emphasise that the source–filter theory of voice production, outlined at the beginning of the chapter, provides an excellent framework for studies that aim to determine the information content of vocal signals in mammal species. We also highlight how playback experiments have provided novel and often foundational insights into the function of static cues in terrestrial mammal vocalisations and human speech, and describe species with anatomical adaptations that allow them to produce calls with unexpectedly high or low frequency components (F0 and formants), effectively manipulating vocal cues to socially relevant traits, such as body size and dominance rank.

It is our hope that this chapter spurs and guides future research. In particular, playback studies and re-synthesis techniques should be used to shed light on the function of F0 variation in nonhuman terrestrial mammals. In addition, the recent discovery of previously undocumented vocal adaptations in the koala (*Phascolarctos cinereus*) and the North American elk (*Cervus canadensis*) indicate that there is still much to be learned about how mammals produce vocal signals (Charlton et al. 2013a; Reby et al. 2016). Indeed, while the causal link between vocal production and acoustic variation in humans is well established (Titze 1994), the biomechanical, and especially the physiological determinants of acoustic variation in the vocalisations of many nonhuman mammals remain unknown. Whereas in humans, research is now needed to gauge the extent to which men and women volitionally modulate the F0 and formants of their voices to exaggerate various traits, particularly across real-life social contexts (Pisanski et al. 2016a). Future studies that combine detailed examination of vocal anatomy and the acoustic structure of vocal signals with physiological measures are now required to address this knowledge gap.

References

Abitbol J, Abitbol P, Abitbol B (1999) Sex hormones and the female voice. J Voice 13:424–446

Apicella C, Feinberg D (2009) Voice pitch alters mate-choice-relevant perception in hunter–gatherers. Proc R Soc B 276(1659):1077–1082

Bachorowski J, Owren M (1999) . . . talker sex and individual talker identity are present in a short vowel segment produced in running . . . . J Acoust Soc Am 106:1054

Bachorowski JA, Smoski MJ, Owren MJ (2001) The acoustic features of human laughter. J Acoust Soc Am 110(3):1581e1597

Barelli C, Mundry R, Heistermann M, Hammerschmidt K (2013) Cues to androgens and quality in male gibbon songs. PLoS One 8:e82748

Beckford NS, Schain D, Roor SR, Schanbacher B (1985) Androgen stimulation and laryngeal development. Ann Otol Rhinol Laryngol 94:634–640

Benitez M, le Roux A, Fischer J, Beehner J, Bergman T (2016) Acoustic and temporal variation in gelada (*Theropithecus gelada*) loud calls advertise male quality. Int J Primatol 37:1–18
Boidron L, Boudenia K, Avena C, Boucheix J-M, Aucouturier J-J (2016) Emergency medical triage decisions are swayed by computer-manipulated cues of physical dominance in caller's voice. Sci Rep 6:30219
Borkowska B, Pawlowski B (2011) Female voice frequency in the context of dominance and attractiveness perception. Anim Behav 82:55–59
Bowling DL, Garcia M, Dunn JC, Ruprecht R, Stewart A, Frommolt KH, Fitch WT (2017) Body size and vocalization in primates and carnivores. Sci Rep 7:41070
Bradbury J (1977) Lek mating behavior in the hammer–headed bat. Z Tierpsychol 45:225–255
Bradbury J, Vehrencamp SL (2011) Principles of animal communication. Sinear Associates Inc, Sunderland, MA
Briefer E, McElligott AG (2011) Mutual mother offspring vocal recognition in an ungulate hider species (*Capra hircus*). Anim Cogn 14:585–598
Bryant G, Haselton M (2009) Vocal cues of ovulation in human females. Biol Lett 5:12–15
Cartei V, Bond R, Reby D (2014) What makes a voice masculine: physiological and acoustical correlates of women's ratings of men's vocal masculinity. Horm Behav 66:569–576
Carter GG, Skowronski MD, Faure PA, Fenton B (2008) Antiphonal calling allows individual discrimination in white-winged vampire bats. Anim Behav 76:1343–1355
Charlton B (2014) Vocal distinctiveness in the harsh coughs of southern hairy-nosed wombats (*Lasiorhinus latifrons*). Acta Acust Acust 100:719–723
Charlton B (2015) The acoustic structure and information content of female koala vocal signals. PLoS One 10:e0138670
Charlton B, Reby D (2016) The evolution of acoustic size exaggeration in terrestrial mammals. Nat Commun 7:12739
Charlton B, Reby D, McComb K (2007a) Female perception of size-related formant shifts in red deer, *Cervus elaphus*. Anim Behav 74:707–714
Charlton B, Reby D, McComb K (2007b) Female red deer prefer the roars of larger males. Biol Lett 3:382–385
Charlton B, Reby D, McComb K (2008) Effect of combined source (F0) and filter (formant) variation on red deer hind responses to male roars. J Acoust Soc Am 123:2936–2943
Charlton B, Zhang Z, Snyder R (2009a) Vocal cues to identity and relatedness in giant pandas (*Ailuropoda melanoleuca*). J Acoust Soc Am 126:2721–2732
Charlton B, Zhihe Z, Snyder R (2009b) The information content of giant panda, *Ailuropoda melanoleuca*, bleats: acoustic cues to sex, age and size. Anim Behav 78:893–898
Charlton B, Zhang Z, Snyder R (2010) Giant pandas perceive and attend to formant frequency variation in male bleats. Anim Behav 79:1221–1227
Charlton B, Ellis W, McKinnon A, Brumm J, Nilsson K, Fitch W (2011a) Perception of male caller identity in koalas (*Phascolarctos cinereus*): acoustic analysis and playback experiments. PLoS One 6:e20329
Charlton B, Ellis W, McKinnon A, Cowin G, Brumm J, Nilsson K, Fitch W (2011b) Cues to body size in the formant spacing of male koala (*Phascolarctos cinereus*) bellows: honesty in an exaggerated trait. J Exp Biol 214:3414–3422
Charlton B, Keating JL, Kersey D, Rengui L, Huang Y, Swaisgood RR (2011c) Vocal cues to male androgen levels in giant pandas. Biol Lett 7:71–74
Charlton B, Ellis W, Brumm J, Nilsson K, Fitch W (2012a) Female koalas prefer bellows in which lower formants indicate larger males. Anim Behav 84:1565–1571
Charlton B, Swaisgood R, Zhihe Z, Snyder R (2012b) Giant pandas attend to androgen-related variation in male bleats. Behav Ecol Sociobiol 66:969–974
Charlton B, Ellis WAH, Larkin R, Fitch WT (2012c) Perception of size-related formant information in male koalas (*Phascolarctos cinereus*). Anim Cogn 15:999–1006
Charlton B, Frey R, McKinnon A, Fritsch G, Fitch W, Reby D (2013a) Koalas use a novel vocal organ to produce unusually low-pitched mating calls. Curr Biol 23:R1035–R1036

Charlton B, Taylor A, Reby D (2013b) Are men better than women at acoustic size judgements? Biol Lett 9:20130270

Charlton B, Whisson D, Reby D (2013c) Free-ranging male koalas use size-related variation in formant frequencies to assess rival males. PLoS One 8:e70279

Charlton B, Keating JL, Rengui L, Huang Y, Swaisgood RR (2015) The acoustic structure of male giant panda bleats varies according to intersexual context. J Acoust Soc Am 138:1305–1312

Charlton B, Taylor AM, Reby D (2017) Function and evolution of vibrato-like frequency modulation in mammals. Curr Biol 27:2692–2697

Charrier I, Mathevon N, Jouventin P (2002) How does a fur seal mother recognize the voice of her pup? An experimental study of *Arctocephalus tropicalis*. J Exp Biol 205:603–612

Charrier I, Mathevon N, Jouventin P (2003) Individuality in the voice of fur seal females: an analysis study of the pup attraction call in *Arctocephalus tropicalis*. Mar Mamm Sci 19:161–172

Charrier I, Pitcher BJ, Harcourt R (2009) Vocal recognition of mothers by Australian sea lion pups: individual signature and environmental constraints. Anim Behav 78:1127–1134

Charrier I, Ahonen H, Harcourt R (2011) What makes an Australian sea lion (*Neophoca cinerea*) male's bark threatening? J Comp Psychol 125:385–392

Chiba T, Kajiyama M (1958) The vowel: its nature and structure. Phonetic Society of Japan, Tokyo-Kaiseikan

Dabbs JM, Mallinger A (1999) High testosterone levels predict low voice pitch among men. Personal Individ Differ 27:801–804

Dunn JC, Halenar LB, Davies TG, Cristobal-Azkarate J, Reby D, Sykes D, Dengg S, Fitch WT, Knapp LA (2015) Evolutionary trade-off between vocal tract and testes dimensions in howler monkeys. Curr Biol 25:2839–2844

Fant G (1960) Acoustic theory of speech production. Mouton, The Hague

Faragó T, Pongrácz P, Miklósi Á, Huber L, Virányi Z, Range F (2010) Dogs' expectation about signalers' body size by virtue of their growls. PLoS One 5:e15175

Fedurek P, Slocombe KE, Enigk DK, Thompson ME, Wrangham R, Muller MN (2016) The relationship between testosterone and long-distance calling in wild male chimpanzees. Behav Ecol Sociobiol 70:659–672

Feinberg DR, Jones BC, Little AC, Burt DM, Perrett DI (2005) Manipulations of fundamental and formant frequencies influence the attractiveness of human male voices. Anim Behav 69:561–568

Fischer J, Kitchen DM, Seyfarth RM, Cheney DL (2004) Baboon loud calls advertise male quality: acoustic features and their relation to rank, age, and exhaustion. Behav Ecol Sociobiol 56:140–148

Fischer J, Semple S, Fickenscher G, Jürgens R, Kruse E, Heistermann M, Amir O (2011) Do women's voices provide cues of the likelihood of ovulation? The importance of sampling regime. PLoS One 6:e24490

Fitch W (1997) Vocal tract length and formant frequency dispersion correlate with body size in rhesus macaques. J Acoust Soc Am 102(2):1213–1222

Fitch WT, Fritz JB (2006) Rhesus macaques spontaneously perceive formants in conspecific vocalisations. J Acoust Soc Am 120:2132–2141

Fitch WT, Giedd J (1999) Morphology and development of the human vocal tract: a study using magnetic resonance imaging. J Acoust Soc Am 106:1511–1522

Fitch WT, Hauser MD (1995) Vocal production in nonhuman-primates - acoustics, physiology, and functional constraints on honest advertisement. Am J Primatol 37:191–219

Fitch WT, Hauser MD (2002) Unpacking "honesty": generating and extracting information from acoustic signals. In: Megala-Simmons A, Popper A (eds) Animal communication. Springer, Berlin, pp 65–137

Fitch W, Reby D (2001) The descended larynx is not uniquely human. Proc R Soc Lond B Biol Sci 268(1477):1669–1675

Fouquet M, Pisanski K, Mathevon N, Reby D (2016) Seven and up: individual differences in male voice fundamental frequency emerge before puberty and remain stable throughout adulthood. R Soc Open Sci 3:160395

Frey R, Gebler A (2003) The highly specialized vocal tract of the male Mongolian gazelle (*Procapra gutturosa* Pallas, 1777 - *Mammalia, Bovidae*). J Anat 203:451–471

Frey R, Hofmann RR (1996) Evolutionary morphology of the proboscideal nose of Guenther's Dikdik (*Rhyncotragus guetheri* Thomas 1894—*Mammalia, Bovidae*). Zool Anz 235:31–51

Frey R, Volodin I, Volodina E (2007) A nose that roars: anatomical specializations and behavioural features of rutting male saiga. J Anat 211:717–736

Frey R, Gebler A, Olson KA, Odonkhuu D, Fritsch G, Batsaikhan N, Stuermer IW (2008) Mobile larynx in Mongolian gazelle: retraction of the larynx during rutting barks in male Mongolian gazelle (*Procapra gutturosa Pallas*, 1777). J Morphol 269:1223–1237

Frey R, Volodin I, Volodina E, Soldatova NV, Juldaschev ET (2011) Descended and mobile larynx, vocal tract elongation and rutting roars in male goitred gazelles (*Gazella subgutturosa* Güldenstaedt, 1780). J Anat 218:566–585

Gamba M, Colombo C, Giacoma C (2012) Acoustic cues to caller identity in lemurs: a case study. J Ethol 30:191–196

Garcia M, Charlton BD, Wyman MT, Fitch WT, Reby D (2013) Do red deer stags (*Cervus elaphus*) use roar fundamental frequency (F0) to assess rivals? PLoS One 8:e83946

Garcia M, Herbst CT, Bowling DL, Dunn JC, Fitch WT (2017) Acoustic allometry revisited: morphological determinants of fundamental frequency in primate vocal production. Sci Rep 7:10450

Gaulin S, Boster J (1985) Cross-cultural differences in sexual dimorphism: is there any variance to be explained? Ethol Sociobiol 6:219–225

Ghazanfar A, Turesson H, Maier J, Vandinther R, Patterson RD, Logothetis N (2007) Vocal-tract resonances as indexical cues in rhesus monkeys. Curr Biol 17:425–430

Gonzalez J (2004) Formant frequencies and body size of speaker: a weak relationship in adult humans. J Phon 32:277–287

Gourbal BEF, Barthelemy M, Petit G, Gabrion C (2004) Spectrographic analysis of the ultrasonic vocalisations of adult male and female BALB/c mice. Naturwissenschaften 91:1–5

Harris TR, Fitch WT, Goldstein LM, Fashing PJ (2006) Black and white colobus monkey (*Colobus guereza*) roars as a source of both honest and exaggerated information about body mass. Ethology 112:911–920

Hughes SMA-O, Harrison MA (2017) Your cheatin' voice will tell on you: detection of past infidelity from voice. Evol Psychol 15:1474704917711513

Hillenbrand JM, Clark MJ (2009) The role of F0 and formant frequencies in distinguishing the voices of men and women. Atten Percept Psychophys 71:1150–1166

Hodges-Simeon CR, Gaulin SJC, Puts DA (2010) Different vocal parameters predict perceptions of dominance and attractiveness. Hum Nat 21:406–427

Hollien H, Green R, Massey K (1994) Longitudinal research on adolescent voice change in males. J Acoust Soc Am 96:2646–2653

Ives DT, Smith DRR, Patterson RD (2005) Discrimination of speaker size from syllable phrases. J Acoust Soc Am 118:3816

Janik V, Sayigh L, Wells RS (2006) Signature whistle shape conveys identity information to bottlenose dolphins. Proc Natl Acad Sci 103:8293–8297

Kelemen G, Sade J (1960) The vocal organ of the howling monkey (*Alouatta palliata*). J Morphol 107:123–140

Kidjo N, Cargnelutti B, Charlton BD, Wilson C, Reby D (2008) Vocal behaviour in the endangered Corsican deer: description and phylogenetic implications. Bioacoustics 18:159–181

Klemuk S, Riede T, Walsh E, Titze I (2011) Adapted to roar: functional morphology of tiger and lion vocal folds. PLoS One 6:e27029

Knowles KK, Little AC (2016) Vocal fundamental and formant frequencies affect perceptions of speaker cooperativeness. J Exp Psychol 69:1657–1675

Koda H, Murai T, Tuuga A, Goossens B, Nathan SKSS, Stark DJ, Ramirez DAR, Sha JCM, Osman I, Sipangkui R, Seino S, Matsuda I (2018) Nasalization by *Nasalis larvatus*: larger noses audiovisually advertise conspecifics in proboscis monkeys. Sci Adv 4:eaaq0250

Kreiman J, Sidtis D (2011) Foundations of voice studies: an interdisciplinary approach to voice production and perception. Wiley, Hoboken, NJ

Lemasson A, Remeuf K, Trabalon M, Cuir F, Hausberger M (2015) Mares prefer the voices of highly fertile stallions. PLoS One 10:e0118468

Leongómez JD, Mileva VR, Little AC, Roberts SC (2017) Perceived differences in social status between speaker and listener affect the speaker's vocal characteristics. PLoS One 12:e0179407

Levrero F, Mathevon N (2013) Vocal signature in wild infant chimpanzees. Am J Primatol 75:324–332

Levrero F, Mathevon N, Pisanski K, Gustafsson E, Reby D (2018) The pitch of babies' cries predicts their voice pitch at age 5. Biol Lett 14:20180065

Liu N, Jiang Z, Zhang L, Zhong Z, Ping X, Xu H, Li C (2016) Bioacoustic cues and their relations to dominance rank in Père David's deer stags. Anim Prod Sci 56:971–977

Markova D, Richer L, Pangelinan M, Schwartz DH, Leonard G, Perron M, Pike GB, Veillette S, Chakravarty MM, Pausova Z, Paus T (2016) Age- and sex-related variations in vocal-tract morphology and voice acoustics during adolescence. Horm Behav 81:84–96

McComb KE (1991) Female choice for high roaring rates in red deer, *Cervus elaphus*. Anim Behav 41:79–88

McComb K, Reby D, Baker L, Moss C, Sayialel S (2003) Long-distance communication of acoustic cues to social identity in African elephants. Anim Behav 65:317–329

McElligott AG, Birrer M, Vannoni E (2006) Retraction of the mobile descended larynx during groaning enables fallow bucks (*Dama dama*) to lower their formant frequencies. J Zool 270:340–345

Mehta PH, Josephs RA (2010) Testosterone and cortisol jointly regulate dominance: evidence for a dual-hormone hypothesis. Horm Behav 58:898–906

Melendez KV, Feng AS (2010) Communication calls of little brown bats display individual-specific characteristics. J Acoust Soc Am 128:919–923

Montano KJ, Tigue CC, Isenstein SGE, Barclay P, Feinberg DR (2017) Men's voice pitch influences women's trusting behavior. Evol Hum Behav 38:293–297

Moulines E, Charpentier F (1990) Pitch synchronous waveform processing techniques for text-to-speech synthesis using diphones. Speech Comm 9:453–467

O'Connor JJM, Re DE, Feinberg DR (2011) Voice pitch influences perceptions of sexual infidelity. Evol Psychol 9:147470491100900109

O'Connor JJM, Fraccaro PJ, Feinberg DR (2012) The influence of male voice pitch on women's perceptions of relationship investment. J Evol Psychol 10:1–13

Ohala JJ (1984) An ethological perspective on common cross-language utilization of F0 of voice. Phonetica 41:1–16

Owren MJ, Rendall D (2001) Sound on the rebound: bringing form and function back to the forefront in understanding nonhuman primate vocal signaling. Evol Anthropol 10:58–71

Palacios V, Font E, Márquez R (2007) Iberian wolf howls: acoustic structure, individual variation, and a comparison with North American populations. J Mammal 88:606–613

Pavela Banai I, Banai B, Bovan K (2017) Vocal characteristics of presidential candidates can predict the outcome of actual elections. Evol Hum Behav 38:309–314

Pfefferle D, Fischer J (2006) Sounds and size: identification of acoustic variables that reflect body size in hamadryas baboons, *Papio hamadryas*. Anim Behav 72:43–51

Pietraszewski D, Wertz AE, Bryant GA, Wynn K (2017) Three-month-old human infants use vocal cues of body size. Proc R Soc B 284:20170656–20170659

Pisanski K, Feinberg DR (2018) Voice attractiveness. In: Belin P, Fruehholz S (eds) Oxford handbook of voice perception. Oxford University Press, New York

Pisanski K, Rendall D (2011) The prioritization of voice fundamental frequency or formants in listener's assessments of speaker size, masculinity, and attractiveness. J Acoust Soc Am 129:2201–2212
Pisanski K, Fraccaro PJ, Tigue CC, O'Connor JJM, Feinberg DR (2014a) Return to Oz: voice pitch facilitates assessments of men's body size. J Exp Psychol Hum Percept Perform 40:1316–1331
Pisanski K, Fraccaro PJ, Tigue CC, O'Connor JJM, Röder S, Andrews PW, Fink B, DeBruine LM, Jones BC, Feinberg DR (2014b) Vocal indicators of body size in men and women: a meta-analysis. Anim Behav 95:89–99
Pisanski K, Cartei V, McGettigan C, Raine J, Reby D (2016a) Voice modulation: a window into the origins of human vocal control? Trends Cogn Sci 20:04–318
Pisanski K, Oleszkiewicz A, Sorokowska A (2016b) Can blind persons accurately assess body size from the voice? Biol Lett 12:20160063
Pisanski K, Feinberg D, Oleszkiewicz A, Sorokowska A (2017) Voice cues are used in a similar way by blind and sighted adults when assessing women's body size. Sci Rep 7:10329
Pisanski K, Bhardwaj K, Reby D (2018) Women's voice pitch lowers after pregnancy. Evol Hum Behav 39:457–463
Pitcher BJ, Briefer EF, McElligott AG (2015) Intrasexual selection drives sensitivity to pitch, formants and duration in the competitive calls of fallow bucks. BMC Evol Biol 15:149
Puts D, Hodges C, Cárdenas R, Gaulin S (2007) Men's voices as dominance signals: vocal fundamental and formant frequencies influence dominance. Evol Hum Behav 28(5):340–344
Puts DA, Apicella CL, Cardenas RA (2012a) Masculine voices signal men's threat potential in forager and industrial societies. Proc R Soc B 279:601–609
Puts DA, Jones BC, DeBruine LM (2012b) Sexual selection on human faces and voices. J Sex Res 49:227–243
Puts DA, Hill AK, Bailey DH, Walker RS, Rendall D, Wheatley JR, Welling LL, Dawood K, Cárdenas R, Burriss RP, Jablonski NG, Shriver MD, Weiss D, Lameira AR, Apicella CL, Owren MJ, Barelli C, Glenn ME, Ramos-Fernandez G (2016) Sexual selection on male vocal fundamental frequency in humans and other anthropoids. Proc R Soc B 283:20152830
Raine J, Pisanski K, Reby D (2017) Tennis grunts communicate acoustic cues to sex and contest outcome. Anim Behav 130:47–55
Rantala MJ, Moore FR, Skrinda I, Krama T, Kivleniece I, Kecko S, Krams I (2012) Evidence for the stress-linked immunocompetence handicap hypothesis in humans. Nat Commun 3:694
Reby D, McComb K (2003) Anatomical constraints generate honesty: acoustic cues to age and weight in the roars of red deer stags. Anim Behav 65:519–530
Reby D, McComb K, Cargnelutti B, Darwin C, Fitch WT, Clutton-Brock TH (2005) Red deer stags use formants as assessment cues during intrasexual agonistic interactions. Proc R Soc B 272:941–947
Reby D, Andre-Obrecht R, Galinier A, Farinas J, Cargnelutti B (2006) Cepstral coefficients and hidden Markov models reveal idiosyncratic voice characteristics in red deer (*Cervus elaphus*) stags. J Acoust Soc Am 120:4080–4089
Reby D, Charlton BD, Locatelli Y, McComb K (2010) Oestrous red deer hinds prefer male roars with higher fundamental frequencies. Proc R Soc B 277:2747–2753
Reby D, Wyman MT, Frey R, Passilongo D, Gilbert J, Locatelli Y, Charlton BD (2016) Evidence of biphonation and source-filter interactions in the bugles of male North American wapiti (*Cervus canadensis*). J Exp Biol 219:1224–1236
Rendall D (2003) Acoustic correlates of caller identity and affect intensity in the vowel-like grunt vocalizations of baboons. J Acoust Soc Am 113:3390–3402
Rendall D, Owren MJ, Rodman PS (1998) The role of vocal tract filtering in identity cueing in rhesus monkey (*Macaca mulatta*) vocalizations. J Acoust Soc Am 103:602–614
Rendall D, Owren MJ, Weerts E, Hienz RD (2004) Sex differences in the acoustic structure of vowel-like grunt vocalizations in baboons and their perceptual discrimination by baboon listeners. J Acoust Soc Am 115:411–421

Rendall D, Vokey JR, Nemeth C (2007) Lifting the curtain on the Wizard of Oz: biased voice-based impressions of speaker size. J Exp Psychol Hum Percept Perform 33:1208–1219

Riede T, Fitch WT (1999) Vocal tract length and acoustics of vocalization in the domestic dog (*Canis familiaris*). J Exp Biol 202:2859–2867

Ryalls JH, Lieberman P (1982) Fundamental-frequency and vowel perception. J Acoust Soc Am 72:1631–1634

Sanvito S, Galimberti F, Miller EH (2007) Vocal signalling in male southern elephant seals is honest but imprecise. Anim Behav 73:287–299

Schön-Ybarra M (1995) A comparative approach to the non-human primate vocal tract: implications for sound production. In: Zimmermann E, Newman JD (eds) Current topics in primate vocal communication. Plenum Press, NewYork, pp 185–198

Searby A, Jouventin P (2003) Mother-lamb acoustic recognition in sheep: a frequency coding. Proc R Soc B 270:1765–1771

Šebesta P, Kleisner K, Tureček P, Kočnar T, Akoko RM, Třebický V, Havlíček J (2017) Voices of Africa: acoustic predictors of human male vocal attractiveness. Anim Behav 127:205–211

Smith DRR, Patterson RD, Turner R (2005) The processing and perception of size information in speech sounds. J Acoust Soc Am 117:305–318

Stoeger AS, Baotic A (2016) Information content and acoustic structure of male African elephant social rumbles. Sci Rep 6:1–8

Stulp G, Barrett L (2014) Evolutionary perspectives on human height variation. Biol Rev 91:206–234

Suthers RA, James MF (1973) Mechanisms of sound production by echolocating bats. Am Zool 13:1215–1226

Taylor A, Reby D, McComb K (2010) Size communication in domestic dog, *Canis familiaris*, growls. Anim Behav 79:205–210

Taylor A, Reby D, McComb K (2011) Cross modal perception of body size in domestic dogs (*Canis familiaris*). PLoS One 6:e17069

Taylor A, Charlton BD, Reby D (2016) Vocal production by terrestrial mammals: source, filter and function. In: Suthers RA, Fitch WT, Fay RR, Popper A (eds) Vertebrate sound production and acoustic communication. Springer, Berlin, pp 229–259

Titze IR (1994) Principles of voice production. Prentice Hall, Englewood Cliffs, NJ

Titze IR (2011) Vocal fold mass is not a useful quantity for describing F0 in vocalization. J Speech Lang Hear Res 54:520–522

Torriani MV, Vannoni E, McElligott AG (2006) Mother-young recognition in an ungulate hider species: a unidirectional process. Am Nat 168:412–420

Townsend S, Charlton B, Manser M (2014) Acoustic cues to identity and predator context in meerkat barks. Anim Behav 94:143–149

Vannoni E, McElligott AG (2007) Individual acoustic variation in fallow deer (*Dama dama*) common and harsh groans: a source-filter theory perspective. Ethology 113:223–234

Vannoni E, McElligott AG (2008) Low frequency groans indicate larger and more dominant fallow deer (*Dama dama*) males. PLoS One 3:e3113

Weissengruber GE, Forstenpointner G, Peters G, Kübber Heiss A, Fitch WT (2002) Hyoid apparatus and pharynx in the lion (*Panthera leo*), jaguar (*Panthera onca*), tiger (*Panthera tigris*), cheetah (*Acinonyx jubatus*) and domestic cat (*Felis silvestris f. catus*). J Anat 201:195–209

Wyman MT, Mooring MS, McCowan B, Penedo MCT, Reby D, Hart LA (2012) Acoustic cues to size and quality in the vocalizations of male North American bison, *Bison bison*. Anim Behav 84:1381–1391

Zhang J, Reid SA (2017) Aggression in young men high in threat potential increases after hearing low-pitched male voices: two tests of the retaliation-cost model. Evol Hum Behav 38:513–521

Chapter 6
Coding for 'Dynamic' Information: Vocal Expression of Emotional Arousal and Valence in Non-human Animals

Elodie F. Briefer

Abstract Emotions guide behavioural decisions in response to events or stimuli of importance for the organism, and thus, are an important component of an animal's life. Communicating emotions to conspecifics allows, in turn, the regulation of social interactions (e.g. approach and avoidance). The existence of common rules governing vocal expression of affective states across species has been proposed as a function of the motivational state (i.e. intention of behaviour) of the emitter ('motivation-structural rules') and as a function the two main dimensions of emotions, valence (positive versus negative) and arousal (bodily activation). In this chapter, I review the potential for vocalisations to serve as universal non-invasive indicators of animal emotions, by considering the latest evidence for common rules existing across species according to the two dimensions of emotions ('emotional-dimension rules'). Vocal indicators of emotional arousal have been relatively well studied. Cross-species comparison shows that, when arousal increases, vocalisations tend to be louder and are produced at faster rates, with higher frequencies (both source- and filter-related) and a more variable fundamental frequency (F0). In contrast, indicators of valence have only been investigated in a few species. The evidence so far indicates that, compared with negative vocalisations, positive vocalisations tend to be shorter, with a lower and less variable F0. Yet, comparison of vocal indicators of valence between closely related species suggests that these indicators are more species specific than indicators of arousal, which have clearly been conserved throughout evolution. To conclude, I further suggest a new set of rules that could explain the acoustic structure of vocalisations across species, which combine features predicted by the motivation-structural rules, the emotional-dimension rules, and characteristics of the social relationship involving the emitter and receiver.

E. F. Briefer (✉)
Behavioural Ecology Group, Section for Ecology and Evolution, Department of Biology, University of Copenhagen, Copenhagen, Denmark
e-mail: elodie.briefer@bio.ku.dk

T. Aubin, N. Mathevon (eds.), *Coding Strategies in Vertebrate Acoustic Communication*, Animal Signals and Communication 7,
https://doi.org/10.1007/978-3-030-39200-0_6

6.1 Introduction

Emotions are defined as intense, short-lived affective reactions to specific events or stimuli of importance for the organism. Their crucial function is to guide behavioural decisions in response to these triggering events or stimuli (e.g. approach or avoidance). Emotions can be characterised by two main dimensions (dimensional approach, Russell 1980) as follows: valence (positive versus negative) and arousal (bodily activation/excitation). The dimensional approach differs from the 'discrete emotion approach', which focuses on different specific emotional states (e.g. fear, rage, and anxiety). Many possible associations between valence and arousal have been proposed in the literature (e.g. independence, symmetric V-shaped relation, and asymmetric V-shaped relation, Kuppens et al. 2013). A useful model for studying non-human animal emotions, which is also largely supported by the literature on humans, consists of considering emotional arousal as the intensity of bipolar valence, comprising appetitive (positive) and defensive (negative) motivational systems (V-shaped relation between arousal and valence, Fig. 6.1; Bradley et al. 2001; Bradley and Lang 2007; Kuppens et al. 2013)

Although the existence of emotions in animals was already acknowledged by Darwin (1872), scientific interest in this topic only began after a century, during which emotions were still regarded as unobservable processes that could not be objectively studied (Fraser 2009). Since then, and particularly over the past two decades, the interest in animal emotions has considerably increased, mainly for human benefit (e.g. pharmaceutical development), and also to study animal behaviour and improve animal welfare (Proctor et al. 2013). Indeed, animal welfare researchers have emphasised the need to assess the mental health of captive animals

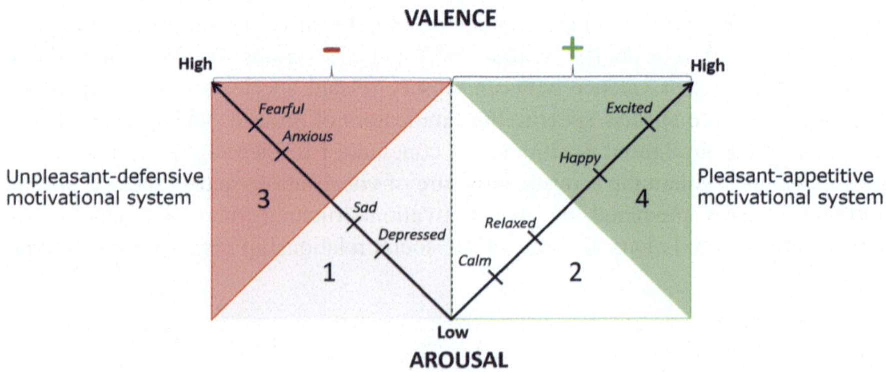

Fig. 6.1 Illustration of the bipolar valence-arousal model, 'arousal as intensity' (V-shaped relation between valence and arousal). Positive emotions are part of the pleasant-appetitive motivational system, while negative emotions are part of the unpleasant-defensive motivational system. Examples of human emotions are shown. Arousal is defined as the intensity of bipolar valence (e.g., Bradley et al. 2001). Quadrants 1, 2, 3, and 4 encompass low-arousal negative emotions, low-arousal positive emotions, high-arousal negative emotions, and high-arousal positive emotions, respectively

in addition to their physical health (Dawkins 1990). Further, they have mentioned the importance of not only avoiding negative emotions such as frustration, suffering and distress, but also promoting positive emotions (Dawkins 1990; Boissy et al. 2007). As a result, new frameworks that offer researchers tools to study emotions have been developed (Désiré et al. 2002; Mendl et al. 2010). Although the application of these frameworks is still in its infancy, they have already led to the establishment of objective indicators of emotions in several species (e.g. sheep, *Ovis aries*, Reefmann et al. 2009c; goats, *Capra hircus*, Briefer et al. 2015b; horses, *Equus caballus*, Briefer et al. 2015a; cattle, *Bos taurus*, Proctor and Carder 2014; and pigs, *Sus scrofa domesticus*, Reimert et al. 2013).

One of the frameworks established by Mendl et al. (2010) to study emotions proposes the assessment of the valence and arousal of animal emotions using the neural, physiological, behavioural, and cognitive changes that accompany emotions. These changes, along with the subjective component (‘feelings’) that cannot be accessed in non-human animals using existing methods, constitute the components of emotions (Dolan 2002; Anderson and Adolphs 2014). Mendl et al.’s framework (2010) has led to the discovery of several neural and physiological indicators (e.g. heart rate and its variability, respiration rate, skin temperature, electrodermal response, neuroendocrine activity, brain activity, Reefmann et al. 2009b; Gygax et al. 2013; Briefer et al. 2015b; Proctor and Carder 2015a; Vögeli et al. 2015), behavioural indicators (e.g. ear and tail postures, visible eye white, movements, vocalisation *rate* and type, Reefmann et al. 2009a; Imfeld-Mueller et al. 2011; Imfeld-Mueller and Hillmann 2012; Reimert et al. 2013; Proctor and Carder 2014, 2015b), as well as cognitive indicators of emotions (e.g. appraisal processes, judgement biases, Veissier et al. 2009; Burman et al. 2011; Doyle et al. 2011; Briefer and McElligott 2013; Briefer Freymond et al. 2014). So far, physiological changes seem to mainly indicate emotional arousal, while neural, behavioural, and cognitive changes additionally indicate emotional valence (Mendl et al. 2009; Briefer et al. 2015a, b). Nonetheless, ideal indicators of emotions that are simultaneously non-invasive, rapid, and provide an accurate indication of the current emotional state of the animal (including both emotional arousal and valence) have not been discovered yet. Such indicators would be highly relevant for several disciplines, including animal welfare, neuroscience, psychopharmacology, and evolutionary zoology.

Vocalisations are promising indicators of animal emotions (Manteuffel et al. 2004). Unlike physiological measurements, which often require manipulation of the animals, either to place a monitor (e.g. heart rate or brain activity monitoring, Gygax et al. 2013) or to collect samples for hormonal analyses (e.g. saliva or blood sampling, Merlot et al. 2011), vocalisations can be recorded non-invasively from a distance, without imposing additional stress on the animals. Using real-time acoustic monitoring, which is being increasingly developed (e.g. see Schön et al. 2004; Moura et al. 2008), vocalisations can provide an almost instantaneous measure of emotions, unlike for example, cognitive biases, which require training (Mendl et al. 2009), or hormonal measurements, which require further analyses (Briefer Freymond et al. 2015). Vocalisations also provide an immediate measure of the current affective state of the animal, similar to physiological measures linked to the sympatho–adreno–medullary (SAM) axis (e.g. heart rate, respiration rate, noradrenaline and adrenaline), as opposed to measures linked to the hypothalamo–pituitary–adrenocortical (HPA) axis (e.g. ACTH and glucocorticoids), which are characterised by a slower response (Romero 2004). This feature might be important for the assessment of short-term affective reactions to specific stimuli (i.e. emotions), as opposed to long-term affective states (i.e. moods, Mendl et al. 2009). In addition, the structure of vocalisations includes some parameters (e.g. *fundamental frequency F0*, *duration*, *amplitude modulation AM*, *frequency modulation FM*, but not overall *amplitude* or *energy distribution*, Gerhardt 1998; see Table 6.1 for a definition of the vocal parameters in italics), which might be less dependent on recording conditions (e.g. distance from the animal, obstacles, movements and body/head orientation, and weather conditions), compared to other non-invasive measurements like infrared thermography (Cilulko et al. 2013). The disadvantage of vocal indicators of emotions, however, is that vocalisations are not produced in every emotional state (e.g. low-arousal situations, depression) and could be restricted to social contexts. Thus, it might be necessary to combine measures related to the structure of vocalisations, with measures of the rate and type of vocalisations produced, as well as other measures (e.g. behaviour), in order to obtain a complete evaluation of the affective state of animals (Manteuffel et al. 2004).

In addition to the above-mentioned advantages of using vocalisations as indicators of emotions (i.e. non-invasive, rapid, referring to the current emotional state, and partially resistant to recording conditions), vocal indicators are also potentially reliable ('honest', Zahavi 1982). These indicators are mainly produced involuntarily, as a result of changes in tension and action of muscles used for voice production (phonation, resonance and articulation), as well as changes in respiration and salivation, caused by somatic and autonomic nervous system activation (Scherer 1986). Additional effects of HPA activation on vocalisations are also expected, since increased glucocorticoid levels have been related to changes in the acoustic structure of the human voice (Puts et al. 2006; Giddens et al. 2013; Taylor et al. 2016b), pig grunts (Schrader and Todt 1998), yellow-bellied marmot (*Marmota flaviventris*) alarm calls (Blumstein and Chi 2012), and zebra finch (*Taeniopygia guttata*) calls (Perez et al. 2012). Consequently, vocal expression of emotions has been observed across species (Briefer 2012), including humans, which can control vocal production (Scherer et al. 2001; Goudbeek and Scherer 2010). In fact, vocalisations of non-human animals are expected to more accurately reflect emotions compared with human voice, since most species seem to have lower voluntary control on vocal output compared to humans (Jürgens 2009). For instance, although voluntary alteration of the overall amplitude or duration of vocalisations through a certain control over the respiratory system seems to be a common ability in the animal kingdom, voluntary modification of sound frequencies, which involves control over the vocal apparatus muscles, is more rare (Janik and Slater 1997). Therefore, attempts by emitters to conceal their emotions by changing the structure of vocalisations are either not possible in some species, or would require a deviation from adaptive changes to vocal production caused by emotions (i.e. activation of the

Table 6.1 Vocal parameters listed in this chapted in italic and their description

Parameter	Description
Duration	Total duration of the vocalisation/elements in the vocalisation
Rate	Number of vocalisations/elements produced per time unit
Interval	Mean silence duration between two vocalisations/elements
Fundamental frequency or F0	Fundamental frequency, lowest frequency of the vocalisation
F0 contour	F0 values across the vocalisation (includes mean, start, end, minimum and maximum values)
F0 range	Difference between the maximum and minimum F0
Jitter	Cycle-to-cycle frequency variation of F0
Shimmer	Cycle-to-cycle amplitude variation of F0
Amplitude	Amount of energy in the vocalisation
Amplitude contour	Amplitude values across the vocalisation (includes mean, start, end, minimum and maximum values)
Amplitude modulation or 'AM'	Variation in amplitude relative to the total duration
Frequency spectrum	Amplitude as a function of frequency
Energy distribution	Distribution of energy in the spectrum (e.g. energy quartiles, amount of energy in various part of the spectrum, ratio between harmonics/F0)
Peak frequency	Frequency value with the maximum amplitude
Dominant frequency band (FB*n*)	High amplitude frequency band (FB1, 2, 3, etc.) across the vocalisation (includes mean, start, end, minimum and maximum values)
Frequency range	Frequency range in the spectrum (e.g. difference between energy quartiles)
Frequency modulation or 'FM'	Dominant frequency or F0 variability/modulation across the call
Formant	Concentration of acoustic energy around particular frequencies in the vocalisation wave, which corresponds to a resonance of the vocal tract
F1, 2, 3, 4, etc.	First, second, third, fourth, etc. formants
F*n* contour	Values of formant frequencies (F1, 2, 3, etc.) across the vocalisation (includes mean, start, end, minimum and maximum values)
Non-linear phenomena	Complex intrusions into the normal spectral structure (e.g. subharmonics, deterministic chaos, biphonation, frequency jumps)
Spectral noise	Proportion of noise in the vocalisation, where the harmonic structure is not clear or cannot be detected (e.g. chaos)
Wiener entropy	Ratio of the geometric mean to the arithmetic mean of the spectrum (0: pure tone; 1: random noise)
Harmonicity	Also called 'harmonic-to-noise ratio'; ratio of amplitude peaks of detectable harmonics to noise threshold (higher values indicate more tonal vocalisations)

somatic and autonomic nervous systems and the HPA, Scherer 1981; Zahavi 1982). Changes in the structure of non-human animals as a function of emotional arousal have been relatively well studied in negative contexts (e.g. castration, pain, hunger, stress). Studies on vocal indicators of arousal in positive contexts, as well as indicators of valence, however, are lacking (Briefer 2012). Thus, further studies are required to evaluate if vocalisations can also accurately inform us about the positive arousal and valence experienced by animals.

Indicators of emotions should ideally be valid across species. This would allow us to assess the emotions of any domestic or wild species using the same set of indicators, and thus, the same tools (e.g. equipment and analysis software). Cross-species indicators would also enable the direct comparison of emotional reactions to certain stimuli between species. Darwin (1872) already described between-species similarities in expression of emotions, including vocal and facial expressions. Since then, the existence of common rules governing the acoustic structure of vocalisations as a function of the motivational (i.e. tendencies in behaviour) or emotional state of the emitter have been proposed and tested by several authors. Collias (1960), Morton (1977), and August and Anderson (1987) contributed to establishing or refining the motivation-structural rules, by comparing the vocalisations of numerous species of mammals and birds. According to these rules, fearful or appeasing contexts are typically characterised by high-frequency, tonal sounds, while aggressive contexts are indicated by low-frequency, loud, and noisy sounds with a wide bandwidth (but see Kelly et al. 2017, which shows that the absolute interspecific value of emotional signals should not be overestimated). August and Anderson (1987) additionally proposed that friendly contexts could be associated with more soft, low-frequency, amplitude-modulated and rhythmic sounds, as has been described in Carnivora ('purr', Peters 1984, 2002). Common changes across species in the way they communicate emotional arousal, hunger, pain, and stress, as well as predator-related urgency, provides strong evidence suggesting that vocal indicators of arousal have been conserved throughout evolution (Scheiner and Fischer 2011; Briefer 2012; Zimmermann et al. 2013). Vocal indicators of valence, however, have not been investigated in enough species to be able to draw a clear picture of their evolution (Briefer 2012). Nonetheless, recent literature that does compare indicators valence in closely related species suggests that they might be more species-specific, unlike indicators of emotional arousal (Maigrot et al. 2017). Therefore, although vocal indicators of emotional arousal have the potential to be valid across species, this might not be the case for indicators of valence.

In this chapter, I will compile the latest evidence for common rules existing across species according to the two dimensions of emotions ('emotional-dimension rules'). I will mainly focus on mammals, while also referring to birds and other classes when relevant. I will first discuss the potential proximate mechanisms underlying the expression of emotions, followed by a review of existing evidence, along with a description of methods used to assess this phenomenon. Since the evidence concerning the arousal dimension of emotions is much stronger than that concerning the valence dimension, I will start describing the findings related to arousal before valence. The most commonly used vocal parameters, which will be mentioned throughout the text in italics, are described in Table 6.1. Vocal parameters that appear in the text, but are not outlined in Table 6.1, will be accompanied by a description in the text.

6.2 Mechanisms: Vocal Expression of Emotions

The 'source–filter' theory of voice production (Fant 1960), which was originally developed to study human speech production, has been applied to non-human animal communication and led to considerable advances in the field (see reviews in Taylor and Reby 2010; Taylor et al. 2016a; see also Chap. 5). This theory stipulates that mammal vocalisations are generated by vibrations of the vocal folds (source, determining *F0*), which are triggered by the airflow originating at the lungs. The sound produced at the source is subsequently filtered by the supra-laryngeal vocal tract (filter; pharynx, oral, and nasal cavities), resulting in amplified frequencies called '*formants*' (Titze 1994). Source-related parameters mainly depend on the anatomy and physiology of the larynx, while filter-related parameters are determined by the anatomy and physiology of the supra-laryngeal vocal tract (Titze 1994). Therefore, *F0* is affected by changes in the action and/or tension of the muscles of the larynx, while *formants* depend on the vocal tract shape and length. More precisely, the shape of the vocal tract determines the frequency of the lower *formants*, while its length, which can be increased through retraction of the larynx, determines the frequency of the highest *formants* (Fant 1960). In addition, changes in respiration or the action and/or tension of the respiratory muscles can affect the *duration* and *F0* by determining the subglottal pressure (i.e. pressure generated by the lungs beneath the larynx). By linking the acoustic parameters of vocalisations to their mode of production (caller's anatomical or physiological attributes), the source–filter theory allows us to understand why acoustic variation occurs. Further, it can help to make predictions about how vocalisations should change according to the factors under investigation (e.g. size, dominance rank, hormonal status, and emotions, Taylor and Reby 2010; see also Chap. 5 in this volume). Parameters that are flexible and can change in a short time frame should be of particular attention when studying emotions.

According to the 'polyvagal hypothesis' (Porges 1995), the mammalian nervous system evolved to promote social interactions and social bonds in safe environments. This theory is based on the observation of a parallel phylogenetic shift in mammals between the neural regulation of the autonomic nervous system (i.e. the emergence of a myelinated vagal system regulating supradiaphragmatic organs), and the emergence of an integrated social engagement system that consists of traits that notably optimise both the production (diaphragm to coordinate vocalisations with respiratory effort and volume) and reception of vocalisations (detached middle ear bones). These phylogenetic transitions were accompanied by anatomical and neurophysiological connections between the regulation of the heart and bronchi through the myelinated vagus and striated muscles of the face and head, including the middle ear and facial muscles, and the pharynx and larynx (Porges 1995; Porges and Lewis 2010). In addition to this association, structures involved in the social engagement system also interact with the HPA axis, neuropeptides (i.e. oxytocin and vasopressin), and the immune system, providing a mechanical and functional rationale for the presence of indicators of emotions in mammal vocalisations (Porges 2001; Porges and Lewis 2010).

Emotions can act at the level of the brain, larynx, vocal tract or lungs to influence vocalisations. At the level of the brain, emotions can impact on vocalisations through the vocal control pathways, which have afferent connections with higher sensory, emotional and homeostatic brain areas (see reviews in Jürgens and Ploog 1981; Jürgens 2002, 2009). As a result, natural sounding species-specific vocalisations can be elicited by electrical stimulation or pharmacological activation of various emotion-related brain areas (e.g. hypothalamus, amygdala, anterior cingulate cortex, and periaqueductal gray, Jürgens 2009). Some of these areas (like the anterior cingulate cortex and periaqueductal gray) are primary vocalisation substrate that directly control the readiness to vocalise, while others (like the hypothalamus and amygdala) trigger vocalisations secondarily (i.e. indirectly) following stimulus-induced motivational changes (Jürgens 1976; Jürgens and Ploog 1981).

At the level of the lungs and the vocal apparatus, the effect of emotions on vocalisations can be predicted from emotion-related physiological changes to these structures. In addition, facial expressions accompanying emotions (e.g. lip raising or curling, and mouth opening) could affect the structure of vocalisations, since the mouth and lips constitute the opening of the vocal tract and impact on *formant* frequencies (Scherer 1986). However, given that scientific interest in facial expressions of non-human animals only recently started (e.g. Facial Action Coding System, 'FACS', Waller et al. 2012; Caeiro et al. 2013; Wathan et al. 2015; 'Grimace scale', Sotocinal et al. 2011; Dalla Costa et al. 2014), it is still too early to predict the effect of facial expressions of emotional arousal and valence on vocalisations. I will thus only discuss the potential effect that physiological changes to the lungs and vocal apparatus (i.e. activation of the SAM and HPA axis) have on vocal output (Scherer 1986; Briefer 2012; Giddens et al. 2013). These predictions are mostly valid for emotional arousal, since so far, no clear physiological indicators of valence have been discovered (i.e. valence-specific physiological response that are independent of emotional arousal).

At the level of the lungs, as emotional arousal rises, increased respiration rate will raise *vocalisations/element rate*, therefore, decreasing the *interval* between vocalisations/elements. In addition, increased lung volume following bronchodilation and increased lung pressure should induce higher *amplitude* and higher *F0* by raising the sub-glottal pressure. At the level of the larynx, the net effect of an isometric increase in muscle tension should result in an increase in *F0*, as well as *jitter* and *shimmer* up to a certain point (Scherer 1979). Cardiovascular-induced alterations in vocal-fold biomechanics and heartbeat-related modulation of the subglottal pressure are expected to trigger further increase in *jitter* and *shimmer* (Orlikoff and Baken 1989). The vocal folds are a primary target of steroid hormones, including androgen, oestrogen, and glucocorticoids (Mukudai et al. 2015). However, studies that have investigated the direct relationship between vocal parameters and glucocorticoids are sparse, both in humans and other species. In pigs, an increase in cortisol levels induced by social isolation is negatively correlated with the rate of grunts produced, and positively correlated with the centre frequency (median of the call spectrum) of these calls (Schrader and Todt 1998). Faecal glucocorticoid metabolites revealed that yellow-bellied marmot calls became less noisy (measured using *Wiener entropy*) as the level of glucocorticoids increased (Blumstein and Chi 2012). In zebra finches, a rise in corticosterone levels, which was induced by social isolation or oral administration of corticosterone, resulted in longer calls with higher frequencies (*dominant frequency* and *energy distribution*, Perez et al. 2012). Other studies that investigated the relationship between vocal parameters and glucocorticoids did not find any effect (Lemasson et al. 2015; Mausbach et al. 2017). In humans, cortisol levels positively correlated with *F0* (Pisanski et al. 2016), and with the deviation from the overall average *F0* when under stress (Taylor et al. 2016b). Further, this study did report that *F0* decreased from the average value during a social task and increased during a problem-solving task when the stress response (cortisol) was higher (Taylor et al. 2016b). Therefore, we might expect an increase in *F0* with cortisol levels, but this could depend on the type of stressor. Finally, at the level of the vocal tract, inhibition of salivation as a result of emotional arousal results in higher *formant* frequencies, while pharyngeal constriction and tension of the vocal tract walls result in a shift in *energy distribution* towards higher frequencies (see review in Scherer 1986). Vocalisations might additionally depend on postures and movements that accompany emotions (Zahavi 1982; Scherer 1986; Scheiner and Fischer 2011).

In humans, vocal expressions of discrete emotions (e.g. anger, panic, fear, elation, happiness, and boredom) have been predicted based on physiological changes, postures, and facial expressions, which accompany the evaluation of stimuli or events that generate emotions ('appraisal theories', Scherer 1984; Lazarus 1991). According to appraisal theories, discrete emotions result from the outcome of an organism's evaluation (appraisal) of the situation in which it finds itself, in terms of its significance for survival and well-being. This evaluation involves a series of stimulus evaluation checks (SEC), which are expected to occur in the following sequential order: relevance of the situation (novelty, goal relevance, and intrinsic pleasantness); implication of the situation (outcome probability, discrepancy from expectation, conduciveness, and urgency); coping potential (agent and intentionality, control, power of action, and adjustment); and the normative significance (compatibility with internal and external standards; 'component process model', Scherer 2009). For example, happiness is associated with low suddenness, high intrinsic pleasantness, and very high outcome probability, while fear is associated with high suddenness, low intrinsic pleasantness, and high outcome probability. Since specific physiological changes, postures, and facial expressions can be related to each SEC outcome, predictions regarding the specific effects of these outcomes, and therefore of discrete emotions, on vocal parameters have been established based on speech production theories (Fant 1960; Scherer 1986). As a result of these predictions, hedonic valence, which refers in the component process model to positive or negative outcomes of both intrinsic pleasantness and conduciveness (i.e. conducive vs obstructive) checks, affects the 'wideness' of the voice. Positive outcomes are expected to induce a 'wide voice', while negative outcomes are thought to result in a 'narrow voice' (Scherer 1986; Laukka et al. 2005). A 'wide voice' (positive outcomes) is distinguished by a shift in *energy distribution* towards low frequencies, a decrease in the frequency of first *formant* (*F1 contour*), a slightly broader *F1* bandwidth, and velopharyngeal nasality. In contrast, a 'narrow voice' (negative outcomes) is characterised by a shift in *energy distribution* towards high frequencies, an increase in *F1 contour*, a decrease in the contour of the second and third *formants*, a narrower *F1* bandwidth, and laryngopharyngeal nasality (Scherer 2009). If vocal expression of emotions has been indeed conserved throughout evolution as suggested by Darwin (1872), we might expect similar patterns in non-human animals. Although appraisal theories have been applied to animals (Désiré et al. 2002; Veissier et al. 2009), the structure of vocalisations produced during SECs has, to my knowledge, never been investigated.

To conclude, clear predictions can be made regarding the effect of the physiological response linked to emotional arousal on non-human vocal production, and thus, on the structure of elicited vocalisations. Some of these predictions could also be valid for birds and crocodilians, to which the source–filter theory has been applied, revealing the presence of formants in their vocalisations (birds, Nowicki 1987; Fitch 1999; Fitch and Kelley 2000; Budka and Osiejuk 2013; crocodilians, Reber et al. 2015, 2017). In contrast, the mechanisms behind the production of vocal indicators of valence are more difficult to predict. Deciphering the mechanisms underlying vocal expression of emotional valence requires further research, and could benefit from integrating predictions about the effects that movements, postures, and facial expressions have on vocal production (e.g. FACS). When studying the relationship between neural, physiological, or behavioural indicators of emotions and vocal production mechanisms, careful interpretations regarding the causes and effects should be drawn, since a correlation between an indicator of emotions and vocalisations could be explained in three different ways as follows: (1) direct effect—the indicator of emotion (e.g. hormone) could have a direct effect on the vocal apparatus (e.g. vocal folds or tract), which would then induce a modification of the vocalisations; (2) indirect effect—the indicator of emotion could have an indirect effect on the vocal apparatus (i.e. secondary reaction), because it affects vocal control pathways in the brain, which in turn have an impact on the vocal apparatus and structure of vocalisations; or (3) mere coincidence—both the indicator and the vocal apparatus could be affected by the emotion independently, resulting in a correlation between the indicator and vocalisations, which results from a mere coincidence (e.g. Jürgens 1976).

6.3 Evidence: Vocal Expression of Emotions

Vocal expression of emotions in non-human animals has been mainly studied with the aim of finding welfare indicators (e.g. stress, level of need, and pain, Manteuffel et al. 2004), which were, until recently, tailored to identify and avoid negative situations (Boissy et al. 2007). Another aim of investigating vocal expression of emotions has been to identify variation in the type or structure of vocalisations that are produced in response to a predator threat (alarm vocalisations, Klump and Shalter 1984). These studies highlighted three categories of alarm calls as follows: 'urgency-based' alarm calls that vary as a function of the speed and distance of the approaching predator, independently of its type (e.g. Sciuridae, Blumstein and Armitage 1997); 'functionally referential' alarm calls that vary with the predator type, independently of the urgency of the situation (e.g. vervet monkey, *Chlorocebus pygerythrus*, Seyfarth et al. 1980); and alarm calls that vary according to both the urgency level and predator type (e.g. meerkats, *Suricata suricatta* Manser et al. 2002). Since functionally referential alarm calls resemble human semantic communication, in that they constitute signs referring to objects or events other than the producer's emotions, the nature of alarm calls (i.e. urgency-based, functionally referential, or both) has been investigated in a wide range of species, in order to elucidate the evolutionary route of human language (Townsend and Manser 2013). As a result of these focuses of research, the evidence on vocal expression of emotions is strongly biased towards the expression of emotional arousal during negative situations, rather than emotional arousal during positive situations and vocal expression of emotional valence. In my previous review (Briefer 2012), of the 58 studies found on the topic, only four studies focussed on emotional valence exclusively, while eight additional studies discussed both valence and arousal. The remaining studies ($n = 46$) investigated the effect of emotional arousal, or other contexts related to emotional arousal (stress, hunger, pain, and urgency), on vocalisations (including both *rate*, types of vocalisations produced, and vocal structure).

Vocal expression of emotions has been mainly investigated using experimental studies, by placing the animals in various situations, which are of opposing valence (e.g. social separation and reunion, Briefer et al. 2015a) or induce different levels of emotional arousal (e.g. low arousal: separation of offspring from the mother; high arousal: separation plus handling, Scheumann et al. 2012). Alternatively, vocal indicators of emotions have been highlighted without manipulating the animals using observational studies, by recording the animals during naturally occurring situations of opposing valence (e.g. agonistic and affiliative interactions, Soltis et al. 2011), or that are characterised by different levels of arousal (e.g. presence of a predator at different distances, Manser 2001; see Sects. 6.3.1 and 6.3.2, and Briefer 2012, for more details on the contexts). The situations during which data are recorded should be adapted to each species, since the contexts eliciting vocalisations can differ widely. Ideally, studies aimed at finding indicators of emotional valence should compare vocalisations produced during several positive and negative situations of similar arousal. Alternatively, in cases where situations of opposite valence also trigger different levels of emotional arousal, the variance in vocal parameters, which is linked to arousal, should be controlled for using an indicator of this dimension (e.g. physiological indicators such as heart rate, Briefer et al. 2015b; or behavioural indicators such as locomotion, Maigrot et al. 2017). Similarly, studies aimed at deciphering indicators of emotional arousal should compare vocalisations produced during situations that are characterised by various levels of arousal (e.g. low, medium, and high), but trigger the same valence. Another approach consists of studying situations that vary both in valence and arousal, and testing which emotional dimension is better at explaining the variance in each vocal parameter (e.g. using model selection procedure based on the Akaike's information criterion, Briefer et al. 2015a, b). In addition, the situations that are used should either be of the same context (e.g. food context: anticipation for food and food frustration Briefer et al. 2015b; social context: social separation and reunion, Briefer et al. 2015a), or include several contexts for each valence or arousal level. Similarly, the situations used should either trigger the same call type, which often occurs when considering the same context, or include several call types for each valence or arousal level. In cases where several contexts or call types are included in the analyses, vocal parameters that vary between valence or with arousal in a similar way across contexts or call types (e.g. increase or decrease) can be considered as good indicators of these dimensions (Maigrot et al. 2017).

The underlying emotional state of the animals during vocal production can be verified using neural, physiological, behavioural, and/or cognitive indicators previously described in the literature for each studied species (Mendl et al. 2010). Since indicators of valence have only been investigated in a few species, however, this will often not be possible for the valence dimension. In the absence of such indicators, the valence of situations can be previously validated using, for example, approach/avoidance behaviour (e.g. preference tests, see Kirkden and Pajor 2006). Alternatively, it can be inferred from knowledge of the behaviour of the species and the function of emotions. For instance, positive emotions result from encounters with rewarding stimuli, which enhance fitness and trigger approach behaviour towards the reward (e.g. affiliative interactions, food reward). In contrast, negative emotions result from encounters with punishing stimuli, which threaten fitness, and trigger avoidance behaviour (e.g. agonistic interactions, social isolation, Mendl et al. 2010). In the arousal-as-intensity version of the two dimensional model of emotions (Fig. 6.1), valence is a bipolar factor (positive versus negative), while emotional arousal can take any value within the species range (Bradley et al. 2001). If this model is used, valence will therefore either be positive or negative, while arousal should ideally be scored as a continuous factor (e.g. using heart rate), which also accounts for individual variation in emotional reactions.

In the rest of this section, I will summarise the findings on vocal indicators of emotional arousal and valence, which I previously reviewed (Briefer 2012), while adding more recent studies on the topic, with the aim of identifying common rule governing changes in the acoustic structure of vocalisations as a function of emotions ('emotional-dimension' rules). I will mainly focus on non-human mammals, with a few bird examples, and studies that analysed the structure of vocalisations (i.e. not only of *rate*, or type of calls produced). As part of these examples, I will describe the findings of a project carried out by my colleagues and myself on vocal correlates of emotional valence in ungulates, including closely related domestic and wild species (goats; domestic horses; Przewalski's horses, *Equus przewalskii*; domestic pigs; and wild boars, *Sus scrofa*).

6.3.1 Evidence for Vocal Expression of Emotional Arousal

As mentioned in the previous section, vocal expression of emotional arousal has been mainly studied during negative situations. The most common experimental situations that are used include partial or full social isolation (e.g. Schrader and Todt 1993; Monticelli et al. 2004; Siebert et al. 2011), human approach test (e.g. Marchant et al. 2001; Gogoleva et al. 2010a, b), and routine farm procedures (e.g. castration and branding, Weary et al. 1998; Watts and Stookey 1999; Puppe et al. 2005; see review in Briefer 2012). More recent studies have used the following: various handling and naturally occurring situations in pandas (*Ailuropoda melanoleuca*, Stoeger et al. 2012); separation from the mother and handling by an experimenter in kittens (*Felis catus*, Scheumann et al. 2012); manual squeezing and isolation in relation to age, body weight, and health in piglets (Illmann et al. 2013); a back test in pigs (Linhart et al. 2015); social isolation, anticipation for food, and food frustration in goats (Briefer et al. 2015b); and social separation and reunion with one or several group members in horses (Briefer et al. 2015a). Observational studies mainly assessed the level of emotional arousal according to predator threat (e.g. Blumstein and Arnold 1995; Manser 2001), the intensity of agonistic interactions (e.g. Gouzoules and Gouzoules 1989; Siebert and Parr 2003; Bastian and Schmidt 2008), and more recently, group composition in humpback whales (*Megaptera novaeangliae*, in stable versus affiliating groups, Dunlop 2017).

Since the valence and arousal dimensions might not be fully independent of each other (Kuppens et al. 2013), emotional arousal could have a different effect on vocalisations within positive contexts, compared to negative ones. However, to my knowledge, the only studies that investigated how the structure of vocalisations changes with emotional arousal within neutral or positive situations were conducted by Rendall (2003), Gogoleva et al. (2010a), Soltis et al. (2011), and Collins et al. (2011). Rendall (2003) investigated grunts produced by chacma baboons (*Papio cynocephalus ursinus*) in a move context (i.e. when the group is about to initiate a move; likely neutral) and infant context (i.e. usually friendly, positive social interaction during which the emitter approaches a mother with an infant and attempts to touch or handle her baby), and identified two discrete arousal conditions (high versus low) within each context using behavioural indicators. Gogoleva et al. (2010a) tested how tame and aggressive strains of silver foxes (*Vulpes vulpes*) react to a human approaching and retrieving, thus, triggering various levels of emotional arousal. The situation was assumed to be positive for the tame strain and negative for the aggressive strain. Soltis et al. (2011) studied the rumbles of African elephants (*Loxodonta africana*) produced in neutral (minimal social activity), negative (dominance interaction), and positive (affiliative interaction) social contexts. Finally, Collins et al. (2011) investigated calls produced during mother–pup interactions characterised by different levels of valence and arousal (i.e. reunion, isolation, and nursing) in Weddell seals (*Leptonychotes weddellii*).

The 54 studies on emotional arousal reviewed by Briefer (2012) revealed strong evidence for an increase in *rate*, *F0 contour*, *F0 range*, *amplitude contour*, *energy distribution* (towards higher frequencies), *peak frequency* and *formant contour,* and a decrease in *interval* with arousal (5–21 studies supported these changes, with a maximum of 2 studies with opposing shifts). The most noticeable changes were the consistent increase in *F0 contour* (n = 21 studies), *amplitude contour* (n = 12 studies), and *peak frequency* (n = 11 studies). In addition, there was some unanimous evidence for an increase in *jitter* (n = 3 studies) and the time of the peak frequency (i.e. time point of *peak frequency* relative to the total *duration*; n = 4 studies). It also appears that vocalisations became noisier (increase in *spectral noise* and decrease in *Wiener entropy* and *harmonicity*), although 2 of the 13 studies that measured noise demonstrated the opposite trend. Zimmermann et al. (2013), who carried out a similar review (n = 39 studies) focussing on *duration*, *rate*, *F0 contour*, and *harmonicity*, also found strong support for an increase in *rate*, and some support for an increase in *F0 contour* and *duration* with emotional arousal. The more recent studies, which were mentioned above (Scheumann et al. 2012; Stoeger et al. 2012; Illmann et al. 2013; Briefer et al. 2015a, b; Linhart et al. 2015; Dunlop 2017), demonstrated similar changes for most parameters including increases in *rate*, frequencies (*F0 contour*, *peak frequency*, *energy distribution*, and/or *formants*), *frequency range* and noise (i.e. the increase in *spectral noise* and decrease in *harmonicity*), with emotional arousal. Surprisingly however, Scheumann et al. (2012) found a decrease in *F0 contour* with arousal in domestic kittens, and Briefer et al. (2015b) demonstrated a decrease in *F1 contour* in goats that could be linked to opening of the mouth with arousal. In addition, Linhart et al. (2015) found that some parameters that changed with arousal differed between call types. Notably, although *harmonicity* decreased with arousal in pig grunts, indicating a raise in noise as found in other species, it increased with arousal in screams. Studies investigating vocal expression of emotional arousal in positive situations (Rendall 2003; Gogoleva et al. 2010a; Soltis et al. 2011) found that arousal-related changes in these contexts were consistent with those mentioned above (i.e. increase in *rate*, *duration*, *F0 contour*, *F0 range*, and *amplitude contour*; decrease in *interval*), except for a decrease in *F2 contour* in baboon's grunts (Rendall 2003), and a decrease in *peak frequency* in the vocalisations of tame silver foxes (Gogoleva et al. 2010a). Although the decrease in *F2 contour* with emotional arousal in the baboons' grunts was similar between the two contexts investigated (infant and move, Rendall 2003), and consistent with previous literature on the same species (Meise et al. 2011), the decrease in *peak frequency* in silver foxes' vocalisations was in contrast to the increase in this parameter with arousal during the negative situation (Gogoleva et al. 2010a).

In birds (zebra finches), social isolation and oral administration of corticosterone resulted in longer calls with a higher frequencies (*dominant frequency* and *energy distribution*, Perez et al. 2012), which is similar to what has been described in mammals. However, compared to spontaneous songs (i.e. not induced by a playback simulating territorial intrusion), provoked songs of territorial birds have lower frequencies (e.g. *peak frequency* in skylarks, *Alauda arvensis*, Geberzahn and Aubin 2014) and *amplitude contour* ('soft songs', see review in Searcy and Beecher 2009 for a review), and are produced at faster *rates* (e.g. banded wren, *Thryophilus pleurostictus*, Vehrencamp et al. 2013) during the widely studied context of territorial aggression. Low frequencies in aggressive contexts (but not low amplitudes) conform to the motivation-structural rules (Morton 1977; Ehret 2006). Therefore, it could be that high-arousal aggressive vocalisations, unlike high-arousal fearful or positive ones, are characterised by low frequencies instead of high ones. Low-frequency voices (deep respiration, low F0, and high amplitude; 'full voice') are also predicted in humans following a confident outcome of the coping potential stimulus evaluation check (Scherer 1986). Future studies on vocal expression of arousal might benefit from differentiating between vocalisations produced by the aggressor and victim during agonistic interactions.

6.3.2 Evidence for Vocal Expression of Emotional Valence

As mentioned previously (Sect. 6.3), studies investigating vocal expression of valence are scarce (only 8/58 studies reviewed in Briefer 2012). However, in recent years, the focus of animal welfare research has shifted from negative to positive affective states, after Boissy et al. (2007) highlighted the importance of not only minimising negative emotions, but also promoting positive emotions to improve animal welfare (Proctor et al. 2013). This shift in research interests resulted in an increase in studies aimed at finding indicators of valence that would allow us to assess if an animal is experiencing a positive or negative emotion. Vocal indicators of valence have been studied using, for example, observations of affiliative versus abusive infant handling in rhesus monkeys (*Macaca mulatta*, Jovanovic and Gouzoules 2001), various appetitive versus aversive stimuli in rats (see review in Knutson et al. 2002), grooming by an experimenter versus isolation and threat-induced contexts in grey mouse lemurs (*Microcebus murinus*, Scheumann et al. 2007), difference in reaction to humans between tame and aggressive strains in silver foxes (Gogoleva et al. 2010a), playing versus disturbance/aggression (unfamiliar human intrusion) or bone guarding in dogs (*Canis familiaris*, Yin and McCowan 2004; Taylor et al. 2009; Faragó et al. 2010), playing versus disturbance/aggression in pigs (Chan 2011), positive versus negative human interactions in cats (Yeon et al. 2011), and observations of affiliative versus agonistic interactions in elephants (Soltis et al. 2011). More recent studies include various farm-related positive (e.g. nursing and reunion) and negative (e.g. castration and crushing) situations in pigs (Tallet et al. 2013), various naturally occurring situations of positive (feed), neutral (travel and rest), and negative (agonistic interactions and alarm) valence in bonobos (*Pan paniscus*, Clay et al. 2015), and positive versus negative conditioning in pigs (Leliveld et al. 2016).

Since vocal indicators of valence have only been studied in very few species, identifying common rules governing the structure of vocalisations in positive and negative contexts is challenging. In addition to the paucity of studies on the topic, most of them did not assess emotional arousal, and thus, we cannot rule out the possibility that the changes in vocal parameters observed between the positive and negative context(s) are in fact related to arousal (Briefer 2012). Compared to negative emotions, positive emotions are often of lower arousal (Boissy et al. 2007). Therefore, vocal changes occuring between negative and positive situations are often likely to result from a decrease in emotional arousal (see Sect. 6.3.1 for a list of expected changes). Furthermore, even in cases when the arousal of two situations of opposite valence is expected to be similar, slight changes in emotional arousal can occur and affect vocalisations. A good example is our study on goats, where social isolation was assumed a priori to induce higher arousal than anticipation for food or food frustration (Briefer et al. 2015b). However, heart rate data revealed that, surprisingly, this context induced a similar arousal level as the control condition (undisturbed in a familiar pen). This was the case at least on the day of the recordings, which was preceded by 3 days of habituation in order to minimise stress linked to novelty. This example demonstrates that it is difficult to assume the arousal triggered by a situation without relying on physiological or behavioural data. However, relying on physiological or behavioural indicators of arousal can also be problematic in some cases; for example, in cases such as 'freezing', where the animal stops moving to evaluate a stimulus (e.g. presence of a predator), which is accompanied by a decrease in heart rate (i.e. 'alarm bradycardia', e.g. Jacobsen 1979), but likely not a decrease in cognitive and potentially subjective arousal (or in the overall emotional response). In general, assessing emotions using measures corresponding to several components (neurophysiological, behavioural, and/or cognitive) should increase accuracy.

Changes in emotional valence have been suggested to result in a change in the type of call produced, while changes in emotional arousal are more likely to result in the modification of the acoustic structure of a given call type (Manser 2010). Although this might be true for most cases, such as the negative 22-kHz and positive 50-kHz rat ultrasonic vocalisations (Knutson et al. 2002; Burgdorf and Moskal 2009), purring-like sounds in Carnivora and other mammalian orders (Peters 1984, 2002), and laughter-like sounds in apes (Ross et al. 2009), there are some exceptions to the rule. For instance, contact calls are typically produced in a wide range of positive and negative situations (e.g. African elephant rumbles, Soltis et al. 2011; bonobos peeps, Clay et al. 2015; goat bleats, Briefer et al. 2015b; horse whinnies, Briefer et al. 2015a; and pig grunts, Briefer et al. 2019). Other examples include pig barks, which are produced both in play (positive) and alarm (negative) contexts (Chan 2011), or dog barks (Yin and McCowan 2004) and growls (Taylor et al. 2009), which can occur both in play and aggression contexts. The acoustic structure of all these call types varies significantly with valence.

The literature reviewed by Briefer (2012) revealed only one parameter, i.e. *duration*, which could indicate emotional valence across species. *Duration* was the only parameter that was measured in more than one study and consistently decreased from negative to positive contexts. Other acoustic parameters either differed between species or had only been investigated in one study. For example, both shifts towards higher frequencies (*F0*, *peak frequency*, and *energy distribution*; Yin and McCowan 2004; Taylor et al. 2009) and towards lower frequencies (Fichtel et al. 2001; Jovanovic and Gouzoules 2001; Scheumann et al. 2007) from negative to positive contexts have been identified. More recent studies found a decrease in *duration* (Tallet et al. 2013), and an increase in *energy distribution*, in the maximum frequency observed in the spectrum and in *frequency range* (bandwidth) in pigs (Leliveld et al. 2016), and a decrease in *F0* in bonobos from negative to positive contexts (Clay et al. 2015). In addition, Clay et al. (2015) found that *duration* tended to be longer in positive, compared to negative contexts, which does not corroborate with previous studies (Briefer 2012). Although Clay et al. (2015) and Tallet et al. (2013) did not include any measure of emotional arousal, Leliveld et al. (2016) collected physiological and behavioural data. In this latter study, physiological data (heart rate and heart rate variability) suggested that the positive and negative situations (positive and negative conditioning) induced similar arousal levels (Leliveld et al. 2016). Yet, the *energy distribution, frequency range*, and maximum frequency were all higher in the positive situation, compared to the negative one, suggesting that these vocal indicators accurately indicated emotional valence and not arousal.

In addition to the above-mentioned studies on vocal expression of valence, my colleagues and I conducted a project that aimed at investigating the evolution of vocal expression of emotional valence in ungulates, including domestic species (goats, horses, and pigs) and closely related wild species (Przewalski's horses and wild boars). Although we originally tried to use similar methods in all species to allow direct comparisons, the contexts of recordings and vocal parameters that were measured had to be adapted to each species. Further, our methods of analyses improved throughout the project. All species were recorded in positive and negative contexts adapted to each, which were experimental in domestic species and naturally occurring in wild species, since the latter were studied in zoos and could not be manipulated. The goats were recorded during anticipation for food versus isolation and food frustration (Briefer et al. 2015b), the horses during social reunion versus separation (Briefer et al. 2015a), the pigs during a positive situation in which they were in pair in the same pen with food and water versus isolation in a barren pen (Briefer et al. 2019), and the Przewalski's horses and wild boars during anticipation for food and affiliative interactions versus agonistic interactions and social separation (for Przewalski's horses only, Maigrot et al. 2017, 2018). All these studies focussed on within-call type and within individual changes between positive and negative contexts. To this aim, we specifically analysed goat bleats, horse and Przewalski's horse whinnies, and pig and wild boar grunts. For the wild species, we additionally compared the proportion of the most common call types (Przewalski's horse whinnies, squeals and nickers; wild boar squeals, screams and grunts) associated with negative versus positive contexts, and tested if vocal expression of valence differed between call types. To control for changes in vocalisations that could be due to emotional arousal and not valence, we assessed this dimension based on heart rate in the domestic species, and locomotion in the wild species. In pigs, Przewalski's horses and wild boars, we controlled for arousal by including its indicator (heart rate or locomotion) in the linear-mixed effects models aimed at testing the effect of valence on vocal structure. In goats and domestic horses, we tested which dimension (i.e. arousal or valence) explained the majority of the variance seen in each parameter. As a result, the best indicators of valence highlighted by our studies were defined in the wild species, as vocal parameters that showed similar changes from negative to positive context across call types, after controlling for movement. In goats and horses, the best valence indicators were parameters whose variance was better explained (or only explained) by valence rather than arousal. In pigs, these were the parameters that loaded the most on principal component (PC) scores that were extracted from a principal component analysis (PCA), and were affected by the valence after controlling for heart rate.

In goats, we found that positive bleats had a more stable *F0* (lower *F0 range* and FM extent, i.e. mean peak-to-peak variation of each F0 modulation) compared to negative bleats (Briefer et al. 2015b). In horses, we discovered that whinnies were composed of two fundamental frequencies along with their respective harmonics, suggesting biphonation. The lower fundamental frequency '*F0*' and higher fundamental frequency 'G0' ranged from 52 to 1050 Hz and from 493 to 3012 Hz, respectively (Briefer et al. 2015a). Interestingly, our analyses revealed that G0 and *duration* indicated the emotional valence of the emitter, while *F0* and *energy distribution* indicated emotional arousal (i.e. segregation of information about valence and arousal, Briefer et al. 2015a). Positive whinnies were characterised by a lower G0 and shorter *duration*, compared to negative whinnies (Briefer et al. 2015a). Przewalski's horse whinnies showed a similar structure, with a *F0* and G0 at 34–790 Hz and 683–2340 Hz, respectively. However, indicators of emotional valence in this species differed noticeably from those identified in domestic horses. Indeed, Przewalski's horses' best indicators of valence were G0 range (i.e. difference between minimum and maximum G0), *energy distribution* and AM rate (i.e. number of complete cycles of *AM* per second), which all decreased from negative to positive vocalisations (across call types; whinnies, squeals and nickers). Further, AM extent (i.e. mean peak-to-peak variation of each *AM*) and the time of maximum *F0* (i.e. time point of the maximum *F0* value relative to the total duration) increased from negative to positive vocalisations (Maigrot et al. 2017). Similar differences were found between domestic pigs and wild boars. In pigs, positive grunts were shorter, with a lower *harmonicity* and *F0 contour*, and had a shorter range of *F3* and higher *formants*, compared to negative grunts (Briefer et al. 2019). Although positive wild boar vocalisations (including squeals, screams and grunts) were shorter and had a shorter *F3* range than negative ones, they had, unlike pig grunts, a lower *energy distribution* and *formants*, in addition to a lower AM rate (Maigrot et al. 2018). Therefore, our project revealed some common indicators of valence across our studied species (e.g. *duration* in horses, wild boars, and pigs; fundamental frequency in horses (G0) and pigs (*F0*); and fundamental frequency range in goats (*F0 range*) and Przewalski's horses (G0 range)). However, striking differences between closely related species, such as the absence of common indicators between domestic and Przewalski's horses, and opposite changes in *formants* between domestic pigs and wild boars, were also noted.

6.4 Discussion and Conclusion

The majority of the studies reviewed in this chapter demonstrated that vocalisations were louder and were produced at faster *rates*, with higher frequencies (both source- and filter-related), and a more variable *F0*, when arousal increased. These alterations correspond closely to the effects of the physiological changes linked to an increase in arousal on the acoustic structure of vocalisations. This suggests that vocal expression of emotional arousal has been conserved throughout evolution in mammals (and maybe birds), as a result of shared anatomical and physiological circuits involved in autonomic regulation, and the production and perception of vocalisations (Porges and Lewis 2010). Some arousal-related changes to the structure of vocalisations might however depend on the type of call (e.g. *harmonicity*, Linhart et al. 2015), the valence (e.g. *peak frequency*, Gogoleva et al. 2010a), the context of production, or other confounding factors (e.g. behaviour of the emitter, aggressor versus victim during agonistic interactions, distance between emitter and receiver). Concerning vocal expression of emotional valence, the evidence gathered by Briefer (2012), together with more recent studies mentioned in this chapter, revealed that a decrease in *duration* from negative to positive contexts is supported by eight studies (Briefer et al. 2015a, 2019; Maigrot et al. 2018; Fichtel et al. 2001; Brudzynski 2007; Taylor et al. 2009; Faragó et al. 2010; Tallet et al. 2013), with only one study demonstrating the opposite trend (not significant, Clay et al. 2015). A decrease in *F0* from negative to positive contexts was found in five studies (Briefer et al. 2015a, 2019; Fichtel et al. 2001; Soltis et al. 2011; Clay et al. 2015), with two studies demonstrating the opposite (in dogs, Taylor et al. 2009; Faragó et al. 2010). Finally, *F0 range* decreased from negative to positive contexts in the three species in which it was measured (Soltis et al. 2011; Briefer et al. 2015b; Maigrot et al. 2017). The rest of the parameters were either only measured in one study, or were found to change in different directions between studies. Therefore, there is now growing evidence for positive vocalisations being shorter, and likely, with lower and less variable *F0*, than negative vocalisations. Such cross-species indicators might be especially useful when differentiating between positive and negative emotions experienced by captive animals, in order to improve their welfare by minimising negative situations and promoting positive ones (Boissy et al. 2007). However, these changes seem to differ from findings in humans; further, as my project on ungulates revealed, vocal expression of valence can vary widely between closely related species. It could therefore be the case that vocal expression of valence is more species specific, unlike vocal expression of emotional arousal, which is well conserved throughout evolution. Further studies that simultaneously compare vocal expression of emotional valence and arousal in closely related species would be particularly useful to test this hypothesis.

The changes mentioned above do not fully correlate with the motivation-structural rules, which predicts aggressive vocalisations to be noisy and with low frequencies, fearful/appeasing vocalisations to be tonal and with high frequencies, and friendly vocalisations to have low amplitudes and frequencies (Morton 1977;

August and Anderson 1987). Friendly vocalisations, which can be assumed to reflect positive-low arousal emotions, are indeed predicted both by the motivation-structural rules and the emotional-dimension rules to have low frequencies (August and Anderson 1987; Peters 2002; Briefer 2012; this chapter). However, aggressive and fearful/appeasing vocalisations, which are predicted to largely differ in structure by the motivation-structural rules in terms of frequencies and spectral noise, can both be assumed to reflect negative-high arousal emotions. The emotional-dimension rules would predict both kinds of vocalisations to be high in frequency, whereas the motivation-structural rules suggests that aggressive and fearful/appeasing vocalisations are low and high in frequency, respectively. The discrepancy between these two sets of common rules could be solved if we combine them and describe rules according to: (a) the valence and arousal dimensions, along with (b) the characteristics of the social relationships indicating whether the emitter is a subordinate that reacts to a threat or not, and (c) the function of the vocalisation in terms of whether it is aimed at inducing approach or retreat from conspecifics. The amplitude of vocalisations and their rate could indicate the emotional arousal, with higher amplitude, faster rate vocalisations indicating higher arousal emotions. The duration of vocalisations could indicate the emotional valence, with positive vocalisations being shorter than negative ones. The frequency of vocalisations could indicate whether the emitter is acting as a subordinate or not in the specific context of the vocalisation, with high frequencies indicating subordination, and low frequencies indicating dominance or absence of dominance relationship (e.g. friendly vocalisations). Finally, the presence of spectral noise could indicate the function of the vocalisation, with noisy vocalisations aimed at inducing a retreat, and tonal vocalisations an approach. These combined sets of rules might be able to explain to a large extent the acoustic features of vocalisations and could be investigated further.

Acknowledgements I thank the Swiss National Science Foundation for funding the project on vocal expression of emotions in domestic and wild ungulates mentioned in this chapter (project no. PZ00P3_148200) and the ETH Zürich for the possibility to spend time on reading for and writing this text.

References

Anderson DJ, Adolphs R (2014) A framework for studying emotions across species. Cell 157:187–200

August PV, Anderson JGT (1987) Mammal sounds and motivation–structural rules: a test of the hypothesis. J Mammal 68:1–9

Bastian A, Schmidt S (2008) Affect cues in vocalizations of the bat, *Megaderma lyra*, during agonistic interactions. J Acoust Soc Am 124:598–608

Blumstein DT, Armitage KB (1997) Alarm calling in yellow-bellied marmots: I. The meaning of situationally variable alarm calls. Anim Behav 53:143–171

Blumstein DT, Arnold W (1995) Situational specificity in Alpine-marmot alarm communication. Ethology 100:1–13

Blumstein DT, Chi YY (2012) Scared and less noisy: glucocorticoids are associated with alarm call entropy. Biol Lett 8:189–192

Boissy A, Manteuffel G, Jensen MB et al (2007) Assessment of positive emotions in animals to improve their welfare. Physiol Behav 92:375–397

Bradley MM, Lang PJ (2007) The International Affective Picture System (IAPS) in the study of emotion and attention. In: Coan JA, Allen JBB (eds) Handbook of emotion elicitation and assessment. Oxford University Press, New York, pp 29–46

Bradley M, Codispoti M, Cuthbert B, Lang P (2001) Emotion and motivation I: defensive and appetitive reactions in picture processing. Emot Wash DC 1:276–298

Briefer EF (2012) Vocal expression of emotions in mammals: mechanisms of production and evidence. J Zool 288:1–20

Briefer EF, McElligott AG (2013) Rescued goats at a sanctuary display positive mood after former neglect. Appl Anim Behav Sci 146:45–55

Briefer EF, Maigrot A-L, Mandel R et al (2015a) Segregation of information about emotional arousal and valence in horse whinnies. Sci Rep 4:9989

Briefer EF, Tettamanti F, McElligott AG (2015b) Emotions in goats: mapping physiological, behavioural and vocal profiles. Anim Behav 99:131–143

Briefer EF, Vizier E, Gygax L, Hillmann E (2019) Expression of emotional valence in pig closed-mouth grunts: involvement of both source- and filter-related parameters. J Acoust Soc Am 145:2895

Briefer Freymond S, Briefer EF, Zollinger A et al (2014) Behaviour of horses in a judgment bias test associated with positive or negative reinforcement. Appl Anim Behav Sci 158:34–45

Briefer Freymond S, Bardou D, Briefer EF et al (2015) The physiological consequences of crib-biting in horses in response to an ACTH challenge test. Physiol Behav 151:121–128

Brudzynski SM (2007) Ultrasonic calls of rats as indicator variables of negative or positive states: acetylcholine–dopamine interaction and acoustic coding. Behav Brain Res 182:261–273

Budka M, Osiejuk TS (2013) Formant frequencies are acoustic cues to caller discrimination and are a weak indicator of the body size of corncrake males. Ethology 119:960–969

Burgdorf J, Moskal JR (2009) Frequency modulated 50 kHz ultrasonic vocalizations reflect a positive emotional state in the rat: neural substrates and therapeutic implications. In: Brudzynski SM (ed) Handbook of mammalian vocalization—an integrative neuroscience approach. Academic, London, pp 209–214

Burman O, McGowan R, Mendl M et al (2011) Using judgement bias to measure positive affective state in dogs. Appl Anim Behav Sci 132:160–168

Caeiro CC, Waller BM, Zimmermann E et al (2013) OrangFACS: a muscle-based facial movement coding system for orangutans (Pongo spp.). Int J Primatol 34:115–129

Chan WY (2011) The meaning of barks: vocal communication of fearful and playful affective states in pigs. PhD thesis, Washington State University

Cilulko J, Janiszewski P, Bogdaszewski M, Szczygielska E (2013) Infrared thermal imaging in studies of wild animals. Eur J Wildl Res 59:17–23

Clay Z, Archbold J, Zuberbühler K (2015) Functional flexibility in wild bonobo vocal behaviour. PeerJ 3:e1124

Collias NE (1960) An ecological and functional classification of animals sounds. In: Lanyon WE, Tavolga WN (eds) Animal sounds and communication. American Institute of Biological Science, Washington, DC, pp 368–391

Collins KT, McGreevy PD, Wheatley KE, Harcourt RG (2011) The influence of behavioural context on Weddell seal (*Leptonychotes weddellii*) airborne mother-pup vocalisation. Behav Proc 87:286–290

Dalla Costa E, Minero M, Lebelt D et al (2014) Development of the Horse Grimace Scale (HGS) as a pain assessment tool in horses undergoing routine castration. PLoS One 9:e92281

Darwin C (1872) The expressions of emotions in man and animals. John Murray, London

Dawkins MS (1990) From an animal's point of view: motivation, fitness, and animal welfare. Behav Brain Sci 13:1–61

Désiré L, Boissy A, Veissier I (2002) Emotions in farm animals: a new approach to animal welfare in applied ethology. Behav Proc 60:165–180

Dolan RJ (2002) Emotion, cognition, and behavior. Science 298:1191

Doyle RE, Hinch GN, Fisher AD et al (2011) Administration of serotonin inhibitor p-Chlorophenylalanine induces pessimistic-like judgement bias in sheep. Psychoneuroendocrinology 36:279–288

Dunlop RA (2017) Potential motivational information encoded within humpback whale non-song vocal sounds. J Acoust Soc Am 141:2204–2213

Ehret G (2006) Common rules of communication sound perception. In: Kanwal J, Ehret G (eds) Behaviour and neurodynamics for auditory communication. Cambridge University Press, Cambridge, pp 85–114

Fant G (1960) Acoustic theory of speech production. Mouton, The Hague

Faragó T, Pongrácz P, Range F et al (2010) "The bone is mine": affective and referential aspects of dog growls. Anim Behav 79:917–925

Fichtel C, Hammerschmidt K, Jürgens U (2001) On the vocal expression of emotion. A multi-parametric analysis of different states of aversion in the squirrel monkey. Behaviour 138:97–116

Fitch WT (1999) Acoustic exaggeration of size in birds via tracheal elongation: comparative and theoretical analyses. J Zool 248:31–48

Fitch WT, Kelley JP (2000) Perception of vocal tract resonances by whooping cranes *Grus americana*. Ethology 106:559–574

Fraser D (2009) Animal behaviour, animal welfare and the scientific study of affect. Spec Issue Anim Suff Welf 118:108–117

Geberzahn N, Aubin T (2014) How a songbird with a continuous singing style modulates its song when territorially challenged. Behav Ecol Sociobiol 68:1–12

Gerhardt HC (1998) Acoustic signals of animals: recording, field measurements, analysis and description. In: Hopp SL, Owren MJ, Evans CS (eds) Animal acoustic communication: sound analysis and research methods. Springer, Berlin, pp 1–25

Giddens CL, Barron KW, Byrd-Craven J et al (2013) Vocal indices of stress: a review. J Voice 27:390.e21–390.e29

Gogoleva S, Svetlana S, Volodin I et al (2010a) Sign and strength of emotional arousal: vocal correlates of positive and negative attitudes to humans in silver foxes (*Vulpes vulpes*). Behaviour 147:1713–1736

Gogoleva S, Volodina E, Volodin I et al (2010b) The gradual vocal responses to human-provoked discomfort in farmed silver foxes. Acta Ethol 13:75–85

Goudbeek M, Scherer KR (2010) Beyond arousal: valence and potency/control cues in the vocal expression of emotion. J Acoust Soc Am 128:1322–1336

Gouzoules H, Gouzoules S (1989) Design features and developmental modification of pigtail macaque, *Macaca nemestrina*, agonistic screams. Anim Behav 37:383–401

Gygax L, Reefmann N, Wolf M, Langbein J (2013) Prefrontal cortex activity, sympatho-vagal reaction and behaviour distinguish between situations of feed reward and frustration in dwarf goats. Behav Brain Res 239:104–114

Illmann G, Hammerschmidt K, Špinka M, Tallet C (2013) Calling by domestic piglets during simulated crushing and isolation: a signal of need? PLoS One 8:e83529

Imfeld-Mueller S, Hillmann E (2012) Anticipation of a food ball increases short-term activity levels in growing pigs. Appl Anim Behav Sci 137:23–29

Imfeld-Mueller S, Van Wezemael L, Stauffacher M et al (2011) Do pigs distinguish between situations of different emotional valences during anticipation? Appl Anim Behav Sci 131:86–93

Jacobsen NK (1979) Alarm bradycardia in white-tailed deer fawns (*Odocoileus virginianus*). J Mammal 60:343–349

Janik VM, Slater PJB (1997) Vocal learning in mammals. Adv Stud Behav 26:59–99

Jovanovic T, Gouzoules H (2001) Effects of nonmaternal restraint on the vocalizations of infant rhesus monkeys (*Macaca mulatta*). Am J Prim 53:33–45

Jürgens U (1976) Reinforcing concomitants of electrically elicited vocalizations. Exp Brain Res 26:203–214

Jürgens U (2002) Neural pathways underlying vocal control. Neurosci Biobehav Rev 26:235–258

Jürgens U (2009) The neural control of vocalization in mammals: a review. J Voice 23:1–10

Jürgens U, Ploog D (1981) On the neural control of mammalian vocalization. Trends Neurosci 4:135–137

Kelly T, Reby D, Levréro F, Keenan S, Gustafson E, Koutseff A, Mathevon N (2017) Adult human perception of distress in the cries of bonobo, chimpanzee, and human infants. Biol J Linn Soc 120:919–930

Kirkden RD, Pajor EA (2006) Using preference, motivation and aversion tests to ask scientific questions about animals' feelings. Appl Anim Behav Sci 100:29–47

Klump GM, Shalter MD (1984) Acoustic behaviour of birds and mammals in the predator context; I. Factors affecting the structure of alarm signals. II. The functional significance and evolution of alarm signals. Z Für Tierpsychol 66:189–226

Knutson B, Burgdorf J, Panksepp J (2002) Ultrasonic vocalizations as indices of affective states in rats. Psychol Bull 128:961–977

Kuppens P, Tuerlinckx F, Russell JA, Barrett LF (2013) The relation between valence and arousal in subjective experience. Psychol Bull 139:917–940

Laukka P, Juslin P, Bresin R (2005) A dimensional approach to vocal expression of emotion. Cogn Emot 19:633–653

Lazarus RS (1991) Progress on a cognitive-motivational-relational theory of emotion. Am Psychol 46:819

Leliveld LMC, Düpjan S, Tuchscherer A, Puppe B (2016) Behavioural and physiological measures indicate subtle variations in the emotional valence of young pigs. Physiol Behav 157:116–124

Lemasson A, Remeuf K, Trabalon M et al (2015) Mares prefer the voices of highly fertile stallions. PLoS One 10:e0118468

Linhart P, Ratcliffe VF, Reby D, Špinka M (2015) Expression of emotional arousal in two different piglet call types. PLoS One 10:e0135414

Maigrot A, Hillmann E, Callista A, Briefer EF (2017) Vocal expression of emotional valence in Przewalski's horses. Sci Rep 18:8779

Maigrot AL, Hillmann E, Briefer EF (2018) Encoding of emotional valence in wild boars (Sus scrofa) calls. Animals 8:85

Manser MB (2001) The acoustic structure of suricates' alarm calls varies with predator type and the level of response urgency. Proc R Soc B 268:2315–2324

Manser MB (2010) The generation of functionally referential and motivational vocal signals in mammals. In: Brudzynski SM (ed) Handbook of mammalian vocalization—an integrative neuroscience approach. Academic, London, pp 477–486

Manser MB, Seyfarth RM, Cheney DL (2002) Suricate alarm calls signal predator class and urgency. Trends Cogn Sci 6:55–57

Manteuffel G, Puppe B, Schön P-C (2004) Vocalization of farm animals as a measure of welfare. Appl Anim Behav Sci 88:163–182

Marchant JN, Whittaker X, Broom DM (2001) Vocalisations of the adult female domestic pig during a standard human approach test and their relationships with behavioural and heart rate measures. Appl Anim Behav Sci 72:23–39

Mausbach J, Braga Goncalves I, Heistermann M et al (2017) Meerkat close calling patterns are linked to sex, social category, season and wind, but not fecal glucocorticoid metabolite concentrations. PLoS One 12:e0175371

Meise K, Keller C, Cowlishaw G, Fischer J (2011) Sources of acoustic variation: implications for production specificity and call categorization in chacma baboon (*Papio ursinus*) grunts. J Acoust Soc Am 129:1631–1641

Mendl M, Burman OHP, Parker RMA, Paul ES (2009) Cognitive bias as an indicator of animal emotion and welfare: emerging evidence and underlying mechanisms. Appl Anim Behav Sci 118:161–181

Mendl M, Burman OHP, Paul ES (2010) An integrative and functional framework for the study of animal emotion and mood. Proc R Soc B 277:2895–2904
Merlot E, Mounier AM, Prunier A (2011) Endocrine response of gilts to various common stressors: a comparison of indicators and methods of analysis. Physiol Behav 102:259–265
Monticelli PF, Tokumaru RS, Ades C (2004) Isolation induced changes in Guinea Pig *Cavia porcellus* pup distress whistles. Acad Bras Cienc 76:368–372
Morton ES (1977) On the occurrence and significance of motivation-structural rules in some bird and mammal sounds. Am Nat 111:855–869
Moura DJ, Silva WT, Naas IA et al (2008) Real time computer stress monitoring of piglets using vocalization analysis. Smart Sens Precis Livest Farming 64:11–18
Mukudai S, Matsuda KI, Nishio T et al (2015) Differential responses to steroid hormones in fibroblasts from the vocal fold, trachea, and esophagus. Endocrinology 156:1000–1009
Nowicki S (1987) Vocal tract resonances in oscine bird sound production: evidence from birdsongs in a helium atmosphere. Nature 325:53–55
Orlikoff RF, Baken RJ (1989) The effect of the heartbeat on vocal fundamental frequency perturbation. J Speech Lang Hear Res 32:576–582
Perez EC, Elie JE, Soulage CO et al (2012) The acoustic expression of stress in a songbird: does corticosterone drive isolation-induced modifications of zebra finch calls? Horm Behav 61:573–581
Peters G (1984) On the structure of friendly close range vocalizations in terrestrial carnivores (Mammalia: Carnivora: Fissipedia). Z Für Säugetierkd 49:157–182
Peters G (2002) Purring and similar vocalizations in mammals. Mammal Rev 32:245–271
Pisanski K, Nowak J, Sorokowski P (2016) Individual differences in cortisol stress response predict increases in voice pitch during exam stress. Physiol Behav 163:234–238
Porges SW (1995) Orienting in a defensive world: mammalian modifications of our evolutionary heritage. A polyvagal theory. Psychophysiology 32:301–318
Porges SW (2001) The polyvagal theory: phylogenetic substrates of a social nervous system. Int J Psychophysiol 42:123–146
Porges SW, Lewis GF (2010) The polyvagal hypothesis: common mechanisms mediating autonomic regulation, vocalizations and listening. In: Brudzynski SM (ed) Handbook of mammalian vocalization—an integrative neuroscience approach. Academic, London, pp 255–264
Proctor HS, Carder G (2014) Can ear postures reliably measure the positive emotional state of cows? Appl Anim Behav Sci 161:20–27
Proctor HS, Carder G (2015a) Nasal temperatures in dairy cows are influenced by positive emotional state. Physiol Behav 138:340–344
Proctor HS, Carder G (2015b) Measuring positive emotions in cows: do visible eye whites tell us anything? Physiol Behav 147:1–6
Proctor SH, Carder G, Cornish RA (2013) Searching for animal sentience: a systematic review of the scientific literature. Animals 3:882–906
Puppe B, Schön P-C, Tuchscherer A, Manteuffel G (2005) Castration-induced vocalisation in domestic piglets, *Sus scrofa*: complex and specific alterations of the vocal quality. Appl Anim Behav Sci 95:67–78
Puts DA, Gaulin SJC, Verdolini K (2006) Dominance and the evolution of sexual dimorphism in human voice pitch. Evol Hum Behav 27:283–296
Reber SA, Nishimura T, Janisch J et al (2015) A Chinese alligator in heliox: formant frequencies in a crocodilian. J Exp Biol 218:2442
Reber SA, Janisch J, Torregrosa K et al (2017) Formants provide honest acoustic cues to body size in American alligators. Sci Rep 7:1816
Reefmann N, Bütikofer Kaszàs F, Wechsler B, Gygax L (2009a) Ear and tail postures as indicators of emotional valence in sheep. Appl Anim Behav Sci 118:199–207
Reefmann N, Bütikofer Kaszàs F, Wechsler B, Gygax L (2009b) Physiological expression of emotional reactions in sheep. Physiol Behav 98:235–241

Reefmann N, Wechsler B, Gygax L (2009c) Behavioural and physiological assessment of positive and negative emotion in sheep. Anim Behav 78:651–659
Reimert I, Bolhuis JE, Rodenburg TB (2013) Indicators of positive and negative emotions and emotional contagion in pigs. Physiol Behav 17:45–50
Rendall D (2003) Acoustic correlates of caller identity and affect intensity in the vowel-like grunt vocalizations of baboons. J Acoust Soc Am 113:3390–3402
Romero LM (2004) Physiological stress in ecology: lessons from biomedical research. Trends Ecol Evol 19:249–255
Ross MD, Owren MJ, Zimmermann E (2009) Reconstructing the evolution of laughter in great apes and humans. Curr Biol 19:1106–1111
Russell J (1980) A circumplex model of affect. J Pers Soc Psychol 39:1161–1178
Scheiner E, Fischer J (2011) Emotion expression—the evolutionary heritage in the human voice. In: Welsch W, Singer W, Wunder A (eds) Interdisciplinary anthropology: the continuing evolution of man. Springer, Heidelberg, pp 105–130
Scherer KR (1979) Nonlinguistic vocal indicators of emotion and psychopathology. In: Izard CE (ed) Emotions in personality and psychopathology. Plenum, New York, pp 493–529
Scherer RK (1981) Vocal indicators of stress. In: Darby J (ed) The evaluation of speech in psychiatry. Grune and Stratton, New York, pp 171–187
Scherer KR (1984) On the nature and function of emotion: a component process approach. In: Scherer KR, Ekman P (eds) Approaches to emotion. Erlbaum, Hillsdale, NJ, pp 293–317
Scherer KR (1986) Vocal affect expression: a review and a model for future research. Psychol Bull 99:143–165
Scherer KR (2009) The dynamic architecture of emotion: evidence for the component process model. Cogn Emot 23:1307–1351
Scherer KR, Banse R, Wallbott HG (2001) Emotion inferences from vocal expression correlate across languages and cultures. J Cross Cult Psychol 32:76–92
Scheumann M, Zimmermann E, Deichsel G (2007) Context-specific calls signal infants' needs in a strepsirrhine primate, the gray mouse lemur (*Microcebus murinus*). Dev Psychobiol 49:708–718
Scheumann M, Roser A-E, Konerding W et al (2012) Vocal correlates of sender-identity and arousal in the isolation calls of domestic kitten (*Felis silvestris catus*). Front Zool 9:36
Schön P-C, Puppe B, Manteuffel G (2004) Automated recording of stress vocalization as a tool to document impaired welfare in pigs. Anim Welf 13:105–110
Schrader L, Todt D (1993) Contact call parameters covary with social context in common marmosets, *Callithrix j. jacchus*. Anim Behav 46:1026–1028
Schrader L, Todt D (1998) Vocal quality is correlated with levels of stress hormones in domestic pigs. Ethology 104:859–876
Searcy WA, Beecher MD (2009) Song as an aggressive signal in songbirds. Anim Behav 78:1281–1292
Seyfarth RM, Cheney DL, Marler P (1980) Vervet monkey alarm calls: semantic communication in a free-ranging primate. Anim Behav 28:1070–1094
Siebert ER, Parr LA (2003) A structural and contextual analysis of chimpanzee screams. Ann N Acad Sci 1000:104–109
Siebert K, Langbein J, Schön P-C et al (2011) Degree of social isolation affects behavioural and vocal response patterns in dwarf goats (*Capra hircus*). Appl Anim Behav Sci 131:53–62
Soltis J, Blowers TE, Savage A (2011) Measuring positive and negative affect in the voiced sounds of African elephants (*Loxodonta africana*). J Acoust Soc Am 129:1059–1066
Sotocinal SG, Sorge RE, Zaloum A et al (2011) The Rat Grimace Scale: a partially automated method for quantifying pain in the laboratory rat via facial expressions. Mol Pain 7:55
Stoeger AS, Baotic A, Li D, Charlton BD (2012) Acoustic features indicate arousal in infant giant panda vocalisations. Ethology 118:896–905
Tallet C, Linhart P, Policht R et al (2013) Encoding of situations in the vocal repertoire of piglets (Sus scrofa): a comparison of discrete and graded classifications. PLoS One 8:e71841

Taylor AM, Reby D (2010) The contribution of source-filter theory to mammal vocal communication research. J Zool 280:221–236
Taylor AM, Reby D, McComb K (2009) Context-related variation in the vocal growling behaviour of the domestic dog (*Canis familiaris*). Ethology 115:905–915
Taylor AM, Charlton BD, Reby D (2016a) Vocal production by terrestrial mammals: source, filter, and function. In: Suthers RA, Fitch WT, Fay RR, Popper AN (eds) Vertebrate sound production and acoustic communication. Springer, Cham, pp 229–259
Taylor CJ, Freeman L, Olguin Olguin D, Kim T (2016b) Deviation in voice pitch as a measure of physiological stress response to group processes. Adv Group Process 33:211–242
Titze IR (1994) Principles of voice production. Prentice Hall, Englewood Cliffs, NJ
Townsend SW, Manser MB (2013) Functionally referential communication in mammals: the past, present and the future. Ethology 119:1–11
Vehrencamp SL, Yantachka J, Hall ML, de Kort SR (2013) Trill performance components vary with age, season, and motivation in the banded wren. Behav Ecol Sociobiol 67:409–419
Veissier I, Boissy A, Désiré L, Greiveldinger L (2009) Animals' emotions: studies in sheep using appraisal theories. Anim Welf 18:347–354
Vögeli S, Wolf M, Wechsler B, Gygax L (2015) Housing conditions influence cortical and behavioural reactions of sheep in response to videos showing social interactions of different valence. Behav Brain Res 284:69–76
Waller BM, Lembeck M, Kuchenbuch P et al (2012) GibbonFACS: a muscle-based facial movement coding system for hylobatids. Int J Primatol 33:809–821
Wathan J, Burrows AM, Waller BM, McComb K (2015) EquiFACS: the equine facial action coding system. PLoS One 10:e0131738
Watts JM, Stookey JM (1999) Effects of restraint and branding on rates and acoustic parameters of vocalization in beef cattle. Appl Anim Behav Sci 62:125–135
Weary DM, Braithwaite LA, Fraser D (1998) Vocal response to pain in piglets. Appl Anim Behav Sci 56:161–172
Yeon SC, Kim YK, Park SJ et al (2011) Differences between vocalization evoked by social stimuli in feral cats and house cats. Behav Proc 87:183–189
Yin S, McCowan B (2004) Barking in domestic dogs: context specificity and individual identification. Anim Behav 68:343–355
Zahavi A (1982) The pattern of vocal signals and the information they convey. Behaviour 80:1–8
Zimmermann E, Leliveld LMC, Schehka S (2013) Towards the evolutionary roots of affective prosody in human acoustic communication: a comparative approach to mammalian voices. In: Altenmüller E, Schmidt S, Zimmermann E (eds) Evolution of emotional communication: from sounds in nonhuman mammals to speech and music in man. Oxford University Press, Oxford, pp 116–132

Chapter 7
Seasonal Hormone Fluctuations and Song Structure of Birds

Manfred Gahr

Abstract Traditionally, it is assumed that singing of birds is a male-typical testosterone-dependent behavior. In this review I point out that singing outside the breeding season is common in males of many songbird species while their testosterone levels are low. Further, females of many tropical and temperate songbird species sing in various context in- or outside of the breeding season but testosterone levels of singing females are low in most cases. These findings question the testosterone-hypothesis of song production of songbirds. However, a key problem is the interpretation of "high" versus "low" testosterone levels, which would require a basic understanding of the quantitative interaction of testosterone and its estrogenic derivate with their specific receptors, as well as the dynamic abundance of these receptors in song controlling systems.

Bird songs are thought to mainly function in the realm of female mate attraction and defense of breeding territories against other males (Searcy and Andersson 1986; Kroodsma and Byers 1991; Catchpole and Slater 1995). Individual variation in song characteristics does affect reproductive success through mate choice and male–male competition (Andersson 1994). In relation, there is evidence of sexual selection for song traits such as song rate, repertoire size, structure of song motor units (syllables), and speed of syllable repetitions (for review: Podos et al. 2004). Current theory predicts that when senders and receivers have different evolutionary interests, as in sexual selection, signals must be subject to some costly constraint to constitute stable, honest indicators of quality (Grafen 1990). Individual variation in vocalizations will therefore depend on the condition of the adult male and its developmental history (Rowe and Houle 1996), but the costs of vocalizations are not well understood (Gil and Gahr 2002).

M. Gahr (✉)
Department of Behavioural Neurobiology, Max Planck Institute for Ornithology, Seewiesen, Germany
e-mail: gahr@orn.mpg.de

T. Aubin, N. Mathevon (eds.), *Coding Strategies in Vertebrate Acoustic Communication*, Animal Signals and Communication 7,
https://doi.org/10.1007/978-3-030-39200-0_7

Many temperate and tropical species, that breed seasonally, show seasonal singing activities, although there are surprisingly few quantitative data of circannual singing of bird species. In male songbirds, we can distinguish three categories (Fig. 7.1a): (MI.) males' singing activity is restricted to the breeding period; (MII.) males sing in the breeding season, and show a second period of singing activity in their post-molt nonbreeding period; (MIII.) males sing year-round. In the second and third cases the uttered song types and or the song pattern might change during the course of the year despite continuous singing activity. Although singing of females of temperate zone species is rare or not well documented, females of some temperate and of many tropical songbird species sing regularly. We can distinguish the following categories (Fig. 7.1b): (FI.) females' singing activity is restricted to the breeding period; (FII.) Females sing in the breeding season, are quiet during the autumnal molt and show again singing activity in their post-molt nonbreeding period; (FIII.) Females sing year-round; (FIV.) Females sing only in the nonbreeding season; (FV.) Females never sing. Clearly, these classifications must be an oversimplification in light of ca 4500 songbird species, but are useful for the discussion of (neuro)endocrine control of singing of males and females.

Due to the functions of male song for mate attraction and territorial defense, it is generally assumed that singing depends on the gonadal hormone testosterone. This predicts that males sing primarily in the breeding season and that females are not singing. However, as stated before, males of some temperate species and of many tropical species are known to sing outside of the breeding season and females of many tropical species sing regularly. In this review, I discuss the evidence for (1) seasonal singing activity, (2) seasonality of song structure, (3) testosterone dependency of song activity and song structure, (4) neural mechanisms of seasonal singing activity, and (5) neural mechanisms of hormone-dependent seasonal song structure. Based on the present survey of seasonal singing of songbirds, in the conclusion (6) I discuss the concept of singing as a testosterone-sensitive male sexual behavior. Although seasonal changes in singing/vocal activity are common in species of most avian orders (e.g., woodpeckers: Tremain et al. 2008; Malacarne et al. 1991) and seasonal changes in song pattern, too, are likely widespread (e.g., Galliformes: Fusani et al. 1993; suboscine-Passeriformes: Robertson et al. 2009), my review is focused on seasonal singing of songbirds (oscine Passeriformes). Songbirds are particularly interesting for this review, since they represent about half of today's bird species and since all songbirds have a homologous brain circuit that controls the song pattern, which facilitates species comparisons of underlying mechanisms.

7.1 Seasonal Singing Activity and Song Type Usage

7.1.1 Seasonal Singing Activity (Fig. 7.1)

In males of most temperate zone species singing activity varies during the breeding cycle being highest during mating and/or territory establishment and declines toward

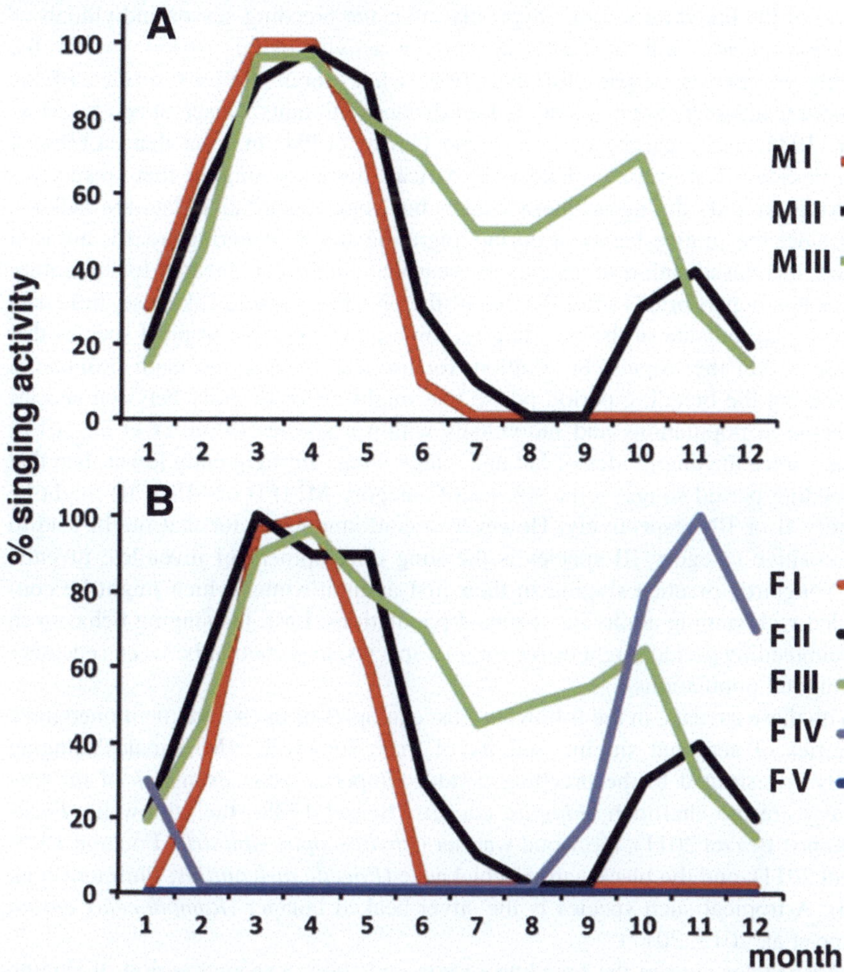

Fig. 7.1 Schematic representations of seasonal singing activity of male (**a**) and female (**b**) songbirds. In **a**, schematic drawings of three types of males' seasonal singing activities are depicted. The timing is based on a hypothetical average breeding season of the Northern Hemisphere: (MI.) Males' singing activity is restricted to the breeding period, (MII.) Males sing in the breeding season, and show a second period of singing activity in their post-molt nonbreeding period, (MIII.) Males sing year-round. In the second and third case the uttered song types and or the song pattern might change during the course of the year despite continuous singing activity. In **b**, schematic drawings of five types of females' seasonal singing activities are depicted: (FI.) Females' singing activity is restricted to the breeding period, (FII.) Females sing in the breeding season, are quite during the autumnal molt and show again singing activity in their post-molt nonbreeding period, (FIII.) Females sing year-round, (FIV.) Females sing only in the nonbreeding season, (FV.) Females never sing. % Singing activity is normalized to the month with highest singing activity of a species type. Month [1–12] represents January to December. (Modified after: Cox 1944; Immelmann 1969; Ritchison 1983; Langmore 1998; Schwabl 1992; Brunton and Li 2006; Price et al. 2008; Bezzel 2011; York et al. 2016; Gahr, unpublished observations)

the end of the breeding season. Typically, after the breeding season individuals of temperate species molt and show no singing activity in this period. In fall, the majority of species prepare for migration, while about 40% of species of the Northern hemisphere are resident or short-distance migrants. Observations by Alexander (1935) and a survey of Cramp and Perrins (1994) indicate that ca 60% of Paleartic-South-European or Palearctic-African migrants sing in their respective wintering grounds during northern winter, but good quantitative data are lacking. Thus, since the singing behavior during migration and in wintering areas is not well studied, the classification of species as "seasonal singing" is bleared for migratory species that depart briefly after the end of the breeding season. Likewise, little data about singing outside of the breeding season are available for tropical species that migrate within the tropics. In addition, for tropical species, the main problem is determining the breeding period per se that might differ strongly between species and between populations and individuals within a species (Stouffer et al. 2013). Further, since the amplitudes of fall and winter songs are frequently lower than that of breeding period songs, some species of category MI (FI) or MII (FII) might be category II or III, respectively. However, a confounding factor that might lead to false-positive category III species is the song development of juveniles; juvenile male songbirds practice singing in their first autumn/winter which might be confounded with singing adults in un-ringed populations. Last, the singing behavior in the nonbreeding period might differ within a species, in particular between sedentary and migrant populations.

With these caveats, in the following I list examples of the before mentioned three categories of seasonal singing activity of male songbirds: (MI.) males' singing activity is restricted to the breeding period: temperate zone examples of the first category are the chaffinch (*Fingilla coelebs*; Bezzel 1988), the willow tit (*Parus montanus*; Bezzel 2011), the wood warbler (*Phylloscopus sibilatrix*; Fouarge 1968; Bezzel 2011), and the black-capped chickadee (*Poecile atricapillus*; Philmore et al. 2006). A tropical such species is the silver-beaked tanager (*Ramphocelus carbo*; Quispe et al. 2015, 2017).

(MII.) Males sing in the breeding season and show a second period of singing activity in their post-molt nonbreeding period: Examples are the black redstart (*Phoenicurus ochruros*; Apfelbeck et al. 2013), the canary (*Serinus canaria*; Leitner et al. 2001a, b; Voigt and Leitner 2008), the serin (*Serinus serinus*; Bezzel 2011), the European robin (*Erithacus rubecula*; Lack 1943; Bezzel 2011), the garden warbler (*Sylvia borin*; Bezzel 2011), the common chiffchaff (*Phylloscopus collybita*; Bezzel 2011), the song trush (*Turdus philomelos*; Hegelbach and Spaar 2000; Alexander 1935), the great tit (*Parus major*; Bezzel 2011; Van Duyse et al. 2003), the stonechats (*Saxicola torquata*: V. Canoine personal communication), the nightingale (*Luscinia megarhynchos*: Kunc et al. 2006; Alexander 1935), the marsh warbler (*Acrocephalus palustris*: Kelsey 1988), and the great reed warbler (*Acrocephalus arundinaceus*; Sorensen et al. 2016). In these species, the postbreeding period of singing varies: some species start in fall and are able to sing throughout winter [e.g., the European robin (Lack 1943; Bezzel 2011) and black redstart (Apfelbeck et al. 2012; Bezzel 2011)] while others start singing only after their fall migration at the wintering grounds [e.g., the great reed warbler (Sorensen et al. 2016)]. Further, interindividual differences in singing activity seem much more pronounced in fall and winter singing periods as compared to the reproductive season, to the extent that not all males of a species are singing or at least sing infrequent outside the breeding season (e.g., Alexander 1935; Bezzel 2011; Sorensen et al. 2016).

(MIII.) Males sing year-round: Temperate such species are the winter wren (*Troglodytes troglodytes*: Bezzel 2011), the coal tit (*Parus ater*: Bezzel 2011), the Eurasian nuthatch (*Sitta europaea*; Bezzel 2011), and the white-crowned sparrow (*Zonotrichia leucophrys*: DeWolfe and Baptista 1995). Tropical examples are the white-browed sparrow weaver (York et al. 2016; Voigt and Gahr 2011), the forest weaver (*Ploceus bicolor*: Wickler and Seibt 1980; Schmidl and Gahr unpublished data) and the red-cheeked cordon-bleus (*Uraeginthus bengalus*: Gahr and Güttinger 1986). The zebra finch (*Taeniopygia guttata*), originating of inner Australia, sings year-round, depending on water availability (Zann 1996; Johnson and Rashotte 2002). As stated above, for most tropical species quantitative data covering the whole year are missing.

Even in species that sing only in the breeding season (Category I) large call repertoires might be uttered year-round, e.g., the black-capped chickadee sings in the breeding season (Phillmore et al. 2006) but produces learned contact calls (the chick-a-dee call and the gargle call) year-round (Ficken et al. 1978; Ficken and Weise 1984; Hughes et al. 1998). This indicates a general complication of the categorization of seasonal singing, that is, to distinguish songs from calls in the first place. As the chickadee example indicates even vocal learning does not always help to distinguish songs from calls. In most studies included in this review, authors have used complexity such as being composed of several sequentially uttered sounds, of more than one sound type, and the length of the sounds to distinguish between songs and calls. This structural definition of a song must of course be a species-specific definition and there is not always a clear dichotomy between call and song, e.g., female canaries utter a call, the female-specific trill that is composed of several syllables in a mating context (Amy et al. 2015). This call is more complex than all other calls uttered by female canaries (Mulligan and Olson 1969), more structured than plastic songs uttered by some female canaries (Pesch and Güttinger 1985; Gahr, personal observation) but less complex than male canary songs. However, this canary call is as complex as, e.g., the "chit" songs of female red-winged blackbirds that are produced in a mating context (*Agelaius phoeniceus*) and more complex than the "teer" songs of these females (Beletsky 1983; Yasukawa et al. 1987).

In difference to all male songbirds, females of some species do not sing at all (e.g., the zebra finch: Morris 1954) and if they sing do not always do so in the breeding season (e.g., the European robin: Lack 1943; Hoelzel 1986; Schwabl 1992). Further, in reports of female singing of Northern temperate zone species, only a fraction of all females of a particular species are observed to sing (e.g., Arcese et al. 1988; Sandell and Smith 1997; Schwabl 1992; Bensch and Hasselquist 1992; Pesch and Güttinger 1985; Langmore and Davies 1997). Thus, the seasonal context of female singing seems to be more variable within a species and between species (for reviews: Ritchison 1983; Langmore 1998; Slater and Mann 2004). Like in males, female songs might be part of duets or solo songs, a distinction that is relevant in the frame of seasonality only in case of multiple distinct song types of a species. Further, most examples of seasonal or year-round female singing lack circannual quantitative data. According to the females' singing behavior, species can be classified as follows (Fig. 7.1):

(FI.) Females' singing activity is restricted to the breeding period: Temperate zone examples are the alpine accentors (*Prunella collaris*: Langmore et al. 1996), the black-headed grosbeaks (*Pheuticus melanocephalus*: Ritchison 1983), the American dipper (*Cinclus americanus*: Bakus 1959), the Northern cardinal (*Cardinalis cardinalis*: Ritchison 1986; Laskey 1944). Although females of these and other species sing only in the breeding season, the context is varied, including nest-building, nest-relief, incubation, brooding, feeding chicks, pair-bond maintenance, family-group maintenance, and only rarely courtship and mate attraction (for review Ritchison 1983; Langmore 1998; Slater and Mann 2004). The example for the latter is the alpine accentor (Langmore et al. 1996).

(FII.) Females sing in the breeding season and have a second singing activity in their post-molt nonbreeding period: An example might be the streak-backed oriole (Price et al. 2008), but data covering the entire year would be required.

(FIII.) Females sing year-round: Examples are the New Zealand bellbird (*Anthornis melanura*; Brunton and Li 2006), the superb fairy wrens (*Malurus cyaneus*; Cooney and Cockburn 1995), the white-browed sparrow weaver (York et al. 2016; Voigt and Gahr 2011), and the forest weaver (Wickler and Seibt 1980; Schmidl and Gahr unpublished data), and likely females of many duetting tropical oscines, such as of the wren family (Slater and Mann 2004). As stated above, for most tropical species quantitative data in relation to phenology are missing.

(FIV.) Females sing only outside of the breeding season: Such examples are the canary (Pesch and Güttinger 1985; Ko and Gahr unpublished data), the European robin (European robin; Lack 1943; Schwabl 1992), and the starling (Pavlova et al. 2007). In the latter, presence of males and nest-boxes suppresses singing.

(FV.) Female never sing, e.g., the zebra finch and the Bengalese finch (*Lonchura striata domestica*) (Morris 1954; Immelmann 1969).

7.1.2 Seasonal Song Type Usage

In some species males produce several song types of which one or several are restricted to the breeding season. The species-typical songs uttered in the breeding season are frequently called full song (also called primary song or crystallized song). Here we use the term full song for songs loudly uttered in the breeding season or year-round. A well-studied such species is the colony-living white-browed sparrow weaver in which the dominant males sing two types of full song: the so-called morning song throughout the breeding season and the duet song throughout the year (Voigt et al. 2006; York et al. 2016) (Fig. 7.2a). The duet song is shared between the dominant male, all subordinate males and the females of a colony (Fig. 7.2b). Duet song, which extends into chorus song if more than two individuals join, is uttered year-round. Solo song and duet song of the dominant male white-browed sparrow weaver differ in the temporal organization and in syllables' structures (Voigt et al. 2006).

There is no species known in which females sing multiple song types with at least one type restricted to the reproductive context. A potential such case might be the courtship songs of cordon bleus, in which males and females combine singing with other vocal and with visual displays (Gahr and Güttinger 1986; Ota et al. 2015).

However, the courtship song per se is similar to the song produced outside of the breeding season that might serve a pair-maintaining function. Song types that might be restricted to certain seasons are the so-called soft songs (also called quiet song, twitter song, whisper song) of males of certain species, which are low amplitude songs. According to calculations of Anderson et al. (2008), the full songs might be perceived by humans as about 4–8 louder than the soft songs of the same species, e.g., European robins (Lack 1943; Dabelsteen et al. 1997) and European blackbirds (*Turdus merula*; Dabelsteen 1984; Dabelsteen and Pedersen 1990) utter soft songs in

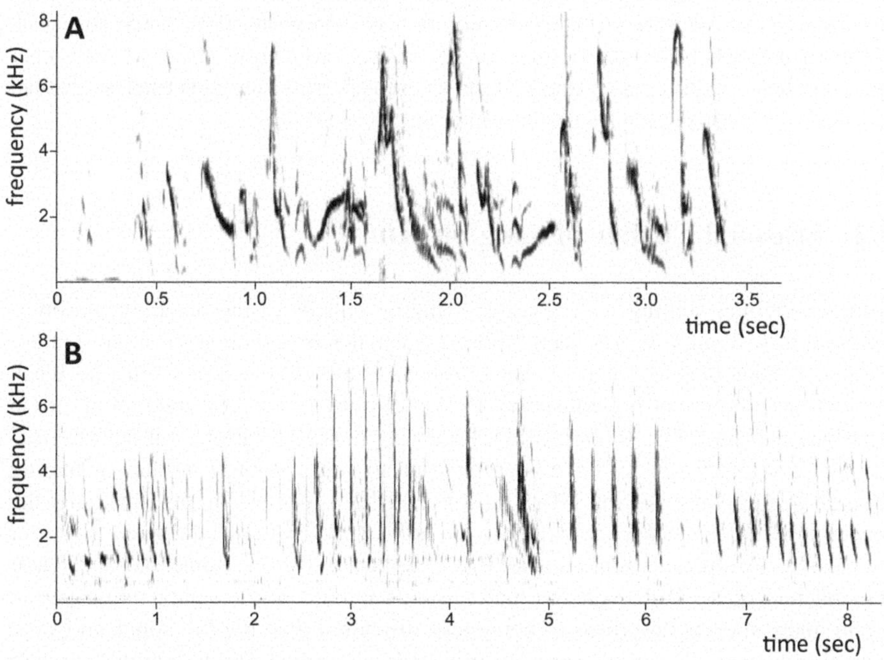

Fig. 7.2 White-browed sparrow weavers possess an extraordinary vocal communication system with two completely different types of song. Sonagrams from field recordings showing a duet song (**a**, similar in structure to chorus song) and a sequence from morning (solo) song (**b**) of white-browed sparrow weavers. While all group members engage in duet and chorus singing, morning song is only produced by the dominant male of the group. Note that time scales are different

winter. Likewise, the dunnock (*Prunella modularis*: Snow 1988), the alpine accentor (*Prunella collaris*: Langmore et al. 1996), the dark-eyed junco (*Junco hyemalis*: Titus 1998), the whitethroat (*Sylvia communis*: Balsby 2000), the swamp sparrow (*Melospiza georgiana*: Ballentine et al. 2008), and the song sparrow (*Melospiza melodia*: Anderson et al. 2008) produce soft songs that are only audible over a few meters. In case of the song sparrow, males broadcasted loudly their full song and two forms of soft songs, one similar in structure to the full song but with low amplitude and one that is unstructured, variable and sang with low amplitude (Anderson et al. 2007, 2008). The same males produce these songs types within short-time windows during the breeding season (Anderson et al. 2007, 2008). Likewise, male red-wings (Lampe 1991), European blackbirds (Dabelsteen 1984), and dark-eyed juncos (Titus 1998) produce soft songs that are described as being different in temporal organization, including more high-pitched elements and/or being more variable than the full songs during the breeding season. However, since soft songs are difficult to hear and to record over distance, it is in most species unclear if these songs represent a year-round song type that is uttered in the breeding season next to the loudly advertised full song, or a song type that is restricted to the nonbreeding season. Further, in most cases the low audibility of soft songs resulted in the lack of suitable recordings that would allow qualifying the soft songs as different or similar in structure to the full songs (Dabelsteen et al. 1998; Morton 2000; Anderson et al. 2008). It is unlikely that soft songs simply reflect production errors of the singer since the less structured soft songs of the swamp sparrows are nevertheless distinct reoccurring motor sequences (Anderson et al. 2007, 2008).

7.2 Seasonal Change of Song Structure

Autumnal/winter singing, i.e., "audible" singing outside of the breeding season as opposed to soft singing, has been reported for males of a number of nonmigratory temperate zone species such as the European robin, the winter wren, the black redstart, and the great tit (Armstrong 1955; Schwabl 1992; Apfelbeck et al. 2012; Van Duyse et al. 2003; Bezzel 2011). Likewise, males of various but not all long-distance migratory species have been reported to sing regularly on their wintering grounds (Cramp and Perrins 1994). In some of these species singing outside of the breeding period songs appear similar to the full song (e.g., the winter wren: Kreutzer 1973; the European robin: Lack 1943; Schwabl 1992; Ramenda and Gahr unpublished data). However, the males' songs uttered in the nonbreeding period are in many species described as being less structured than the full songs and were termed subsongs. The term subsong originally described quite singing (i.e., soft songs) (Nicholson and Koch 1936) but was later used to label songs that were less structured than the typical full songs broadcasted during the breeding season (Lister 1953a). Nowadays, the term subsong is frequently used to describe the unstructured songs of juveniles uttered during song development. To avoid confusion, in this review we use the term "plastic song" for songs of adults uttered outside of the

Table 7.1 Seasonal change of song structures of species studied quantitatively in the wild

Song structure	Species	References
Syllable repertoire	Wild canary	Leitner et al. (2001a, b)
	Mockingbird	Howard (1974), Logan (1983)
	Tui	Hill et al. (2015)
	Starling	Van Hout et al. (2012)
Syllable repetition rate	Great reed warbler	Sorensen et al. (2016), Wegrzyn and Leniowski (2009)
Syllable stereotypy	Wild canary	Leitner et al. (2001a, b)
Phrase length	Song sparrow	Smith et al. (1997)
	Black redstart	Apfelbeck et al. (2013)
	Mockingbird	Howard (1974), Logan (1983)
	Starling	Van Hout et al. (2009)
Song length	Wild canary	Leitner et al. (2001a, b)
	Great reed warbler	Sorensen et al. (2016)
Song stereotypy	Wild canary	Leitner et al. (2001a, b)
	Song sparrow	Smith et al. (1997)

"Phrase" refers to larger subunits of the song that are composed of sequences of several syllables or repetitions of the same syllable or both

breeding season, i.e., autumnal/wintering songs that are structurally different from full songs.

Very few data are available in which the song pattern of the breeding period and the nonbreeding period within a species were compared statistically, and even fewer data are available, in which the song of the same individual singer uttered in different seasons were compared (Table 7.1). The best-studied example of seasonal song structure is the canary (Nottebohm et al. 1986; Leitner et al. 2001a, b; Voigt and Leitner 2008). Wild canaries form nonmigratory island populations at the Canary Islands, the Azores and Madeira. Both wild and domesticated canaries sing a less structured version of their full songs in autumn (Fig. 7.3). Besides a lower song amplitude and singing in groups, the males change their syllable repertoire partially, the syllable repetition rate, the stereotypy of syllables, the stereotypy of syllable sequences, and the song length (Nottebohm et al. 1986; Leitner et al. 2001a, b; Voigt and Leitner 2008). Other temperate zone examples likely singing plastic songs in autumn and winter are the song thrush, the European blackbird, the mistle trush, the marsh warbler, the willow warbler, and the chaffinch (Thorpe 1958, 1961; Thorpe and Pilcher 1958; Kelsey 1988; Sorensen 2014). In fact, Thorpe states: "we may safely guess that as further material comes to hand, subsong will be found to be a very widespread phenomenon amongst song birds" (Thorpe 1961, p. 70). Similar, Lister (1953b) reported plastic song for many species of the Indian subcontinent. However, about 60 years later, the data situation is still mainly descriptive if not anecdotic:

Including the canary, quantitative comparisons of songs and plastic songs are available for only seven species (Fig. 7.4): In the song sparrow, song parts (trills) were composed of more syllables and syllables of the entire song were more

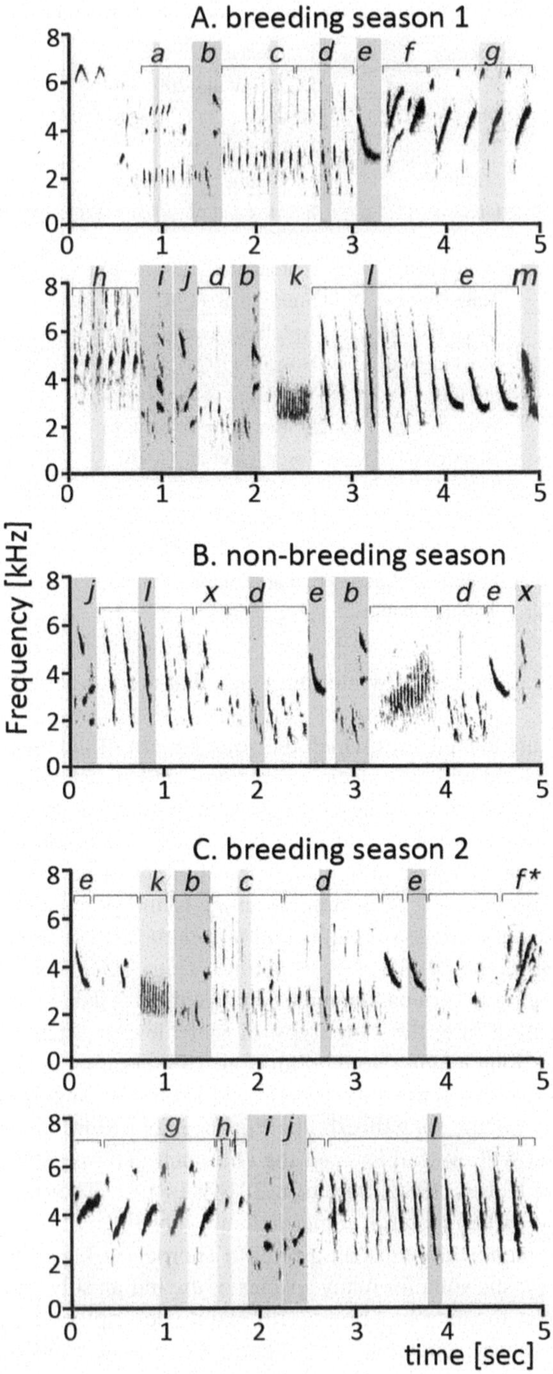

Fig. 7.3 Seasonal changes of the song pattern of wild canary males. Full songs of the breeding season (**a**) differ from plastic songs of the autumn (**b**) but are similar to the full songs uttered in the next breeding season (**c**). The repertoire composition, the number of unrepeated syllables and the

stereotyped in the full songs as compared to plastic songs (Smith et al. 1997). In black redstarts, full songs had more elements in two subparts (A and C) than songs of the nonbreeding period while other structural song parameters did not differ significantly between (Apfelbeck et al. 2013). In mockingbirds (*Mimus polyglottos*) the full song phrases are composed of significantly more repetitions of a given syllable type and the same individuals use partially different syllable repertoires in the spring breeding versus the autumnal nonbreeding season (Howard 1974; Logan 1983). In the New Zealand honeyeater, the tui (*Prosthemadera novaeseelandiae*), in the breeding season males' songs contained significantly greater proportions of trill components compared with songs in the nonbreeding season (Hill et al. 2015). In European starlings, full songs are characterized by longer song bouts and a larger repertoire (Van Hout et al. 2009). Further, although many species seem to sing during migration and on the wintering grounds (see above), only the great reed warbler (Sorensen et al. 2016; Wegrzyn and Leniowski 2009) was studied in detail in both periods. In difference to the song of the breeding season, songs of males wintering in Zambia were longer and less repetitive (Sorensen et al. 2016). In the willow warbler (*Phylloscopus trochilus*), songs recorded at the wintering grounds in Zambia seem to be somewhat longer with more variable syllable structure as compared to songs of breeding birds in Europe (Sorensen 2014), but this has not been analyzed statistically.

In summary, seasonality of the song structure has been studied quantitatively only in very few species (Table 7.1). Nevertheless, anecdotic reports on the seasonal occurrence of plastic songs suggest that it is a broad phenomenon occurring in various songbird families (e.g., Acrocephalidae; Fingillidae; Meliphagidae; Mimidae; Passerellidae; Phylloscopidae; Sturnidae; Turdidae). Seasonal structural changes such as song length, syllable repetition rates, syllable consistency, and repertoire composition are species specific (Table 7.1). However, individuals of some species

Fig. 7.3 (continued) song length change seasonally in wild canaries (Leitner et al. 2001a, b; Voigt and Leitner 2008); in domesticated canaries the syllable repetition rate changes in addition (Nottebohm et al. 1986; Voigt and Leitner 2008). Shown are spectrograms of song segments recorded from the same male in the field in April 1996 (breeding season 1, **a**), in November 1996 (nonbreeding season, **b**), and again in April 1997 (breeding season 2, **c**). Each syllable is a stereotyped set of one to several continuous sounds each with a typical frequency modulation. To illustrate this, some syllables are labeled with gray, green, or red. In a sequence, a syllable is repeated one to several times. Syllables that the bird produced year-round are labeled gray, those that are produced only in the breeding seasons are labeled red, those that were found only in one period are indicated green. A sequence of syllables from the 1996 breeding season (**a**) was marked with letters. Syllables that were produced in the following nonbreeding (**b**) and breeding season (**c**) are marked with the letters used in (**a**). About 75% of syllables such as b, d, e, i, j, l had correlation coefficients between 0.85 and 0.95 and are, therefore, produced year-round. About 25% of the syllables were produced seasonally. Half of those syllables (red) reappeared in the following year, i.e., are produced annually. Correlation coefficients for such syllables (for example c, g, h, k) were between 0.86 and 0.93. Certain syllables such as "a" and "x" (green) were not found in the subsequent seasons. "f" (**a**) and "f*" (**c**) are related syllables but with a correlation coefficients below 0.75 (after Leitner et al. 2001a, b)

species	song rate	song length	segment length	repertoire size	stereo-typy	syllable rate	frequency range	references
zebra finch	x	x	-	-	-	-	x	2,4,6,18,28
Bengalese finch	x	-	-	-	n.a.	-	n.a.	19
black redstart	-	-	-	-	n.a.	x	x	1
blue tit	-	n.a.	-	-	-	n.a.	n.a.	11,12
blue-headed vireo	x	n.a.	n.a.	n.a.	n.a.	n.a.	n.a.	27
canary	x	x	x	-	x	x	n.a.	3,5,8
dark-eyed junco	x							9
house finch	x	n.a.	n.a.	n.a.	n.a.	n.a.	n.a.	23
pied flycatcher	x	n.a.	n.a.	n.a.	n.a.	n.a.	n.a.	20
red-winged blackbird	x	n.a.	n.a.	-	-	n.a.	n.a.	7.29
song sparrow	x/-	n.a.	x	-	x	-	n.a.	15,21,22,30
starling	x/-	-	n.a.	x	n.a.	-	n.a.	16,25,26
white crowned sparrow	x	n.a.	n.a.	n.a.	x	n.a.	n.a.	10, 14
white-throated sparrow	x	n.a.	n.a.	n.a.	n.a.	n.a.	n.a.	13
chaffinch	x	n.a.	n.a.	n.a.	n.a.	n.a.	n.a.	17,24

Fig. 7.4 Testosterone dependency of singing activity (song rate) and song structure (song length, segment length, syllable repertoire size, stereotypy, syllable rate, frequency range) of songbirds. Classifications are based on published studies that employed castration, hormone treatment and/or inhibiting of endogenous hormone production and androgen and estrogen receptor activity. Red: testosterone sensitive. Green: Estrogen sensitive. Yellow: Testosterone and estrogen sensitive. N. A. = not analyzed in the respective publication. References: 1 = Apfelbeck et al. 2012; 2 = Arnold 1975; 3 = Boseret et al. 2006; 4 = Cynx et al. 2005; 5 = Fusani et al. 2003; 6 = Harding et al. 1983; 7 = Harding et al. 1988; 8 = Heid et al. 1985; 9 = Ketterson et al. 1992; 10 = Konishi 1965; 11 = Kunc et al. 2006; 12 = Kurvers et al. 2008; 13 = Maney et al. 2009; 14 = Meitzen et al. 2007a; 15 = Nowicki and Ball 1989; 16 = Pinxten et al. 2002; 17 = Poulsen 1951; 18 = Pröve 1974; 19 = Ritschard et al. 2011; 20 = Silverin 1980; 21 = Smith et al. 1997; 22 = Soma et al. 1999; 23 = Strand et al. 2008; 24 = Thorpe 1958; 25 = Van Hout et al. 2009; 26 = Van Hout et al. 2012; 27 = Van Roo 2004; 28 = Walters et al. 1991; 29 = Weatherhead et al. 1993; 30 = Wingfield 1994 (From Gahr 2014)

such as the winter wren and the skylark (*Alauda arvensis*) change song parameters dependending on the social context (Camacho-Schlenker et al. 2011; Geberzahn and Aubin 2014), which might be confounded with seasonal changes. Further, some observations of seasonal song structures might be due to juveniles that undergo song development in their first autumn and winter.

Another indication of seasonal changes in song structure is the occurrence of adult song learning. Adult learning of new song syllables likely affects overall song structure, particular in species that sing stereotyped songs in the breeding season such as canaries or song sparrow, in which changes of the syllable repertoire might affect syllable sequencing, song stereotypy, and or song length (Nottebohm and Nottebohm 1978; Leitner et al. 2001a, b; Nordby et al. 2001). Well-documented examples of changes of adult repertoires are the song sparrow (Nordby et al. 2001), the starling (Adret-Hausberger et al. 1990; Eens et al. 1992), the brown-headed cowbird (*Molothrus ater*; O'Loghlen and Rothstein 2002), and the mockingbird (Derrickson 1987). In light of seasonal structural changes, it is not relevant if adult song learning is de novo learning or altered syllable usage of material learned as juveniles.

Structural seasonal changes of the song of adult female songbirds have not been reported.

7.3 Testosterone-Dependent Seasonal Singing

7.3.1 Testosterone-Dependent Singing Activity

The gonadal dependency of singing of male birds was known for centuries based on the castration of roosters. Berthold with his seminal testis transplantation experiments was the first to recognize that a gonadal hormone, now known as testosterone, is controlling vocal activity and vocal structure (Berthold 1849). Likewise, in songbirds, seasonal singing activity was found to correlate with the hormonal activity of the testicles (Bullough 1942; Davis 1958; Armstrong 1973; Wingfield and Farner 1975). Testosterone is the main androgen released by the testicles and can be converted by the enzyme 5α-reductase into the androgen 5α-dihydrotestosterone (both androgens bind to the androgen receptor) and into the estrogen 17β-estradiol (a potent ligand of both estrogen receptor α [ERα] and estrogen receptor β [ERβ]) via the enzyme aromatase, which occurs in various tissues including the brain (Schlinger and Arnold 1991; Saldanha et al. 2000; Fusani et al. 2000, 2001; Schlinger and Remage-Healey 2012).

In temperate zones, the annual increase in day length triggers seasonal gonadal growth, which results in increased levels of circulating testosterone, an important initiator of reproductive behaviors. Supporting data of testosterone-dependent vocal activity (Fig. 7.4) come from hormone treatments of species of several avian orders including songbirds (e.g., Heid et al. 1985; Dloniak and Deviche 2001; Van Hout et al. 2009), suboscine passerines (Kroodsma 1985), parrots (Brockway 1968; Nespor et al. 1996), galliformes species (Marler et al. 1962; Andrew 1963; Chiba and Hosokawa 2006; Beani et al. 2000; Fusani et al. 1994), night herons (Noble and Wurm 1940), doves (Bennett 1940), and gulls (Terkel et al. 1976; Groothuis and Meeuwissen 1992). In particular, in all testosterone treatment experiments, male birds increased their singing activities (for review Gahr 2014). There is so far no report that testosterone treatment would fail to affect singing activity of birds. Further, males of MI species have maximal testosterone levels at the beginning of the breeding/singing period (e.g., Schwabl 1992; Fusani 2008; Goymann and Landys 2011; Apfelbeck et al. 2017).

In contrast to the testosterone hypothesis of singing, male birds of category MII and MIII sing outside of the breeding season when testicles are regressed (Armstrong 1973) and the levels of circulating testosterone are in many cases very low, around or below 100 pg per mL blood plasma (canary: Leitner et al. 2001a, b; Voigt and Leitner 2008/black redstart: Apfelbeck et al. 2013/song sparrow: Smith et al. 1997; Wingfield and Hahn 1994; but Soma et al. 1999/great tit: van Duyse et al. 2003/stonechats: Gwinner et al. 1994; Goymann et al. 2006). Because these values are close to the detection limits of most testosterone assays, (Wingfield and Farner 1975;

Leitner et al. 2001a, b; Gwinner et al. 1994; Goymann et al. 2006; Apfelbeck et al. 2013) the real blood testosterone concentration might even be lower. Since the detection of low amounts of testosterone is particularly sensitive to the sample volume and assay procedures, species differences of low testosterone levels might in part be procedural. In the breeding seasons, the amounts of circulating testosterone are in average 10–200 times higher: In the canary (Leitner et al. 2001a, b; Voigt and Leitner 2008), the great tit (Röhss and Silverin 1983; Van Duyse et al. 2003), the willow tit (Silverin et al. 1986), the black redstart (Apfelbeck et al. 2013), the European robin (Schwabl 1992), the mockingbird (Logan and Wingfield 1995), and the stonechats (Gwinner et al. 1994; Goymann et al. 2006) testosterone is high in the breeding singing season but low in the nonbreeding singing periods. It is unclear if these low levels of testosterone are relevant for the control of singing activity outside of the breeding season. In relation, inhibition of the androgen receptor did not affect singing outside of the breeding periods in male black redstarts and European robins (Schwabl and Kriner 1991; Apfelbeck et al. 2012, 2013).

Thus, testosterone data of seasonal (category MII) and year-round (category MIII) singing males support the notion that elevated levels are important for initiating singing in the breeding season but are not supportive for a testosterone-sensitive mechanism underlying autumnal and wintery singing. Further, although the overall relationship between testosterone and high singing activities is well established in the breeding season, there is mixed evidence of how individual variation in testosterone levels relate to individual variation in singing activity (e.g., Ketterson et al. 1992 but Saino and Moller 1995).

The effect of testosterone on song performance might involve androgenic and estrogenic action of testosterone and its metabolites. In the male zebra finch, aromatase inhibitor reduced the amount of directed (presumably courtship related) singing but not that of undirected singing (Walters et al. 1991). In great tits, estrogen might play a role for singing activity but this pathway was tested together with the androgenic control of singing (Van Duyse et al. 2005). However, singing rate was not estrogen sensitive in other species tested in this way, the canary (Fusani et al. 2003; Rybak and Gahr 2004), the black redstart (Apfelbeck et al. 2012), and the house finch (*Carpodacus mexicanus*: Strand et al. 2008).

As stated above, females of many species sing too. Testosterone is produced in the ovary as a precursor of estrogens (Johnson 1990). In relation, testosterone is highest in the chicken hen about 6–8 h before ovulation (Etches and Cheng 1981; Robinson et al. 1988). However, maximal concentrations of circulating testosterone (range: 50 pg–3.8 ng/mL) of female birds are four- to tenfold lower than in male conspecifics during the breeding season (Ketterson et al. 2005), and are around or below the detection limits in the nonbreeding periods (Ketterson et al. 2005). Concentrations of circulating estrogens in female songbirds are highest during the nest-building and the egg-laying period with concentration between 100 and 700 pg per mL blood plasma and are low throughout the rest of the year (Silverin et al. 1986; Wingfield and Farner 1978; Dawson 1983; Schwabl et al. 1980; Schwabl 1992; Gwinner et al. 1994).

In support of testosterone dependency of female singing, females that sing periodically or circumstantially are thought to secrete higher levels of testosterone in these periods. Such examples might be the female song sparrow (*Melospizia melodia*) (Arcese et al. 1988), the European robin (Schwabl 1992), the blue-capped cordon blues (Geberzahn and Gahr 2011), and female dunnocks, in which competition for male reproductive investment elevates testosterone (Langmore et al. 2002). Further, testosterone treatment induced singing in all adult female birds that differentiated the necessary song controlling brain structures during development (for reviews: Gahr 2007, 2014). However, these experimental concentrations of circulating testosterone are in average 10–500 times higher than those measured in untreated females and are higher than those found in males during the breeding season (e.g., Kriner and Schwabl 1991; Fusani et al. 2003; Hartog et al. 2009). On the other hand, female dark-eyed juncos have relatively high testosterone levels but do not sing regularly (Ketterson et al. 2005). Unfortunately, there are only little hormone data of females that spontaneously singing when circulating levels of testosterone are expected being low (e.g., female white-browed sparrow weaver: Wingfield et al. 1991 but York et al. 2016). Although estrogens have been shown to mediate some of the activational action of testosterone on vocalizations (Fusani et al. 2003; Fusani and Gahr 2006), these actions are due to estrogens derived from testosterone in the brain. Unfortunately, there are no estrogen data of females that were singing during nest-building/egg-laying period. Female European robins singing in winter had low to undetectable levels of estradiol (Schwabl 1992).

As stated above for the male birds, it is unclear if the low testosterone levels are relevant for the control of female singing behavior.

Thus, in summary, elevated levels of circulating testosterone correlate with males' singing activity in the breeding season but not with males' singing outside of the breeding season and not with the females' singing activities. A major problem of behavior–hormone correlations is the temporal resolution of data sampling, which is dense (e.g., daily) concerning the singing but sparse (weakly or biweekly or even less frequent) in case of the blood hormone measurements. In addition, testosterone production might undergo circadian changes. For these reasons short periods of high or elevated testosterone concentration are easily missed. Further, birds need to be caught for blood sampling. Thus, in most longitudinal field studies that cover one or more seasons, the singing activity data and testosterone data originate frequently from different individuals. Correlations of this type are particularly worrisome if the data sets originate from entirely different populations and years (e.g., Rost 1990, 1992) and should be avoided. Last, unfortunately the current testosterone assays cannot resolve low blood hormone concentrations. Thus, it needs to be seen if low levels of circulating testosterone are meaningful for song control.

7.3.2 *Testosterone-Dependent Seasonal Song Structure*

Vocalizations of some species (e.g., the canary, starling) change in structure after castration of adult males (Berthold 1849; Heid et al. 1985; Van Hout et al. 2009), can be reinstated with testosterone treatment after castration in adulthood (Heid et al. 1985), and can be induced in adult females by testosterone treatment (e.g., Shoemaker 1939; Leonard 1939; Konishi 1965; Kern and King 1972; Vallet et al. 1996). However, little attention has been paid to verify if testosterone-induced vocalizations of females are indeed "male-typical," e.g., testosterone-treated female canaries sing male-like songs (Shoemaker 1939; Leonard 1939), which in average are composed of much fewer syllables compared to male canaries (Hartley and Suthers 1990; Fusani et al. 2003). However, some testosterone-treated female canaries produce songs with sexual release quality (Vallet et al. 1996). Likewise, in autumn, testosterone-induced songs of male chaffinches were reported being less structured than the full songs (Thorpe 1958).

In difference to overall testosterone sensitivity of singing activity, song structure such as song length, song fragment length (e.g., motif, tour, phrase), song unit repertoire (element, syllable, song type), song unit stereotypy, song unit repetition rates, or the frequency range are sensitive to testosterone treatment in a species-specific manner (Fig. 7.4; Gahr 2014). Thus, there are large species differences to the extent of which the song pattern is testosterone-sensitive, from little in the zebra finch to stark in the canary. In relation, acute levels of testosterone do not correlate with song structure such as repertoire size in the red-winged blackbird (*Agelaius phoeniceus*) (Weatherhead et al. 1993) or the canary. This might be due to fact that song structure depends on the organizational effect of testosterone, that is, there is a delay between testosterone's effect on neuronal and neural endophenotypes and the occurrence of the overt phenotypes such as the song. In general, the relation between individual changes in testosterone levels and changes in song structure as well as interindividual differences in testosterone levels and such differences in song structure are vague.

The only example of brain-derived estrogens that control the song structure comes from the adult canary; estrogens are required to sing songs with high syllable-repetition rates (Fusani et al. 2003; Rybak and Gahr 2004; Fusani and Gahr 2006), a feature that is important for the sexual quality of canaries' songs (Kreutzer and Vallet 1991).

7.4 Neural Mechanisms of Seasonal Singing Activity

Singing activity seems not controlled directly by the neural vocal control system that controls the structure of the song as detailed in the next paragraph. Even male canaries with bilaterally lesioned HVC (an important song control region of the songbird brain, see Fig. 7.5), try to sing although with no or little audible outcome

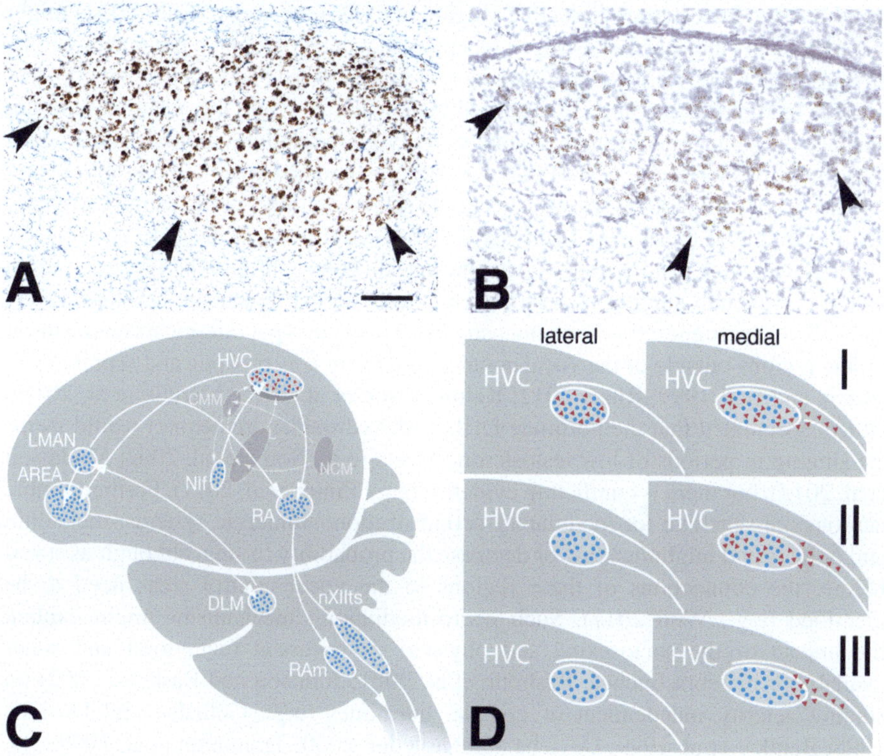

Fig. 7.5 Distribution of androgen receptors (AR) and estrogen receptors (ERα) in the vocal control system. In **a**, we depict the expression of AR mRNA in the HVC of a male canary and in **b**, the ERα mRNA in the HVC of a great tit (*Parus major*) of the reproductive season. The mRNA expressing cells (brown) were labeled with a nonradioactive in situ hybridization method. In **c**, we show the distribution of AR (blue dots) and of ERα (red triangles) in the areas of a schematic vocal control system of songbirds. Some thalamic brain areas that appear important for coordination of the left and right vocal control network are omitted (see Wild 1997). Note that ER expression in vocal areas is limited to HVC and differs strongly between species (see **d** and Table 7.1). In **d**, we represent the distribution of AR and ERα in the lateral and medial part of the HVC: in type I, ERα is expressed throughout the entire HVC, in type II ERα is expressed in the medial HVC but not or very low in the lateral part, in type III ERα expression is low even in the medial part of HVC. In all songbirds, AR is expressed throughout HVC and ERα is found ventromedial to HVC. In Area X, AR is abundant only in some individuals. *Abbreviations*: Area X; DLM, nucleus dorsolateralis anterior, pars medialis; DM, dorsomedial nucleus of the midbrain nucleus intercollicularis; HVC, proper name; Field L; lMAN, lateral magnocellular nucleus of the anterior nidopallium; mMAN, medial magnocellular nucleus of the anterior nidopallium; NC, caudal nidopallium; NIF, nucleus interface of the nidopallium; nXIIts, tracheosyringeal portion of the nucleus hypoglossus; RA, robust nucleus of the arcopallium; RAm, nucleus retroambigualis; rVRG, rostro-ventral respiratory group (after Frankl-Vilches and Gahr 2017)

(Nottebohm et al. 1976). In a non-songbird, the adult male ring dove (*Streptopelia risoria*), estrogens derived from testosterone in the preoptic area are important to stimulate sexual vocal displays (Hutchison and Steimer 1984). Likewise, in songbirds the testosterone-inducible singing activity (see above; Gahr 2014) might be mediated by androgen or estrogen receptors that are abundant in preoptic neurons: Alward et al. (2013) showed that singing rate was increased by testosterone implants into the medial preoptic area of castrated male canaries while this treatment did not enhance the song structure such as song stereotypy. In relation, implants of estrogens plus dihydrotestosterone near HVC of white-crowned sparrows did not affect song rate (Meitzen et al. 2007a). How testosterone affects cellular mechanisms of preoptic neurons and how those would activate the vocal control areas needs to be seen.

However, such testosterone-dependent mechanisms acting in the preoptic region would not explain singing in seasons when testosterone levels are baseline. Riters and colleagues suggest that song intensity involves neuro-peptidergic mechanisms in limbic regions outside of the vocal control system such as preoptic and septal nuclei (Riters and Ball 1999; Riters 2012; Kelm-Nelson et al. 2013; Merullo et al. 2016). Others speculated that the hormone DHEA (dehydroepiandrosterone) would mediate singing in periods of low testosterone levels (e.g., Soma et al. 2002; Maddison et al. 2012), but there is conflicting evidence (e.g., Pintér et al. 2011). Further, limbic neurons are sensitive for modulatory action of monoaminergic systems, melatonin and opioids and might increase or decrease the probability to sing, although as stated before, the connections of these regions to the vocal control areas need to be identified (e.g., Wild 2017). Such neuro-modulating mechanisms might explain the impact of the sociosexual and physical environment (e.g., food and water availability in zebra finches: Rashotte et al. 2001; Johnson and Rashotte 2002) on singing activity independent of gonadal hormones (e.g., Gulledge and Deviche 1998; Bentley et al. 1999; Deviche and Gulledge 2000; Tramontin et al. 1999).

An alternative possibility of testosterone-sensitive singing activity in times of low testosterone is the increased sensitivity of vocal neurons for testosterone and its androgenic and estrogenic derivatives. Such an example is the vocal neurons of the silver-beaked tanager (*Ramphocelus carbo*), an endemic neotropical songbird from the Amazon region. In these males, seasonal activation of full song correlated with an increased expression of androgen receptors in the HVC, while testosterone levels remained basal for several weeks after onset of song activity (Quispe et al. 2016). Circulating levels of testosterone started to rise later in the breeding season and coincided with consummatory sexual behaviors. This indicates that equatorial silver-beaked tanagers expands the annual period of singing by transiently increasing the sensitivity of the vocal control system to testosterone, a mechanism that avoids potentially detrimental effects of prolonged periods of elevated testosterone (Quispe et al. 2016).

7.5 Neural Mechanisms of Testosterone-Dependent Seasonal Song Structure

As detailed before, in many tropical and some temperate zone birds, the same song pattern (full song) can be heard year-round (e.g., winter wren: Kreutzer 1973; the great tit): while others sing additional song types in the breeding period (e.g., morning song of white-browed sparrow weaver: Voigt et al. 2006; directed song of zebra finches: Pröve 1974; Sossinka and Böhner 1980; Kao and Brainard 2006) or sing plastic songs in the nonbreeding period (e.g., great reed warbler: Sorensen et al. 2016; canary: Nottebohm et al. 1986; Leitner et al. 2001a, b; Voigt and Leitner 2008). Species and sex differences in seasonal stability versus seasonal plasticity might dependent on the testosterone sensitivity of neural vocal control regions, on the sensitivity of the genome to changing testosterone levels, and might depend on species-specific effects of testosterone-induced gene expression on neural circuit formation. Alternatively, seasonal vocal pattern might be influenced by testosterone sensitivity of auditory processing (Henry and Lucas 2009; Velez et al. 2015) in auditory areas that give input to the vocal control system. In relation, androgen receptors (ARs) are expressed in many neurons of the auditory forebrain (Metzdorf et al. 1999). This scenario is not further discussed in the present review.

7.5.1 Seasonal Testosterone Sensitivity of Vocal Control Regions: Androgen and Estrogen Receptors in Vocal Neurons

In songbirds, neural song control is achieved by a chain of interconnected brain areas in the fore-, mid-, and hindbrain (Nottebohm et al. 1976; Wild 1997; Hahnloser et al. 2002; Amador et al. 2013) (Fig. 7.5c). In particular, forebrain vocal control areas such as the HVC are evolutionary novelties of songbirds (Gahr 2000; Petkov and Jarvis 2012). In addition to the control of the song pattern, these areas are active during call-based vocal communication (Ter Maat et al. 2014; Benichov et al. 2016). The forebrain vocal circuit of songbirds connects to general avian vocal areas in the mid- and hindbrain via a projection of archistriatal neurons (the RA, robust nucleus of the arcopallium), in particular to the syringeal motonucleus (nucleus hypoglossus pars tracheosyringealis) and to respiratory pre-motor nuclei (Wild 1997).

One mode of steroid action in the brain is the alteration of gene expression by binding to intracellular steroid receptors that transactivate transcription of target genes in a ligand-dependent manner (Carson-Jurica et al. 1990). The androgen receptor (AR) has a high affinity for the testosterone and 5α-dihydrotestosterone, but not for 5β-dihydrotestosterone (Grino et al. 1990). The two types of estrogen receptors (ERα, ERβ) bind 17β-estradiol with high affinity. The AR gene codes for the AR protein, the ESR1 gene for ERα protein, and the ESR2 gene for ERβ protein. The AR and ERα but not ERβ occur in the vocal control system (see below).

In songbirds, AR mRNA or AR protein were reported for HVC, RA, and lMAN of all species studied (Balthazart et al. 1992; Bernard et al. 1999; Gahr et al. 1998, 2008; Metzdorf et al. 1999; Fusani et al. 2000; Voigt and Gahr 2011; Fraley et al. 2010; Quispe et al. 2016). Since these include species of various songbird families, among which are the basal Maluridae, the Corvidae, the Malaconotidae, and the derived Fringillidae and Thraupidae, the AR expression in HVC, RA and lMAN seems a general characteristic of songbirds (Frankl-Vilches and Gahr 2017). Further, AR mRNA and protein are reported for mMAN (medial magnocellular nucleus of the anterior nidopallium) and NIF (nucleus interfacialis) in canaries and zebra finches (Balthazart et al. 1992; Metzdorf et al. 1999; Fusani et al. 2000), but these areas have not yet been surveyed in other species. Extrapolating from the HVC, RA and lMAN data, we assume that AR expression in mMAN and NIF is also a common feature of songbirds. In Area X of zebra finches and canaries, ARs occur in only some individuals for unknown reasons (Gahr 2004; Kim et al. 2004). In another Estrildid finch, the wild white-rumped munia (*Lonchura striata*) and its domesticated relative the Bengalese finch (*Lonchura striata domestica*), ARs are expressed in a strain-specific pattern in Area X (Wada et al. 2013). In the brainstem, ARs occur in all respiratory-vocal areas and in syringeal motoneurons (Gahr and Wild 1997; Gahr 2000).

Among forebrain vocal areas, ESR1 mRNA and ERα protein is only expressed in HVC and around the dorsal aspect of RA of canaries and zebra finches (Gahr et al. 1993; Metzdorf et al. 1999). Further comparative data are available for ERα expression and protein abundance in the HVC of various species. These data suggest three types of distribution pattern (Frankl-Vilches and Gahr 2017) (Fig. 7.5d): (I) High expression of ERα throughout the entire HVC (e.g., canary, East-African shrike); (II) High expression of ERα only in the medial part of HVC (e.g., the forest weaver, the black redstart); and (III) No expression in the lateral part and low levels of ERα in the medial part of HVC (e.g., zebra finch, Bengalese finch). In all songbird species, large populations of ERα expressing neurons are found ventromedial to HVC aligning the lateral ventricle, an area including the so-called para-HVC (Johnson and Bottjer 1995), but extending much further medial than the latter (Gahr et al. 1993, and unpublished data).

In summary, all forebrain vocal control areas express or have the potential to express AR. Females of species of Maluridae, the Corvidae, the Malaconotidae, and the derived Fringillidae and Thraupidae that develop a vocal system, AR and ERα are expressed in the same areas as in their male conspecifics (Gahr unpublished data). Thus, since the expression pattern between species is very similar, AR and ERα distribution does not explain species differences in the degree of testosterone sensitivity of song features. In difference, ERα distribution of HVC varies considerably between species and suggests species indifference in estrogen dependency of vocal structures. Further, sexual differences of testosterone-dependent singing are not explained by sex differences in hormone receptor abundance in vocal areas.

Nevertheless, species differences in seasonal dynamics of both AR and ERα expression in vocal control areas might be involved in seasonality of song pattern and neural endophenotypes in a species-specific way (Fusani et al. 2000; Fraley et al. 2010). In the canary, AR expression in HVC is similar in the breeding and the nonbreeding season and low during the molt while ERα expression is high in the nonbreeding season and at the beginning of the breeding season but lower during the later breeding season and the molt (Gahr and Metzdorf 1997; Fusani et al. 2000). In the white-crowned sparrow, AR expression in HVC is higher during the breeding season as compared to the nonbreeding season while AR expression in other vocal areas is similar in the two seasons (Fraley et al. 2010). In general, AR expression in vocal control areas is rather invariant throughout the year except during the molt (Gahr and Metzdorf 1997; Gahr unpublished data).

7.5.2 Seasonal Changes of Transcription of Vocal Control Neurons

The same set of genes was found in genomes of various songbird species (Frankl-Vilches et al. 2015; Lovell et al. 2014; Warren et al. 2010). These findings suggest that the evolution of species-specific testosterone-sensitive song structures and related endophenotypes of songbirds does not result from the gain and loss of genes, but from the species-specific hormone-sensitive gene regulation (Frankl-Vilches et al. 2015). In contrast to the global similarity of songbird genomes, on the nucleotide level there are considerable species differences as shown above for the AR promoter of chicken and zebra finch. Such species differences can impact binding motifs of the ERα, the so-called estrogen response element (ERE) and of the AR, the so-called androgen response element (ARE) as shown for genes expressed in the HVC of the canary and the zebra finch (Frankl-Vilches et al. 2015). About 35% of the ERE-bearing and about 11% of ARE-bearing genes expressed in HVC of canaries were lacking these sites in the corresponding zebra finch orthologous promoters (Frankl-Vilches et al. 2015). Thus, species-specific evolutionary loss or gain (e.g., through point mutations) of EREs and AREs might underlie a species-specific gene pool that can be regulated by the activation of AR and ERα via testosterone and its androgenic and estrogenic metabolites in HVC.

This species-specific sensitivity of genes for AR and ERα together with species-specific testosterone profiles might lead to species differences in seasonal transcriptomes. In the white-crowned sparrow, seasonal comparisons of HVC and RA showed that gene expression is area-specific and time-specific across different reproductive conditions (Thompson et al. 2012). In the European robin, examination of HVC transcriptomes and histological analyses of song control nuclei showed testosterone-induced differentiation processes related to neuron growth and spacing, angiogenesis and neuron projection morphogenesis. Similar effects were found in female canaries treated with testosterone (Dittrich et al. 2014). In contrast, the expression of genes related to synaptic transmission was not enhanced in the HVC of testosterone treated female robins but was strongly upregulated in female canaries. A comparison of the testosterone-stimulated transcriptomes indicated that brain-derived neurotrophic factor (BDNF) likely functions as a common mediator of the testosterone effects in HVC (Dittrich et al. 2014; Frankl-Vilches et al. 2015). BDNF is an important regulator of neuronal and neural plasticity (Kowiański et al. 2017). However, other modes of testosterone action, notably related to synaptic transmission, appeared to be regulated in a more species-specific manner in the HVC of robins and canaries. Divergent effects of testosterone on the HVC of different species might be related to differences between species in regulatory mechanisms of the singing behavior.

7.5.3 Comparisons of Seasonal Testosterone-Induced Differentiation of Song Control Regions and Song Structure

Comparisons of the neuroanatomy of male songbirds suggest that seasonal changes in song structure parallel seasonal changes in HVC size of several species (Nottebohm 1981; Tramontin and Brenowitz 2000). However, this pattern is not universal and a number of species do not show such a correlation (Gahr 1990, 1997; Brenowitz et al. 1991; Leitner et al. 2001a, b; Reeves et al. 2003; Phillmore et al. 2006). Next to HVC size, testosterone and its estrogenic metabolites affect various parameters of song control neurons of adult songbirds in a song-area-specific manner. These effects including synapse density and dendritic arbores in RA (DeVoogd and Nottebohm 1981), GAP-junctions in HVC (Gahr and Garcia-Segura 1996), spiking activity of RA neurons (Meitzen et al. 2007b, 2009), synapse protein expression in HVC (Voigt et al. 2004), and neurotransmitter content (Ball and Balthazart 2010). Although these neuronal features are closer to neural function than song-area size, their relations to testosterone-dependent vocal pattern remains to be seen.

Next, I discuss the consequences of local manipulation of testosterone and estrogen concentration and of testosterone- and/or estrogen-controlled proteins in particular song control regions. First, testosterone infusion into song areas suggest that local action of steroids in one song control area affects downstream areas (Brenowitz and Lent 2002; Meitzen et al. 2007a). In particular the works of Meitzen et al. (2007a) show that estrogenic and androgenic metabolites of testosterone act in HVC to increase firing rate of RA neurons, which correlates with song stereotypy but not with song performance in adult white-crowned sparrows. Such testosterone-induced firing rates of RA neurons reflect seasonality of RA neurons since spontaneous firing rates of such neurons are much higher in spring compared to autumn in white-crowned and song sparrows (Meitzen et al. 2007b, 2009). Likewise, testosterone infusion into RA of adult canaries increased the firing of RA neurons (Breutel and Gahr unpublished data). How HVC neurons affect the firing rate of RA neurons is not known.

In canaries, songs are composed of syllables that are repeated identically several times (so called tours) before switching to the next syllable. Fusani et al. (2003) induced singing in adult female canaries through testosterone treatment, but inhibited in one group of such females the aromatization of testosterone into estrogens with an aromatase inhibitor. Testosterone-induced development of male-like song in female canaries is accompanied by an increase in the expression and enzymatic activity of aromatase in the telencephalon near HVC (Fusani et al. 2001), the only vocal areas with higher levels of estrogen receptors (Fig. 7.2). After 3–4 weeks of testosterone treatment, females developed a male-like song, with exception that such females sang few different syllables. In correlation with the male-like songs (long sequences of repeated syllables), the HVC size of singing females increased and was different from untreated non-singing control females (Fusani et al. 2003). The estrogen-deprived singing females differed, however, in that they produced syllable sequences with lower repetition rates compared to the non-deprived singing females (Fusani et al. 2003). On the behavioral level, similar results were obtained for male canaries (Rybak and Gahr 2004).

The local action of testosterone on HVC and the song structure has been studied in some detail in canaries. Testosterone induces the production of vascular endothelial growth factor (VEGF) and its receptor (VEGFR2 tyrosine-kinase) in the canary HVC, which in turn leads to an upregulation of BDNF production in HVC endothelial cells (Louissaint et al. 2002). Systemic inhibition of the VEGFR2 tyrosine-kinase was sufficient to block testosterone-induced singing of female canaries, even though testosterone exerts its inductive action on HVC morphology such as increased volume and increased total number of HVC neurons (Hartog et al. 2009). Expression of exogenous BDNF in HVC, induced locally by in situ transfection, reversed this VEGFR2 inhibitor-associated blockage of song development. The VEGFR2-inhibited, BDNF-expressing females developed elaborate male-like full song features such as syllable repertoires and high repetition rates of song syllables (Hartog et al. 2009).

The work of Fusani et al. (2003) and Hartog et al. (2009) suggests that androgen-dependent overall HVC morphology (reflected in HVC volume) is necessary for the production of certain (basic canary pattern such as repetition of identical syllables) song structure of canaries while other (estrogen-dependent) song structures (high repetition rates of syllables) require neuronal properties that are independent of vocal area size. The association/dissociation of brain area size and a particular neuronal phenotype might be a species-specific phenomenon since HVC volumes appear estrogen sensitive in song sparrows (Soma et al. 2004) and white-crowned sparrows (Tramontin et al. 2003 but Baker et al. 1984) but androgen sensitive in the canary (Fusani et al. 2003; Fusani and Gahr 2006) and hormone insensitive in the Mexican house finch (Strand et al. 2008).

7.5.4 Males Versus Females

As stated above, the differences between testosterone sensitivity of singing activity and of song pattern, seasonal or not, of male and female songbirds are not explained by the expression of AR and ERα in vocal control areas. A similar conclusion comes from the comparison of the neuroanatomy of the song system of males and females of several species. In species with singing females sexual dimorphisms are found in a species-specific, area-specific way to a varying degree at all organizational levels of the song control areas, the gross-anatomical, ultrastructural, electrophysiological, biochemical, and molecular (gene-expression) level (Nottebohm and Arnold 1976; Nottebohm 1980; Gahr and Metzdorf 1997; Del Negro and Edeline 2002; Riters and Ball 2002; Nealen 2005; Voigt and Gahr 2011; Duncan et al. 2011). Most notable, even in species such as the forest weaver, the East-African shrike and the Northern cardinal, in which male and female sing identical or nearly identical songs, respectively, there are sex differences in the size and organization of forebrain song control areas (Gahr et al. 1998, 2008; Jawor and MacDougall-Shackleton 2008; Voigt and Gahr 2011). Likewise, there is no correlation between song performance and song system anatomy since song areas are smaller in female streak-backed oriols that have a larger song output than the males of this species (Hall et al. 2010). Further, even in females that produce male typical songs due to testosterone treatment, the neuroanatomy is sexually dimorphic and does not allow relating neural features and hormone-dependent song pattern of males or females (Nottebohm 1980; Fusani et al. 2003; Hartog et al. 2009; Voigt and Leitner 2013).

7.5.5 Testosterone-Independent Seasonal Control of Song Structure

Vocal neurons express receptors for glucocorticoids (Suzuki et al. 2011; Senft et al. 2016), retinoic acid (Denisenko-Nehrbass et al. 2000), and melatonin (Gahr and Kosar 1996; Jansen et al. 2005; Fusani and Gahr 2015). Further, vocal neurons are sensitive for modulatory action of monoaminergic systems (Li and Sakaguchi 1997; Ball and Balthazart 2010) and opioids (Gulledge and Deviche 1998; Dloniak and Deviche 2001). These neurochemical phenotypes suggest that song phenotypes are under the control of multiple environmental signals that directly act on song neurons, bypassing the gonadal hormones or acting in concert with the gonadal hormones. The most direct evidence suggesting non-gonadal plasticity of singing comes from works on adult male zebra finches: melatonin affects their song pattern and the electrophysiology of RA neurons transiently (Jansen et al. 2005; Deregnaucourt et al. 2012). Unfortunately, the impact of melatonin on song and vocal neurons of seasonal singers such as the canary has not been tested.

A common mechanism of song control, dependent and independent of gonadal hormone, might be BDNF-sensitive mechanisms of HVC. As explained above for the canary, BDNF activity in the HVC is important for the production of sexually salient song pattern (Hartog et al. 2009). Whatever upregulates BDNF in HVC might facilitate differentiation of song structure. This notion is supported by the findings that BDNF expression, at least in rodents can be modulated by many factors that reflect environmental conditions, next to sex hormones (Dittrich et al. 1999; Louissaint et al. 2002; Scharfman and MacLusky 2005; Berton et al. 2006). In this scenario, BDNF upregulation in the vocal control system, even in birds with low levels of gonadal hormone production should show some form of singing.

7.6 Conclusion. Seasonal Singing: Who, How, and What for?

Traditionally, it is assumed that singing of songbirds and song-like vocalizations of other bird species such as the crowing of the chicken are male-typical behaviors (Thorpe 1961, pp. 41–48). This idea was particularly sponsored by the observation of many naturalists that male birds of the Northern hemisphere sing intensely during the breeding season but not in fall or early winter (Fry 1916; Cox 1944; Saunders 1947, 1948). However, the reconsideration of old observations (Fry 1916; Alexander 1935) and new studies of species of the Northern and Southern hemisphere (e.g., Kelsey 1988; DeWolfe and Baptista 1995; van Duyse et al. 2003; Kunc et al. 2006; Bezzel 2011; Apfelbeck et al. 2013; Sorensen 2014; Sorensen et al. 2016) clearly demonstrate that singing outside the breeding season is a common behavior of male songbirds while their testosterone levels remain low (e.g., Schwabl 1992; Apfelbeck et al. 2013; York et al. 2016). Further, detailed studies of tropical and temperate species in the last three decades and reconsideration of old studies of Australo-Asian species led to the conclusion that singing of female songbirds is an ancient and widespread behavior (e.g., Gahr and Güttinger 1986; Hoelzel 1986; Ritchison 1986; Odom et al. 2014), while testosterone levels of singing females are low in most cases (for review: Ketterson et al. 2005). These findings question the testosterone hypothesis of song production. However, a key problem is the interpretation of "high" and "low" testosterone levels, which would require a basic understanding of the quantitative interaction of testosterone molecules, androgen receptor (AR) and estrogen receptor (ER) densities, and transcriptional regulation, e.g., would the transient upregulation of AR in HVC or the vocal system in general be a mechanism to maintain song structures in seasons when testosterone production is low? Or do other endocrine systems induce singing in males outside of the breeding season and in females at various annual periods? Or are males and females able to regulate AR and ER dependent mechanisms via non-endocrine pathways? These are just a few possibilities to explain the mismatch of the seasonal profiles of testosterone and of singing behavior. More detailed field studies of hormone production and singing of males and females as well as detailed molecular analysis of hormone-receptor–hormone interactions are needed to elucidate the neuroendocrine control of seasonal singing. Insights into the endocrine control of male singing outside the breeding season and of female singing are essential for understanding the proximate and ultimate mechanisms of seasonal singing.

References

Adret-Hausberger M, Güttinger HR, Merkel FW (1990) Individual life-history and song repertoire changes in a colony of starlings (*Sturnus vulgaris*). Ethology 84:265–280

Alexander HG (1935) A chart of bird song. Br Birds 29:190–198

Alward BA, Balthazart J, Ball GF (2013) Differential effects of global versus local testosterone on singing behavior and its underlying neural substrate. Proc Natl Acad Sci USA 110:19573–19578. https://doi.org/10.1073/pnas.1311371110

Amador A, Perl YS, Mindlin GB, Margoliash D (2013) Elemental gesture dynamics are encoded by song premotor cortical neurons. Nature 495:59–64. https://doi.org/10.1038/nature11967

Amy M, Salvin P, Naguib M, Leboucher G (2015) Female signalling to male song in the domestic canary, *Serinus canaria*. R Soc Open Sci 2:140196. https://doi.org/10.1098/rsos.140196

Anderson RC, Nowicki S, Searcy WA (2007) Soft song in song sparrows: response of males and females to an enigmatic signal. Behav Ecol Sociobiol 61:1267–1274. https://doi.org/10.1007/s00265-007-0357-7

Anderson RC, Searcy WA, Peters S, Nowicki S (2008) Soft song in song sparrows: acoustic structure and implications for signal function. Ethology 114:662–676. https://doi.org/10.1111/j.1439-0310.2008.01518.x

Andersson M (1994) Sexual selection. Princeton University Press, Princeton

Andrew RJ (1963) Effects of testosterone on the behavior of the domestic chick. J Comp Physiol Psychol 56:933–940. https://doi.org/10.1037/h0045105

Apfelbeck B, Kiefer S, Mortega KG, Goymann W, Kipper S (2012) Testosterone affects song modulation during simulated territorial intrusions in male black redstarts (*Phoenicurus ochruros*). PLoS One 7:e52009. https://doi.org/10.1371/journal.pone.0052009

Apfelbeck B, Mortega K, Kiefer S, Kipper S, Vellema M, Villavicencio CP, Gahr M, Goymann W (2013) Associated and disassociated patterns in hormones, song, behavior and brain receptor expression between life-cycle stages in male black redstarts, *Phoenicurus ochruros*. Gen Comp Endocrinol 184:93–102. https://doi.org/10.1016/j.ygcen.2012.11.027

Apfelbeck B, Mortega KG, Flinks H, Illera JC, Helm B (2017) Testosterone, territorial response, and song in seasonally breeding tropical and temperate stonechats. BMC Evol Biol 17:101. https://doi.org/10.1186/s12862-017-0944-9

Arcese P, Stoddard PK, Hiebert SM (1988) The form and function of song in female song sparrows. Condor 90:44–50. https://doi.org/10.2307/1368431

Armstrong EA (1955) The wren. Collins, London

Armstrong EA (1973) A study of bird song, vol 201. Dover, New York

Arnold AP (1975) The effects of castration and androgen replacement on song, courtship, and aggression in zebra finches (*Poephila guttata*). J Exp Zool 191:309–326. https://doi.org/10.1002/jez.1401910302

Baker MC, Bottjer SW, Arnold AP (1984) Sexual dimorphism and lack of seasonal changes in vocal control regions of the white-crowned sparrow brain. Brain Res 295:85–89

Bakus GJ (1959) Observations on the life history of the dipper in Montana. Auk 76:190–207. https://doi.org/10.2307/4081776

Ball GF, Balthazart J (2010) Seasonal and hormonal modulation of neurotransmitter systems in the song control circuit. J Chem Neuroanat 39:82–95. https://doi.org/10.1016/j.jchemneu.2009.08.005

Ballentine B, Searcy WA, Nowicki S (2008) Reliable aggressive signalling in swamp sparrows. Anim Behav 75:693–703. https://doi.org/10.1016/j.anbehav.2007.07.025

Balsby TJS (2000) The function of song in whitethroats *Sylvia communis*. Bioacoustics 11:17–30. https://doi.org/10.1080/09524622.2000.9753447

Balthazart J, Foidart A, Wilson EM, Ball GF (1992) Immunocytochemical localization of androgen receptors in the male songbird and quail brain. J Comp Neurol 317:407–420. https://doi.org/10.1002/cne.90317040

Beani L, Briganti F, Giuseppe C, Concetta L, Dess X, Fulgheri F (2000) Effect of androgens on structure and rate of crowing in the Japanese quail (*Coturnix japonica*). Behaviour 137:417–435

Beletsky LD (1983) Aggressive and pair bond maintenance songs of female red-winged blackbirds (*Agelaius, Phoeniceus*). Z Tierpsychol 62:47–54

Benichov JI, Benezra SE, Vallentin D, Globerson E, Long MA, Tchernichovski O (2016) The forebrain song system mediates predictive call timing in female and male zebra finches. Curr Biol 26:309–318. https://doi.org/10.1016/j.cub.2015.12.037

Bennett MA (1940) The vocal hierarchy in ring doves. II. The effects of testosterone propionate. Ecology 21:148–165

Bensch S, Hasselquist D (1992) Evidence for active female choice in a polygynous warbler. Anim Behav 44:301–311. https://doi.org/10.1016/0003-3472(92)90036-9

Bentley GE, Van't Hof TJ, Ball GF (1999) Seasonal neuroplasticity in the songbird telencephalon: a role for melatonin. Proc Natl Acad Sci USA 96:4674–4679

Bernard DJ, Bentley GE, Balthazart J, Turek FW, Ball GF (1999) Androgen receptor, estrogen receptor alpha, and estrogen receptor beta show distinct patterns of expression in forebrain song control nuclei of European starlings. Endocrinology 140:4633–4643. https://doi.org/10.1210/en.140.10.4633

Berthold AA (1849) Transplantation der Hoden Archiv für Anatomie. Physiologie und Wissenschaftliche Medizin 16:42–46

Berton O, Berton O, McClung CA, DiLeone RJ, Krishnan V, Renthal W, Russo SJ, Graham D, Tsankova NM, Bolanos CA, Rios M, Monteggia LM, Self DW, Nestler EJ (2006) Essential role of BDNF in the mesolimbic dopamine pathway in social defeat stress. Science 311:864–868. https://doi.org/10.1126/science.1120972

Bezzel E (1988) Die Gesangszeiten des Buchfinken (*Fringilla coelebs*). Eine Regionalstudie. J Ornithol 129:71–81. https://doi.org/10.1007/BF01641533

Bezzel E (2011) Gesangsphänologie und Methodenstandards der Brutvogelerfassung – langfristige lokale Erfahrungen. Limicola 25:1–36

Boseret G, Carere C, Ball GF, Balthazart J (2006) Social context affects testosterone-induced singing and the volume of song control nuclei in male canaries (*Serinus canaria*). J Neurobiol 66:1044–1060. https://doi.org/10.1002/neu.20268

Brenowitz EA, Lent K (2002) Act locally and think globally: intracerebral testosterone implants induce seasonal-like growth of adult avian song control circuits. Proc Natl Acad Sci USA 99:12421–12426. https://doi.org/10.1073/pnas.192308799

Brenowitz EA, Nalls B, Wingfield JC, Kroodsma DE (1991) Seasonal-changes in avian song nuclei without seasonal-changes in song repertoire. J Neurosci 11:1367–1374

Brockway BF (1968) Influences of sex hormones on the loud and soft warbling of male budgerigars. Anim Behav 16:5–12

Brunton DH, Li XL (2006) The song structure and seasonal patterns of vocal behavior of male and female bellbirds (*Anthornis melanura*). J Ethol 24:17–25. https://doi.org/10.1007/s10164-005-0155-5

Bullough WS (1942) The reproductive cycles of the British and continental races of the starling (*Sturnus vulgaris L*). Phil Trans R Soc B 231:165–U142

Camacho-Schlenker A, Courvoisier H, Aubin T (2011) Song sharing and singing strategies in the winter wren *Troglodytes troglodytes*. Behav Process 87:260–267

Carson-Jurica MA, Schrader WT, O'Malley BW (1990) Steroid-receptor family - structure and functions. Endocr Rev 11:201–220. https://doi.org/10.1210/edrv-11-2-201

Catchpole CJB, Slater P (1995) Bird song: biological themes and variations. Cambridge University Press, Cambridge, UK. https://doi.org/10.1017/CBO9780511754791

Chiba A, Hosokawa N (2006) Effects of androgens and estrogens on crowings and distress callings in male Japanese quail, *Coturnix japonica*. Horm Behav 49:4–14. https://doi.org/10.1016/j.yhbeh.2005.05.020

Cooney R, Cockburn A (1995) Territorial defense is the major function of female song in the superb fairy-wren, *Malurus-cyaneus*. Anim Behav 49:1635–1647. https://doi.org/10.1016/0003-3472(95)90086-1

Cox PR (1944) A statistical investigation into bird song. Br Birds 38:3–9

Cramp SP, Perrins CM (1994) The birds of the Western Palearctic. Oxford University Press, Oxford

Cynx J, Bean NJ, Rossman I (2005) Testosterone implants alter the frequency range of zebra finch songs. Horm Behav 47:446–451. https://doi.org/10.1016/j.yhbeh.2004.11.018

Dabelsteen T (1984) An analysis of the full song of the blackbird Turdus-Merula with respect to message coding and adaptations for acoustic communication. Ornis Scand 15:227–239. https://doi.org/10.2307/3675931

Dabelsteen T, Pedersen SB (1990) Song and information about aggressive responses of blackbirds, *Turdus-Merula* - evidence from interactive playback experiments with territory owners. Anim Behav 40:1158–1168. https://doi.org/10.1016/S0003-3472(05)80182-4

Dabelsteen T, McGregor PK, Holland J, Tobias JA, Pedersen SB (1997) The signal function of overlapping singing in male robins. Anim Behav 53:249–256

Dabelsteen T, McGregor PK, Lampe HM, Langmore NE, Holland JO (1998) Quiet song in song birds: an overlooked phenomenon. Bioacoustics 9:89–105. https://doi.org/10.1080/09524622.1998.9753385

Davis J (1958) Singing behaviour and the gonad cycle of the rufous-sided towhee. Condor 60:308–336

Dawson A (1983) Plasma gonadal-steroid levels in wild starlings (*Sturnus-vulgaris*) during the annual cycle and in relation to the stages of breeding. Gen Comp Endocrinol 49:286–294. https://doi.org/10.1016/0016-6480(83)90146-6

Del Negro C, Edeline JM (2002) Sex and season influence the proportion of thin spike cells in the canary HVC. Neuroreport 13:2005–2009. https://doi.org/10.1097/00001756-200211150-00003

Denisenko-Nehrbass NI, Jarvis E, Scharff C, Nottebohm F, Mello CV (2000) Site-specific retinoic acid production in the brain of adult songbirds. Neuron 27:359–370. https://doi.org/10.1016/S0896-6273(00)00043-X

Deregnaucourt S, Saar S, Gahr M (2012) Melatonin affects the temporal pattern of vocal signatures in birds. J Pineal Res 53:245–258. https://doi.org/10.1111/j.1600-079X.2012.00993.x

Derrickson KC (1987) Behavioral-correlates of song types of the northern mockingbird (*Mimus-polyglottos*). Ethology 74:21–32

Deviche P, Gulledge CC (2000) Vocal control region sizes of an adult female songbird change seasonally in the absence of detectable circulating testosterone concentrations. J Neurobiol 42:202–211. https://doi.org/10.1002/(Sici)1097-4695(20000205)42:2/202::Aid-Neu4/3.0.Co;2-G

DeVoogd T, Nottebohm F (1981) Gonadal-hormones induce dendritic growth in the adult avian brain. Science 214:202–204. https://doi.org/10.1126/science.7280692

Dewolfe BB, Baptista LF (1995) Singing behavior, song types on their wintering grounds and the question of leap-frog migration in Puget-sound white-crowned sparrows. Condor 97:376–389. https://doi.org/10.2307/1369024

Dittrich F, Feng Y, Metzdorf R, Gahr M (1999) Estrogen-inducible, sex-specific expression of brain-derived neurotrophic factor mRNA in a forebrain song control nucleus of the juvenile zebra finch. Proc Natl Acad Sci USA 96:8241–8246. https://doi.org/10.1073/pnas.96.14.8241

Dittrich F, Ramenda C, Grillitsch D, Frankl-Vilches C, Ko MC, Hertel M, Goymann W, ter Maat A, Gahr M (2014) Regulatory mechanisms of testosterone-stimulated song in the sensorimotor nucleus HVC of female songbirds. BMC Neurosci 15:125. https://doi.org/10.1186/s12868-014-0128-0

Dloniak SM, Deviche P (2001) Effects of testosterone and photoperiodic condition on song production and vocal control region volumes in adult male dark-eyed juncos (*Junco hyemalis*). Horm Behav 39:95–105. https://doi.org/10.1006/hbeh.2000.1621

Duncan KA, Jimenez P, Carruth LL (2011) Distribution and sexually dimorphic expression of steroid receptor coactivator-1 (SRC-1) in the zebra finch brain. Gen Comp Endocrinol 170:408–414. https://doi.org/10.1016/j.ygcen.2010.10.021

Eens M, Pinxten R, Verheyen RF (1992) No overlap in song repertoire size between yearling and older starlings *Sturnus-vulgaris*. Ibis 134:72–76. https://doi.org/10.1111/j.1474-919X.1992.tb07233.x

Etches RJ, Cheng KW (1981) Changes in the plasma concentrations of luteinizing hormone, progesterone, oestradiol and testosterone and in the binding of follicle-stimulating hormone to the theca of follicles during the ovulation cycle of the hen (*Gallus domesticus*). J Endocrinol 91:11–22

Ficken MS, Weise CM (1984) A complex call of the black-capped chickadee (*Parus atricapillus*) I. Microgeographic variation. Auk 101:349–360

Ficken MS, Ficken WR, Witkin S (1978) Vocal repertoire of the black-capped chickadee. Auk 95:34–48. https://doi.org/10.2307/4085493

Fouarge JG (1968) Le Pouillot Siffleur *Phylloscopus sibilatrix*. Le Gerfaut 58:179–368

Fraley GS, Steiner RA, Lent KL, Brenowitz EA (2010) Seasonal changes in androgen receptor mRNA in the brain of the white-crowned sparrow. Gen Comp Endocrinol 166:66–71. https://doi.org/10.1016/j.ygcen.2009.08.001

Frankl-Vilches C, Gahr M (2017) Androgen and estrogen sensitivity of bird song: a comparative view on gene regulatory levels. J Comp Physiol A Neuroethol Sens Neural Behav Physiol 204:113–126. https://doi.org/10.1007/s00359-017-1236-y

Frankl-Vilches C et al (2015) Using the canary genome to decipher the evolution of hormone-sensitive gene regulation in seasonal singing birds. Genome Biol 16:19. https://doi.org/10.1186/s13059-014-0578-9

Fry HJ (1916) A study of the seasonal decline of bird song. Auk 33:28–40. https://doi.org/10.2307/4072114

Fusani L (2008) Testosterone control of male courtship in birds. Horm Behav 54:227–233. https://doi.org/10.1016/j.yhbeh.2008.04.004

Fusani L, Gahr M (2006) Hormonal influence on song structure and organization: the role of estrogen. Neuroscience 138:939–946. https://doi.org/10.1016/j.neuroscience.2005.08.041

Fusani L, Gahr M (2015) Differential expression of melatonin receptor subtypes MelIa, MelIb and MelIc in relation to melatonin binding in the male songbird brain. Brain Behav Evol 85:4–14. https://doi.org/10.1159/000367984

Fusani L, Beani L, Dessì-Fulgheri F (1993) Seasonal individual changes in grey partridge Perdix perdix male call. Mus reg Sci nat Torino 1:173–181

Fusani L, Beani L, Dessifulgheri F (1994) Testosterone affects the acoustic structure of the male call in the grey partridge (*Perdix perdix*). Behaviour 128:301–310. https://doi.org/10.1163/156853994x00307

Fusani L, Van't Hof T, Hutchison JB, Gahr M (2000) Seasonal expression of androgen receptors, estrogen receptors, and aromatase in the canary brain in relation to circulating androgens and estrogens. J Neurobiol 43:254–268. https://doi.org/10.1002/(Sici)1097-4695(20000605)43:3/254::Aid-Neu4/3.0.Co;2-W

Fusani L, Hutchison JB, Gahr M (2001) Testosterone regulates the activity and expression of aromatase in the canary neostriatum. J Neurobiol 49:1–8

Fusani L, Metzdorf R, Hutchison JB, Gahr M (2003) Aromatase inhibition affects testosterone-induced masculinization of song and the neural song system in female canaries. J Neurobiol 54:370–379. https://doi.org/10.1002/neu.10141

Gahr M (1990) Delineation of a brain nucleus - comparisons of cytochemical, hodological, and cytoarchitectural views of the song control nucleus Hvc of the adult canary. J Comp Neurol 294:30–36. https://doi.org/10.1002/cne.902940104

Gahr M (1997) How should brain nuclei be delineated? Consequences for developmental mechanisms and for correlations of area size, neuron numbers and functions of brain nuclei. Trends Neurosci 20:58–62

Gahr M (2000) Neural song control system of hummingbirds: comparison to swifts, vocal learning (songbirds) and nonlearning (suboscines) passerines, and vocal learning (budgerigars) and nonlearning (dove, owl, gull, quail, chicken) nonpasserines. J Comp Neurol 426:182–196

Gahr M (2004) Hormone-dependent neural plasticity in the juvenile and adult song system - what makes a successful male? Ann N Y Acad Sci 1016:684–703. https://doi.org/10.1196/annals.1298.025

Gahr M (2007) Sexual differentiation of the vocal control system of birds. Adv Genet 59:67–105. https://doi.org/10.1016/S0065-2660(07)59003-6

Gahr M (2014) How hormone-sensitive are bird songs and what are the underlying mechanisms? Acta Acustica united with Acustica 100(4):705–718. https://doi.org/10.3813/AAA.918749

Gahr M, Garcia-Segura LM (1996) Testosterone-dependent increase of gap-junctions in HVC neurons of adult female canaries. Brain Res 712:69–73

Gahr M, Güttinger HR (1986) Functional-aspects of singing in male and female *Uraeginthus-bengalus* (Estrildidae). Ethology 72:123–131

Gahr M, Kosar E (1996) Identification, distribution, and developmental changes of a melatonin binding site in the song control system of the zebra finch. J Comp Neurol 367:308–318

Gahr M, Metzdorf R (1997) Distribution and dynamics in the expression of androgen and estrogen receptors in vocal control systems of songbirds. Brain Res Bull 44:509–517. https://doi.org/10.1016/S0361-9230(97)00233-5

Gahr M, Wild JM (1997) Localization of androgen receptor mRNA-containing cells in avian respiratory-vocal nuclei: an in situ hybridization study. J Neurobiol 33:865–876

Gahr M, Güttinger HR, Kroodsma DE (1993) Estrogen-receptors in the avian brain - survey reveals general distribution and forebrain areas unique to songbirds. J Comp Neurol 327:112–122. https://doi.org/10.1002/cne.903270109

Gahr M, Sonnenschein E, Wickler W (1998) Sex difference in the size of the neural song control regions in a dueting songbird with similar song repertoire size of males and females. J Neurosci 18:1124–1131

Gahr M, Metzdorf R, Schmidl D, Wickler W (2008) Bi-directional sexual dimorphisms of the song control nucleus HVC in a songbird with unison song. PLoS One 3:e3073. https://doi.org/10.1371/journal.pone.0003073

Geberzahn N, Aubin T (2014) How a songbird with continuous singing style modulates its song when territorially challenged. Behav Ecol Sociobiol 68:1–12

Geberzahn N, Gahr M (2011) Undirected (solitary) birdsong in female and male blue-capped cordon-bleus (*Uraeginthus cyanocephalus*) and its endocrine correlates. PLoS One 6:e26485. https://doi.org/10.1371/journal.pone.0026485

Gil D, Gahr M (2002) The honesty of bird song: multiple constraints for multiple traits. Trends Ecol Evol 17:133–141. https://doi.org/10.1016/S0169-5347(02)02410-2

Goymann W, Landys MM (2011) Testosterone and year-round territoriality in tropical and non-tropical songbirds. J Avian Biol 42:485–489. https://doi.org/10.1111/j.1600-048X.2011.05464.x

Goymann W, Geue D, Schwabl I, Flinks H, Schmidl D, Schwabl H, Gwinner E (2006) Testosterone and corticosterone during the breeding cycle of equatorial and European stonechats (*Saxicola torquata axillaris* and *S. t. rubicola*). Horm Behav 50:779–785. https://doi.org/10.1016/j.yhbeh.2006.07.002

Grafen A (1990) Biological signals as handicaps. J Theor Biol 144:517–546. https://doi.org/10.1016/S0022-5193(05)80088-8

Grino PB, Griffin JE, Wilson JD (1990) Testosterone at high-concentrations interacts with the human androgen receptor similarly to dihydrotestosterone. Endocrinology 126:1165–1172

Groothuis T, Meeuwissen G (1992) The influence of testosterone on the development and fixation of the form of displays in two age classes of young black-headed gulls. Anim Behav 43:189–208. https://doi.org/10.1016/S0003-3472(05)80215-5

Gulledge CC, Deviche P (1998) Photoperiod and testosterone independently affect vocal control region volumes in adolescent male songbirds. J Neurobiol 36:550–558. https://doi.org/10.1002/(Sici)1097-4695(19980915)36:4/550:Aid-Neu8/3.0.Co;2-V

Gwinner E, Rödl T, Schwabl H (1994) Pair territoriality of wintering stonechats - behavior, function and hormones. Behav Ecol Sociobiol 34:321–327. https://doi.org/10.1007/Bf00197002

Hahnloser RHR, Kozhevnikov AA, Fee MS (2002) An ultra-sparse code underlies the generation of neural sequences in a songbird. Nature 419:65–70. https://doi.org/10.1038/nature00974

Hall ZJ, MacDougall-Shackleton SA, Osorio-Beristain M, Murphy TG (2010) Male bias in the song control system despite female bias in song rate in streak-backed orioles (*Icterus pustulatus*). Brain Behav Evol 76:168–175. https://doi.org/10.1159/000320971

Harding CF, Sheridan K, Walters MJ (1983) Hormonal specificity and activation of sexual-behavior in male zebra finches. Horm Behav 17:111–133. https://doi.org/10.1016/0018-506x(83)90021-1

Harding CF, Walters MJ, Collado D, Sheridan K (1988) Hormonal specificity and activation of social-behavior in male red-winged blackbirds. Horm Behav 22:402–418. https://doi.org/10.1016/0018-506x(88)90011-6

Hartley RS, Suthers RA (1990) Lateralization of syringeal function during song production in the canary. J Neurobiol 21:1236–1248. https://doi.org/10.1002/neu.480210808

Hartog TE et al (2009) Brain-derived neurotrophic factor signaling in the HVC is required for testosterone-induced song of female canaries. J Neurosci 29:15511–15519. https://doi.org/10.1523/JNEUROSCI.2564-09.2009

Hegelbach J, Spaar R (2000) Saisonaler Verlauf der Gesangsaktivität der Singdrossel (*Turdus philomelos*), mit Anmerkungen zum nachbrutzeitlichen Gesangsschub. J Ornithol 141:425–434. https://doi.org/10.1111/j.1439-0361.2000.00035.pp.x

Heid P, Güttinger HR, Prove E (1985) The influence of castration and testosterone replacement on the song architecture of canaries (*Serinus-canaria*). Z Tierpsychol 69:224–236

Henry KS, Lucas JR (2009) Vocally correlated seasonal auditory variation in the house sparrow (*Passer domesticus*). J Exp Biol 212(23):3817–3822. https://doi.org/10.1242/jeb.033035

Hill SD, Amiot C, Ludbrook MR, Ji WH (2015) Seasonal variation in the song structure of Tui (*Prosthemadera novaeseelandiae*). N Z J Ecol 39:110–115

Hoelzel AR (1986) Song characteristics and response to playback of male and female robins *Erithacus-rubecula*. Ibis 128:115–127. https://doi.org/10.1111/j.1474-919X.1986.tb02098.x

Howard RD (1974) The influence of sexual selection and interspecific competition on mockingbird song (*Mimus polyglottos*). Evolution 28:428–438. https://doi.org/10.2307/2407164

Hughes M, Nowicki S, Lohr B (1998) Call learning in black-capped chickadees (*Parus atricapillus*): the role of experience in the development of 'chick-a-dee' calls. Ethology 104:232–249

Hutchison JB, Steimer T (1984) Androgen metabolism in the brain: behavioural correlates. In: de Vries GJ, de Bruin JPC, Uylings HBM, Corner MA (eds) Sex differences in the brain. Elsevier, Amsterdam

Immelmann K (1969) Song development in the zebra finch and other estrildid finches. In: Hinde RA (ed) Bird vocalizations. Cambridge University Press, Cambridge, pp 61–77

Jansen R, Metzdorf R, van der Roest M, Fusani L, ter Maat A, Gahr M (2005) Melatonin affects the temporal organization of the song of the zebra finch. FASEB J 19:848–850. https://doi.org/10.1096/fj.04-2874fje

Jawor JM, MacDougall-Shackleton SA (2008) Seasonal and sex-related variation in song control nuclei in a species with near-monomorphic song, the northern cardinal. Neurosci Lett 443:169–173. https://doi.org/10.1016/j.neulet.2008.07.085

Johnson AL (1990) Steroidogenesis and actions of steroids in the hen ovary. Crit Rev Poult Biol 2:319–346

Johnson F, Bottjer SW (1995) Differential estrogen accumulation among populations of projection neurons in the higher vocal center of male canaries. J Neurobiol 26:87–108. https://doi.org/10.1002/neu.480260108

Johnson F, Rashotte M (2002) Food availability but not cold ambient temperature affects undirected singing in adult male zebra finches. Physiol Behav 76:9–20

Kao MH, Brainard MS (2006) Lesions of an avian basal ganglia circuit prevent context-dependent changes to song variability. J Neurophysiol 96:1441–1455

Kelm-Nelson CA, Stevenson SA, Cordes MA, Riters LV (2013) Modulation of male song by naloxone in the medial preoptic nucleus. Behav Neurosci 127:451–457. https://doi.org/10.1037/a0032329

Kelsey MG (1988) A comparison of the song and territorial behaviour of a long-distance migrant, the Marsh Warbler *Acrocephalus palustris*, in summer and winter. Ibis 131:403–414

Kern MD, King JR (1972) Testosterone-induced singing in female white-crowned sparrows. Condor 74:204–209. https://doi.org/10.2307/1366289

Ketterson ED, Nolan V, Wolf L, Ziegenfus C (1992) Testosterone and avian life histories - effects of experimentally elevated testosterone on behavior and correlates of fitness in the dark-eyed Junco (*Junco hyemalis*). Am Nat 140:980–999. https://doi.org/10.1086/285451

Ketterson ED, Nolan V Jr, Sandell M (2005) Testosterone in females: mediator of adaptive traits, constraint on sexual dimorphism, or both? Am Nat 166(Suppl 4):S85–S98. https://doi.org/10.1086/444602

Kim YH, Perlman WR, Arnold AP (2004) Expression of androgen receptor mRNA in zebra finch song system: developmental regulation by estrogen. J Comp Neurol 469:535–547. https://doi.org/10.1002/cne.11033

Konishi M (1965) The role of auditory feedback in the control of vocalization in the white-crowned sparrow. Z Tierpsychol 22:770–783

Kowiański P, Lietzau G, Czuba E, Waśkow M, Steliga A, Moryś J (2017) BDNF: a key factor with mutipotent impact of brain signaling and synaptic plasticity. Cell Mol Neurobiol 38(3):579–593. https://doi.org/10.1007/s10571-017-0510-4

Kreutzer ML (1973) Contribution à l'étude du chant de proclamation territoriale du troglodyte (*Troglodytes troglodytes*): Stéréotypie, variations géographiques, dialectes, conséquences éthologiques de ces phénomènes. Université Paris V

Kreutzer ML, Vallet EM (1991) Differences in the responses of captive female canaries to variation in conspecific and heterospecific songs. Behaviour 117:106–116. https://doi.org/10.1163/156853991x00148

Kriner E, Schwabl H (1991) Control of winter song and territorial aggression of female robins (*Erithacus-rubecula*) by testosterone. Ethology 87:37–44

Kroodsma DE (1985) Development and use of two song forms by the eastern Phoebe. Wilson Bull 97:21–29

Kroodsma DE, Byers BE (1991) The function(s) of bird song. Am Zool 31:318–328

Kunc HP, Foerster K, Vermeirssen ELM, Kempenaers B (2006) Experimentally elevated plasma testosterone levels do not influence singing behaviour of male blue tits (*Parus caeruleus*) during the early breeding season. Ethology 112:984–992. https://doi.org/10.1111/j.1439-0310.2006.01254.x

Kurvers RHJM, Roberts ML, McWilliams SR, Peters A (2008) Experimental manipulation of testosterone and condition during molt affects activity and vocalizations of male blue tits. Horm Behav 54:263–269. https://doi.org/10.1016/j.yhbeh.2008.03.011

Lack DF (1943) The life of the robin. Witherby HFG, London

Lampe HM (1991) The response of male redwings *Turdus-iliacus* to playback of conspecific songs with or without the terminating Twitter. Ornis Scand 22:137–142. https://doi.org/10.2307/3676544

Langmore NE (1998) Functions of duet and solo songs of female birds. Trends Ecol Evol 13:136–140

Langmore NE, Davies NB (1997) Female dunnocks use vocalizations to compete for males. Anim Behav 53:881–890. https://doi.org/10.1006/anbe.1996.0306

Langmore NE, Davies NB, Hatchwell BJ, Hartley IR (1996) Female song attracts males in the alpine accentor *Prunella collaris*. Proc Roy Soc B 263:141–146. https://doi.org/10.1098/rspb.1996.0022

Langmore NE, Cockrem JF, Candy EJ (2002) Competition for male reproductive investment elevates testosterone levels in female dunnocks, *Prunella modularis*. Proc Roy Soc B 269:2473–2478. https://doi.org/10.1098/rspb.2002.2167

Laskey AR (1944) A study of the cardinal in Tennessee. Wilson Bull 56:27–44

Leitner S, Cornelia V, Gahr M (2001a) Seasonal changes in the song pattern of the non-domesticated Island Canary (*Serinus canaria*), a field study. Behaviour 138:885–904

Leitner S, Voigt C, Garcia-Segura LM, Van't Hof T, Gahr M (2001b) Seasonal activation and inactivation of song motor memories in wild canaries is not reflected in neuroanatomical changes of forebrain song areas. Horm Behav 40:160–168. https://doi.org/10.1006/hbeh.2001.1700

Leonard SL (1939) Induction of singing in female canaries by injections of male hormone. Proc Soc Exp Biol Med 41:229–230

Li R, Sakaguchi H (1997) Cholinergic innervation of the song control nuclei by the ventral paleostriatum in the zebra finch: a double-labeling study with retrograde fluorescent tracers and choline acetyltransferase immunohistochemistry. Brain Res 763(2):239–246

Lister D (1953a) Secondary song: a tentative classification. Br Birds 46:139–143

Lister D (1953b) Secondary song of some Indian birds Bombay. Nat Hist Soc 51:699–706

Logan CA (1983) Biological diversity in avian vocal learning. In: Zeiler MH (ed) Advances in the analysis of behaviour, vol 3. Wiley, Sussex, pp 143–176

Logan CA, Wingfield JC (1995) Hormonal correlates of breeding status, nest construction, and parental care in multiple-brooded northern mockingbirds, *Mimus polyglottos*. Horm Behav 29:12–30. https://doi.org/10.1006/hbeh.1995.1002

Louissaint A Jr, Rao S, Leventhal C, Goldman SA (2002) Coordinated interaction of neurogenesis and angiogenesis in the adult songbird brain. Neuron 34:945–960

Lovell PV et al (2014) Conserved syntenic clusters of protein coding genes are missing in birds. Genome Biol 15:565. https://doi.org/10.1186/s13059-014-0565-1

Maddison CJ, Anderson RC, Prior NH, Taves MD, Soma KK (2012) Soft song during aggressive interactions: seasonal changes and endocrine correlates in song sparrows. Horm Behav 62:455–463. https://doi.org/10.1016/j.yhbeh.2012.08.002

Malacarne G, Cucco M, Camanni S (1991) Coordinated visual-displays and vocal duetting in different ecological situations among Western Palearctic non-passerine birds. Ethol Ecol Evol 3:207–219. https://doi.org/10.1080/08927014.1991.9525369

Maney DL, Lange HS, Raees MQ, Reid AE, Sanford SE (2009) Behavioral phenotypes persist after gonadal steroid manipulation in white-throated sparrows. Horm Behav 55:113–120. https://doi.org/10.1016/j.yhbeh.2008.09.002

Marler P, Kreith M, Willis E (1962) An analysis of testosterone-induced crowing in young domestic cockerels. Anim Behav 10:48–54. https://doi.org/10.1016/0003-3472(62)90130-6

Meitzen J, Moore IT, Lent K, Brenowitz EA, Perkel DJ (2007a) Steroid hormones act transsynaptically within the forebrain to regulate neuronal phenotype and song stereotypy. J Neurosci 27:12045–12057. https://doi.org/10.1523/Jneurosci.3289-07.2007

Meitzen J, Perkel DJ, Brenowitz EA (2007b) Seasonal changes in intrinsic electrophysiological activity of song control neurons in wild song sparrows. J Comp Physiol A 193:677–683. https://doi.org/10.1007/s00359-007-0222-1

Meitzen J, Weaver AL, Brenowitz EA, Perkel DJ (2009) Plastic and stable electrophysiological properties of adult avian forebrain song-control neurons across changing breeding conditions. J Neurosci 29:6558–6567. https://doi.org/10.1523/jneurosci.5571-08.2009

Merullo DP, Angyal CS, Stevenson SA, Riters LV (2016) Song in an affiliative context relates to the neural expression of dopamine- and neurotensin-related genes in male European starlings. Brain Behav Evol 88:81–92. https://doi.org/10.1159/000448191

Metzdorf R, Gahr M, Fusani L (1999) Distribution of aromatase, estrogen receptor, and androgen receptor mRNA in the forebrain of songbirds and nonsongbirds. J Comp Neurol 407:115–129

Morris D (1954) The reproductive behaviour of the zebra finch (*Poephila guttata*), with special reference to pseudofemale behaviour and displacement activities. Behaviour 6:271–322

Morton ES (2000) An evolutionary view of the origins and functions of avian vocal communication. Jpn J Ornitho 49:69–78. https://doi.org/10.3838/jjo.49.69

Mulligan JA, Olson KC (1969) Communication in canary courtship calls. In: Hinde RA (ed) Bird vocalisations. Cambridge University Press, London, UK, pp 165–184

Nealen PM (2005) An interspecific comparison using immunofluorescence reveals that synapse density in the avian song system is related to sex but not to male song repertoire size. Brain Res 1032:50–62. https://doi.org/10.1016/j.brainres.2004.10.059

Nespor AA, Lukazewicz MJ, Dooling RJ, Ball GF (1996) Testosterone induction of male-like vocalizations in female budgerigars (*Melopsittacus undulatus*). Horm Behav 30:162–169. https://doi.org/10.1006/hbeh.1996.0020

Nicholson EM, Koch L (1936) Songs of wild birds. H.F. & G. Witherby, London

Noble GK, Wurm M (1940) The effect of testosterone propionate on the black-crowned night heron. Endocrinology 26:837–850

Nordby JC, Campbell SE, Beecher MD (2001) Late song learning in song sparrows. Anim Behav 61:835–846. https://doi.org/10.1006/anbe.2000.1673

Nottebohm F (1980) Testosterone triggers growth of brain vocal control nuclei in adult female canaries. Brain Res 189:429–436. https://doi.org/10.1016/0006-8993(80)90102-X

Nottebohm F (1981) A brain for all seasons: cyclical anatomical changes in song control nuclei of the canary brain. Science 214:1368–1370

Nottebohm F, Arnold AP (1976) Sexual dimorphism in vocal control areas of the songbird brain. Science 194:211–213

Nottebohm F, Nottebohm ME (1978) Relationship between song repertoire and age in the canary, *Serinus canarius*. Z Tierpsychol 46:298–305. https://doi.org/10.1111/j.1439-0310.1978.tb01451.x

Nottebohm F, Stokes TM, Leonard CM (1976) Central control of song in the canary, *Serinus canarius*. J Comp Neurol 165:457–486. https://doi.org/10.1002/cne.901650405

Nottebohm F, Nottebohm ME, Crane L (1986) Developmental and seasonal changes in canary song and their relation to changes in the anatomy of song-control nuclei. Behav Neural Biol 46:445–471

Nowicki S, Ball GF (1989) Testosterone induction of song in photosensitive and photorefractory male sparrows. Horm Behav 23:514–525. https://doi.org/10.1016/0018-506x(89)90039-1

O'Loghlen AL, Rothstein SI (2002) Ecological effects on song learning: delayed development is widespread in wild populations of brown-headed cowbirds. Anim Behav 63:475–486. https://doi.org/10.1006/anbe.2001.1951

Odom KJ, Hall ML, Riebel K, Omland KE, Langmore NE (2014) Female song is widespread and ancestral in songbirds. Nat Commun 5:3379. https://doi.org/10.1038/ncomms4379

Ota N, Gahr M, Soma M (2015) Tap dancing birds: the multimodal mutual courtship display of males and females in a socially monogamous songbird. Sci Rep 5:16614. https://doi.org/10.1038/srep16614

Pavlova DZ, Pinxten R, Eens M (2007) Seasonal singing patterns and individual consistency in song activity in female European starlings (*Sturnus vulgaris*). Behaviour 144:663–680. https://doi.org/10.1163/156853907781347835

Pesch A, Güttinger H-R (1985) Der Gesang des weiblichen Kanarienvogels. J Ornithol 126:108–110. https://doi.org/10.1007/BF01640450

Petkov CI, Jarvis ED (2012) Birds, primates, and spoken language origins: behavioral phenotypes and neurobiological substrates. Front Evol Neurosci 4:12. https://doi.org/10.3389/fnevo.2012.00012

Phillmore LS, Hoshooley JS, Sherry DF, MacDougall-Shackleton SA (2006) Annual cycle of the black-capped chickadee: seasonality of singing rates and vocal-control brain regions. J Neurobiol 66:1002–1010. https://doi.org/10.1002/neu.20282

Pintér O, Péczely P, Zsebök S, Zelena D (2011) Seasonal changes in courtship behavior, plasma androgen levels and in hypothalamic aromatase immunoreactivity in male free-living European starlings (*Sturnus vulgaris*). Gen Comp Endocrinol 172:151–157. https://doi.org/10.1016/j.ygcen.2011.02.002

Pinxten R, De Ridder E, Balthazart J, Eens M (2002) Context-dependent effects of castration and testosterone treatment on song in male European starlings. Horm Behav 42:307–318. https://doi.org/10.1006/hbeh.2002.1824

Podos J, Huber SK, Taft B (2004) Bird song: the interface of evolution and mechanism. Annu Rev Ecol Evol Syst 35:55–87. https://doi.org/10.1146/annurev.ecolsys.35.021103.105719

Poulsen H (1951) Inheritance and learning in the song of the chaffinch (*Fringilla-coelebs L*). Behaviour 3:216–228. https://doi.org/10.1163/156853951x00278

Price JJ, Yunes-Jiménez L, Osorio-Beristain M, Omland KE, Murphy TG (2008) Sex-role reversal in song? Females sing more frequently than males in the streak-backed oriole. Condor 110:387–392. https://doi.org/10.1525/cond.2008.8430

Pröve E (1974) Der Einfluß von Kastration und Testosteronsubstitution auf das Sexualverhalten männlicher Zebrafinken (*Taeniopygia guttata castanotis Gould*). J Ornithol 115:338–347. https://doi.org/10.1007/BF01644328

Quispe Valdez R, Trappschuh M, Gahr M, Goymann W (2015) Towards more physiological manipulations of hormones in field studies: comparing the release dynamics of three kinds of testosterone implants, silastic tubing, time-release pellets and beeswax. Gen Comp Endocrinol 212:100–105. https://doi.org/10.1016/j.ygcen.2015.01.007

Quispe Valdez R, Sebe F, da Silva ML, Gahr M (2016) Dawn-song onset coincides with increased HVC androgen receptor expression but is decoupled from high circulating testosterone in an equatorial songbird. Physiol Behav 163:334–334. (vol 156, p 1, 2016)

Quispe Valdez R, Brazao Protazio JM, Gahr M (2017) Seasonal singing of a songbird living near the equator correlates with minimal changes in day length. Sci Rep 7:9140. https://doi.org/10.1038/s41598-017-08800-6

Rashotte ME, Sedunova EV, Johnson F, Pastukhov IF (2001) Influence of food and water availability on undirected singing and energetic status in adult male zebra finches (*Taeniopygia guttata*). Physiol Behav 74:533–541. https://doi.org/10.1016/S0031-9384(01)00600-X

Reeves BJ, Beecher MD, Brenowitz EA (2003) Seasonal changes in avian song control circuits do not cause seasonal changes in song discrimination in song sparrows. J Neurobiol 57:119–129. https://doi.org/10.1002/neu.10257

Ritchison G (1983) The function of singing in female black-headed grosbeaks (*Pheucticus-melanocephalus*). Family-group maintenance. Auk 100:105–116

Ritchison G (1986) The singing behavior of female northern cardinal. Condor 88:156–159

Riters LV (2012) The role of motivation and reward neural systems in vocal communication in songbirds. Front Neuroendocrinol 33:194–209. https://doi.org/10.1016/j.yfrne.2012.04.002

Riters LV, Ball GF (1999) Lesions to the medial preoptic area affect singing in the male European starling (*Sturnus vulgaris*). Horm Behav 36:276–286. https://doi.org/10.1006/hbeh.1999.1549

Riters LV, Ball GF (2002) Sex differences in the densities of alpha 2-adrenergic receptors in the song control system, but not the medial preoptic nucleus in zebra finches. J Chem Neuroanat 23:269–277

Ritschard M, Laucht S, Dale J, Brumm H (2011) Enhanced testosterone levels affect singing motivation but not song structure and amplitude in Bengalese finches. Physiol Behav 102:30–35. https://doi.org/10.1016/j.physbeh.2010.10.005

Robertson BA, Fontaine JJ, Loomis E (2009) Seasonal patterns of song structure variation in a suboscine passerine. Wilson J Ornithol 121:815–818. https://doi.org/10.1676/08-155.1

Robinson FE et al (1988) Steroidogenic relationships of gonadotrophin hormones in the ovary of the hen (*Gallus domesticus*). Gen Comp Endocrinol 69:455–466

Röhss M, Silverin B (1983) Seasonal-variations in the ultrastructure of the Leydig-cells and plasma-levels of luteinizing-hormone and steroid-hormones in juvenile and adult male great tits *Parus major*. Ornis Scand 14:202–212. https://doi.org/10.2307/3676154

Rost R (1990) Hormones and behavior - a joint examination of studies on seasonal-variation in song production and plasma-levels of testosterone in the great tit *Parus major*. J Fur Ornithol 131:403–411. https://doi.org/10.1007/Bf01639816

Rost R (1992) Hormones and behavior - a comparison of studies on seasonal-changes in song production and testosterone plasma-levels in the willow-tit *Parus montanus*. Ornis Fennica 69:1–6

Rowe L, Houle D (1996) The lek paradox and the capture of genetic variance by condition dependent traits. Proc Roy Soc B 263:1415–1421. https://doi.org/10.1098/rspb.1996.0207

Rybak F, Gahr M (2004) Modulation by steroid hormones of a "sexy" acoustic signal in an Oscine species, the common canary *Serinus canaria*. Acad Bras Cienc 76:365–367. https://doi.org/10.1590/S0001-37652004000200026

Saino N, Moller AP (1995) Testosterone correlates of mate guarding, singing and aggressive-behavior in male barn swallows, *Hirundo rustica*. Anim Behav 49:465–472. https://doi.org/10.1006/anbe.1995.0060

Saldanha CJ, Tuerk MJ, Kim YH, Fernandes AO, Arnold AP, Schlinger BA (2000) Distribution and regulation of telencephalic aromatase expression in the zebra finch revealed with a specific antibody. J Comp Neurol 423:619–630. https://doi.org/10.1002/1096-9861(20000807)423:4/619:Aid-Cne7/3.0.Co;2-U

Sandell MI, Smith HG (1997) Female aggression in the European starling during the breeding season. Anim Behav 53:13–23. https://doi.org/10.1006/anbe.1996.0274

Saunders AA (1947) The seasons of bird song the beginning of song in spring. Auk 64:97–107. https://doi.org/10.2307/4080066

Saunders AA (1948) The seasons of bird song: the cessation of song after the nesting season. Auk 65:19–30. https://doi.org/10.2307/4080221

Scharfman HE, MacLusky NJ (2005) Similarities between actions of estrogen and BDNF in the hippocampus: coincidence or clue? Trends Neurosci 28:79–85. https://doi.org/10.1016/j.tins.2004.12.005

Schlinger BA, Arnold AP (1991) Brain is the major site of estrogen synthesis in a male songbird. Proc Natl Acad Sci USA 88:4191–4194. https://doi.org/10.1073/pnas.88.10.4191

Schlinger BA, Remage-Healey L (2012) Neurosteroidogenesis: insights from studies of songbirds. J Neuroendocrinol 24:16–21. https://doi.org/10.1111/j.1365-2826.2011.02150.x

Schwabl H (1992) Winter and breeding territorial behavior and levels of reproductive hormones of migratory European robins. Ornis Scand 23:271–276. https://doi.org/10.2307/3676649

Schwabl H, Kriner E (1991) Territorial aggression and song of male European robins (*Erithacus Rubecula*) in autumn and spring - effects of antiandrogen treatment. Horm Behav 25:180–194. https://doi.org/10.1016/0018-506x(91)90049-N

Schwabl H, Wingfield JC, Farner DS (1980) Seasonal variation in plasma levels of luteinizing hormone and steroid hormones in the European blackbird (*Turdus merula*). Vogelwarte - Zeitschrift für Vogelkunde 30:283–294

Searcy WA, Andersson M (1986) Sexual selection and the evolution of song. Annu Rev Ecol Syst 17:507–533. https://doi.org/10.1146/annurev.es.17.110186.002451

Senft RA, Meddle SL, Baugh AT (2016) Distribution and abundance of glucocorticoid and mineralocorticoid receptors throughout the brain of the great tit (*Parus major*). PLoS One 11:e0148516. https://doi.org/10.1371/journal.pone.0148516

Shoemaker HH (1939) Effect of testosterone propionate on behavior of the female canary. Proc Soc Exp Biol Med 41:299–302. https://doi.org/10.3181/00379727-41-10652

Silverin B (1980) Effects of long-acting testosterone treatment on freeliving pied flycatchers, *Ficedula hypoleuca*, during the breeding period. Anim Behav 28:906–912. https://doi.org/10.1016/S0003-3472(80)80152-7

Silverin B, Viebke PA, Westin J (1986) Seasonal-changes in plasma-levels of Lh and gonadal-steroids in free-living willow tits *Parus montanus*. Ornis Scand 17:230–236. https://doi.org/10.2307/3676831

Slater PJB, Mann NI (2004) Why do the females of many bird species sing in the tropics? J Avian Biol 35:289–294. https://doi.org/10.1111/j.0908-8857.2004.03392.x

Smith GT, Brenowitz EA, Beecher MD, Wingfield JC (1997) Seasonal changes in testosterone, neural attributes of song control nuclei, and song structure in wild songbirds. J Neurosci 17:6001–6010

Snow DW (1988) *Prunella modularis* dunnock - social pattern and behaviour; voice. In: Cramp S (ed) Handbook of the birds of Europe, the Middle East and North Africa: the birds of the Western Palearctic, vol 5. Oxford University Press, New York, pp 553–556

Soma KK, Hartman VN, Wingfield JC, Brenowitz EA (1999) Seasonal changes in androgen receptor immunoreactivity in the song nucleus HVC of a wild bird. J Comp Neurol 409:224–236

Soma KK, Wissman AM, Brenowitz EA, Wingfield JC (2002) Dehydroepiandrosterone (DHEA) increases territorial song and the size of an associated brain region in a male songbird. Horm Behav 41:203–212. https://doi.org/10.1006/hbeh.2001.1750

Soma KK, Tramontin AD, Featherstone J, Brenowitz EA (2004) Estrogen contributes to seasonal plasticity of the adult avian song control system. J Neurobiol 58:413–422. https://doi.org/10.1002/neu.10288

Sorensen MC (2014) Singing in Africa: no evidence for a long supposed function of winter song in a migratory songbird. Behav Ecol 25:909–915. https://doi.org/10.1093/beheco/aru058

Sorensen MC, Jenni-Eiermann S, Spottiswoode CN (2016) Why do migratory birds sing on their tropical wintering grounds? Am Nat 187:E65–E76. https://doi.org/10.1086/684681

Sossinka R, Böhner J (1980) Song types in the zebra finch *poephila guttata castanotis*. Z Tierpsychol 53:123–132

Stouffer PC, Johnson EI, Bierregaard RO (2013) Breeding seasonality in central Amazonian rainforest birds. Auk 130:529–540. https://doi.org/10.1525/auk.2013.12179

Strand CR, Ross MS, Weiss SL, Deviche P (2008) Testosterone and social context affect singing behavior but not song control region volumes in adult male songbirds in the fall. Behav Process 78:29–37. https://doi.org/10.1016/j.beproc.2007.12.002

Suzuki K, Matsunaga E, Kobayashi T, Okanoya K (2011) Expression patterns of mineralocorticoid and glucocorticoid receptors in Bengalese finch (*Lonchura Striata Var. Domestica*) brain suggest a relationship between stress hormones and song-system development. Neuroscience 194:72–83. https://doi.org/10.1016/j.neuroscience.2011.07.073

Ter Maat A, Trost L, Sagunsky H, Seltmann S, Gahr M (2014) Zebra finch mates use their forebrain song system in unlearned call communication. PLoS One 9:e109334. https://doi.org/10.1371/journal.pone.0109334

Terkel AS, Moore CL, Beer CG (1976) The effects of testosterone and estrogen on the rate of long-calling vocalization in juvenile laughing gulls, *Larus atricilla*. Horm Behav 7:49–57

Thompson CK et al (2012) Seasonal changes in patterns of gene expression in avian song control brain regions. PLoS One 7:e35119. https://doi.org/10.1371/journal.pone.0035119

Thorpe WH (1958) The learning of song patterns by birds, with especial reference to the song of the chaffinch *fringilla coelebs*. Ibis 100:535–570. https://doi.org/10.1111/j.1474919X.1958.tb07960.x

Thorpe WH (1961) Bird song: the biology of vocal communication and expression in birds. Cambridge University Press, London

Thorpe WH, Pilcher M (1958) The nature and characteristics of subsong. Br Birds 51:509–514

Titus R (1998) Short-range and long-range songs: use of two acoustically distinct song classes by dark-eyed juncos (*Junvo hyemalis*). Auk 115:386–393

Tramontin AD, Brenowitz EA (2000) Seasonal plasticity in the adult brain. Trends Neurosci 23:251–258. https://doi.org/10.1016/S0166-2236(00)01558-7

Tramontin AD, Wingfield JC, Brenowitz EA (1999) Contributions of social cues and photoperiod to seasonal plasticity in the adult avian song control system. J Neurosci 19:476–483

Tramontin AD, Wingfield JC, Brenowitz EA (2003) Androgens and estrogens induce seasonal-like growth of song nuclei in the adult songbird brain. J Neurobiol 57:130–140. https://doi.org/10.1002/neu.10263

Tremain SB, Swiston KA, Mennill DJ (2008) Seasonal variation in acoustic signals of pileated woodpeckers. Wilson J Ornithol 120:499–504

Vallet E, Kreutzer M, Gahr M (1996) Testosterone induces sexual release quality in the song of female canaries. Ethology 102:617–628

Van Duyse E, Pinxten R, Eens M (2003) Seasonal fluctuations in plasma testosterone levels and diurnal song activity in free-living male great tits. Gen Comp Endocrinol 134:1–9. https://doi.org/10.1016/S0016-6480(03)00213-2

Van Duyse E, Pinxten R, Snoeijs T, Eens M (2005) Simultaneous treatment with an aromatase inhibitor and an anti-androgen decreases the likelihood of dawn song in free-living male great tits, *Parus major*. Horm Behav 48:243–251. https://doi.org/10.1016/j.yhbeh.2005.02.013

Van Hout AJM, Eens M, Balthazart J, Pinxten R (2009) Complex modulation of singing behavior by testosterone in an open-ended learner, the European starling. Horm Behav 56:564–573. https://doi.org/10.1016/j.yhbeh.2009.09.010

Van Hout AJ, Pinxten R, Darras VM, Eens M (2012) Testosterone increases repertoire size in an open-ended learner: an experimental study using adult male European starlings (*Sturnus vulgaris*). Horm Behav 62:563–568. https://doi.org/10.1016/j.yhbeh.2012.09.008

Van Roo BL (2004) Exogenous testosterone inhibits several forms of male parental behavior and stimulates song in a monogamous songbird: the blue-headed vireo (*Vireo solitarius*). Horm Behav 46:678–683. https://doi.org/10.1016/j.yhbeh.2004.06.011

Velez A, Gall MD, Lucas JR (2015) Seasonal plasticity in auditory processing of the envelope and temporal fine structure of sounds in three songbirds. Anim Behav 103:53–63. https://doi.org/10.1016/j.anbehav.2015.01.036

Voigt C, Gahr M (2011) Social status affects the degree of sex difference in the songbird brain. PLoS One 6:e20723. https://doi.org/10.1371/journal.pone.0020723

Voigt C, Leitner S (2008) Seasonality in song behaviour revisited: seasonal and annual variants and invariants in the song of the domesticated canary (*Serinus canaria*). Horm Behav 54:373–378. https://doi.org/10.1016/j.yhbeh.2008.05.001

Voigt C, Leitner S (2013) Testosterone-dependency of male solo song in a duetting songbird - evidence from females. Horm Behav 63:122–127. https://doi.org/10.1016/j.yhbeh.2012.10.006

Voigt C, Metzdorf R, Gahr M (2004) Differential expression pattern and steroid hormone sensitivity of SNAP-25 and synaptoporin mRNA in the telencephalic song control nucleus HVC of the zebra finch. J Comp Neurol 475:83–94. https://doi.org/10.1002/cne.20151

Voigt C, Leitner S, Gahr M (2006) Repertoire and structure of duet and solo songs in cooperatively breeding white-browed sparrow weavers. Behaviour 143:159–182. https://doi.org/10.1163/156853906775900739

Wada K et al (2013) Differential androgen receptor expression and DNA methylation state in striatum song nucleus area X between wild and domesticated songbird strains. Eur J Neurosci 38:2600–2610. https://doi.org/10.1111/ejn.12258

Walters MJ, Collado D, Harding CF (1991) Oestrogenic modulation of singing in male zebra finches: differential effects on directed and undirected songs. Anim Behav 42:445–452. https://doi.org/10.1016/S0003-3472(05)80043-0

Warren WC et al (2010) The genome of a songbird. Nature 464:757–762. https://doi.org/10.1038/nature08819

Weatherhead PJ, Metz KJ, Bennett GF, Irwin RE (1993) Parasite faunas, testosterone and secondary sexual traits in male red-winged blackbirds. Behav Ecol Sociobiol 33:13–23. https://doi.org/10.1007/Bf00164342

Wegrzyn E, Leniowski K (2009) Syllable sharing and changes in syllable repertoire size and composition within and between years in the great reed warbler, *Acrocephalus arundinaceus*. J Ornithol 151:255–267

Wickler W, Seibt U (1980) Vocal dueting and the pair bond. Z Tierpsychol 52:217–226. https://doi.org/10.1111/j.1439-0310.1980.tb00713.x

Wild JM (1997) Neural pathways for the control of birdsong production. J Neurobiol 33:653–670. https://doi.org/10.1002/(Sici)1097-4695(19971105)33:5/653:Aid-Neu11/3.0.Co;2-A

Wild JM (2017) The ventromedial hypothalamic nucleus in the zebra finch (*Taeniopygia guttata*): afferent and efferent projections in relation to the control of reproductive behavior. J Comp Neurol 525:2657–2676. https://doi.org/10.1002/cne.24225

Wingfield JC (1994) Regulation of territorial behavior in the sedentary song sparrow, *Melospiza melodia morphna*. Horm Behav 28:1–15. https://doi.org/10.1006/hbeh.1994.1001

Wingfield JC, Farner DS (1975) The determination of five steroids in avian plasma by radioimmunoassay and competitive protein-binding. Steroids 26:311–321

Wingfield JC, Farner DS (1978) The annual cycle in plasma irLH and steroid hormones in feral populations of the white-crowned sparrow, Zonotrichia leucophrys gambelii. Biol Reprod 19:1046–1056

Wingfield JC, Hahn TP (1994) Testosterone and territorial behaviour in sedentary and migratory sparrows. Anim Behav 47:77–89. https://doi.org/10.1006/anbe.1994.1009

Wingfield JC, Hegner RE, Lewis DM (1991) Circulating levels of luteinizing-hormone and steroid-hormones in relation to social-status in the cooperatively breeding white-browed sparrow Weaver, *Plocepasser Mahali*. J Zool 225:43–58

Yasukawa K, Boley RA, Simon SE (1987) Seasonal change in the vocal behavior of female red-winged blackbirds, *Agelaius phoeniceus*. Anim Behav 35:1416–1423. https://doi.org/10.1016/S0003-3472(87)80014-3

York JE, Radford AN, de Vries B, Groothuis TG, Young AJ (2016) Dominance-related seasonal song production is unrelated to circulating testosterone in a subtropical songbird. Gen Comp Endocrinol 233:43–52. https://doi.org/10.1016/j.ygcen.2016.05.011

Zann RA (1996) The zebra finch. Oxford University Press, Oxford

Chapter 8
From Vocal to Neural Encoding: A Transversal Investigation of Information Transmission at Long Distance in Birds

Solveig C. Mouterde

Abstract Acoustic communication in the natural world requires both emitter and receiver to adapt to the loss of information due to the transmission of sound in the environment. At the emitter's end, encoding information into propagation-resistant features may ensure its transmission on ecologically relevant distances. At the receiver's end, making sense of the degraded signal merged with added noise enables animals to produce behaviorally relevant responses. In this chapter, I use a songbird model to investigate both sides of the transmission chain, examining how individual vocal signatures are encoded, degraded and finally discriminated, with a particular emphasis on the receiver's neural encoding of degraded information. The cornerstone of this focal study is the use of naturally degraded vocalizations, combining various aspects of the challenges faced by animals performing auditory scene analysis, in order to address the complexity of real-life ecological constraints. While the individual signature of our songbird model, the zebra finch, is very resistant to propagation-induced degradation, single neurons in the avian auditory cortex have the ability to discriminate highly degraded individual vocal signatures, without prior familiarization or training. In the light of past and recent findings in birds and mammals, I delve further into significant insights uncovered by the research on neural processing and ensemble coding, and discuss the roles of perceptual plasticity and learning in the neural interface between brain and behavior.

Communication is the fabric of animal sociality. It is the foundation on which relationships between mates, rivals, or offspring develop, and population cohesion is maintained. For individuals to communicate properly, both sides of the

S. C. Mouterde (✉)
Equipe de Neuro-Ethologie Sensorielle ENES/Neuro-PSI, Université de Lyon/Saint-Etienne, Saint-Etienne, France

Department of Psychology, University of California, Berkeley, CA, USA

Institut de Recherche Expérimentale et Clinique (IREC), Université catholique de Louvain, Brussels, Belgium
e-mail: solveig.mouterde@uclouvain.be

T. Aubin, N. Mathevon (eds.), *Coding Strategies in Vertebrate Acoustic Communication*, Animal Signals and Communication 7,
https://doi.org/10.1007/978-3-030-39200-0_8

transmission chain must take an active part in the transfer of information. On one end, the emitter sends an acoustical message through a coded signal that can withstand some level of degradation induced by the propagation of sound through the natural environment. At the other end, the receiver(s) make(s) sense of a complex auditory stimulus (containing the potentially degraded initial signal merged with added noise) to extract the relevant information it contains, that is, to decode the signal. In this chapter, we will take a look at the whole picture, using a songbird model to investigate how the information encoded in the original signal is degraded through the environment, and how receivers deal with deciphering this altered information, both on neurological and behavioral levels.

8.1 Active Space and Encoded Information in Animal Vocalizations

In our endeavor, the notion of active space is interesting to consider because it encompasses all aspects of the communication chain. The active space of a signal is the distance from the source (emitter) over which the signal can be perceived by potential receivers (Brenowitz 1982); it is of fundamental importance in acoustic communication as it pertains to the biological relevance of propagated signals (Marler and Slabbekoorn 2004). The active space of a signal depends on its coding by the emitter, which may be the result of an adaptation to the species' lifestyle and environment; it is impacted by the acoustic constraints of the environment in which the signal propagates, which can vary in time (e.g., biotic noise, weather conditions); and it depends as well on the receivers' psychoacoustic abilities, in other words, on their capacity to perceive and decode the signal.

Within a given vocalization, the active space may also differ depending on the type of information being transmitted by the emitter. The white-browed warbler *Basileuterus leucoblepharus* provides a good illustration of this idea: the song of male warblers contains information about their species, as well as their individual identity ("individual signature," Aubin et al. 2004). Researchers found that for this bird living in the dense environment of the Brazilian tropical forest, species-specific information in the male song was encoded in a resistant acoustic feature that propagates at long distance (further than 100 m), while individually specific information degraded rapidly with propagation, restricting individual recognition to neighboring males (Mathevon et al. 2008). In this territorial species, being able to recognize the identity of neighboring males is of critical importance to mediate the male's responses to a song emitted in the vicinity; only a song perceived as being emitted by a stranger will elicit a response. Outside the territorial boundaries however, the transmission of species-specific information in the song is useful to inform potential intruders of the presence of a conspecific, but the added information about individual identity is not strictly necessary. The active space of each type of information in this species' song is thus well adapted to its ecological requirements.

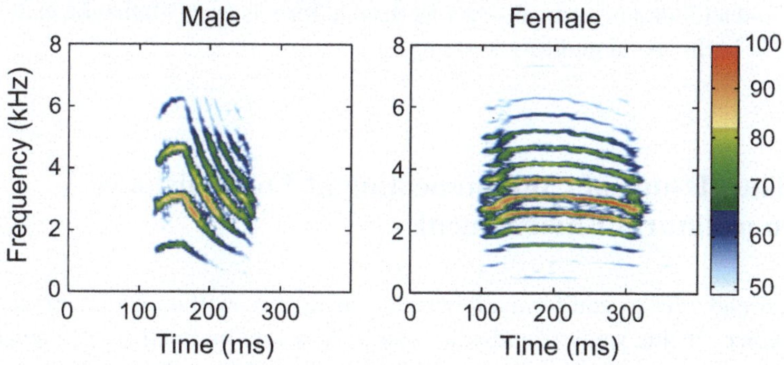

Fig. 8.1 Spectrograms of a male and female zebra finch distance call. The color scale is in relative dB as shown on the color bar, with 100 dB corresponding to the maximum amplitude observed

In other species, individual discrimination at long distance can be of critical importance, e.g., in mated pairs that need to remain within earshot when visual contact has been lost. This is the case for the zebra finch *Taeniopygia guttata*, a small gregarious songbird from subarid regions of Australia, living in large flocks in open grassy country with a scattering of trees and bushes (Zann 1996). Partners form strong pair bonds for life, and because these opportunistic breeders live in a very unpredictable environment, maintaining a strong pair bond between breeding events while living in large fission–fusion groups that are constantly on the move is of utmost importance. In this species, using a strong vocal recognition system could avoid the cost of partners losing each other. Within the zebra finch repertoire, the vocalization used for this purpose is the distance call, which was described by Richard Zann (1984) as a loud call that "will prevent members of a pair getting lost in vegetation but [it] probably serves to help them locate each other in flocks, in which contact between mates may easily be lost, especially when alarmed." The distance call of zebra finches is a complex sound emitted by both sexes, consisting of a harmonic series modulated in frequency as well as amplitude. It is sexually dimorphic (Zann 1984; Vicario et al. 2001), the males' fundamental frequency being higher than the females' (typically 650–1000 Hz vs. 500–600 Hz) as well as usually being shorter and more frequency modulated (Fig. 8.1). It has been shown that the distance call bears an individual signature, and that zebra finches are capable of call-based individual recognition (Zann 1984; Vignal et al. 2004, 2008). Regarding the active space of the individual signature in these calls, it has been estimated in earlier studies, based on naturalistic observations as well as theoretical analyses, to be up to around 100 m (Zann 1996; Lohr et al. 2003).

As a common thread throughout this chapter, I will use the vocal recognition system of zebra finches as a model to study the impact of propagation-induced degradation on information content, and investigate how the birds manage to decode the degraded signals, while working within realistic biological constraints faced by this species. For this purpose, I will focus on the individual identity encoded in distance calls (that is, the individual vocal signature of the calls), and how the fine spectral and temporal acoustical structure needed for this task (Mathevon et al. 2008) is encoded in the avian auditory system.

8.2 The "Issue" of Communicating at Long Distance in a Natural Environment

When a friend talks to you from a few meters away, you will immediately recognize his/her voice. If this same friend calls you from a few hundred meters away, you might hear that it is a man or a woman calling, in any case a human being, but not necessarily that this is your friend. Acoustic signals transmitted over long distances through the atmosphere not only degrade in amplitude, but also in their spectral and temporal structure, which may alter the original information (Wiley and Richards 1982; Forrest 1994). Intensity decreases as a result of spherical spreading, as well as excess attenuation due to the nonhomogeneous environment. Changes in the spectral characteristics of the signal can be due to masking ambient noise, atmospheric absorption, and environmental filtering, whereas changes in the temporal domain may be the result of the reflection, reverberation, and scattering of sound waves (Forrest 1994). These modifications in the signal structure increase with the propagation distance, which makes the transmission of information from emitter to receiver(s) more and more difficult.

As an illustration of the difficulty of this task in the context of individual discrimination, Fig. 8.2 shows the similarity between calls of pairs of different male zebra finches as a function of propagation distance, from 2 to 256 m. The similarity, which is shown here as the spectral correlation between calls, increases with distance along with the propagation-induced signal degradation and the decrease of the signal-to-noise ratio. As an example, and to further illustrate the increasing difficulty of discriminating between individuals at long distance, the spectrograms of the same calls from two different males are shown for various propagation distances (Fig. 8.2). Note however that this propagation-induced degradation of the signal is not solely a hindrance in animal communication, as receivers can use it to locate the emitter and estimate the propagation distance (a process called "ranging"; Naguib 1995; Fotheringham 1997; Holland 1998).

At the receiver's end, the challenge is to interpret this sensory information in the auditory system and display behavioral responses that are adapted to the situation and context. This ability to extract relevant information from a complex environment and to interpret it in terms of behaviorally relevant objects, otherwise known as scene analysis, is universal to species and sensory modalities (Lewicki et al. 2014). This process is especially complex in the auditory context where sounds from various sources and localizations, potentially degraded by transmission through the environment, combine to form a single acoustical signal. While a number of studies have highlighted the proficiency with which animals deal with this difficult task of discriminating what is relevant from what is not in degraded auditory signals

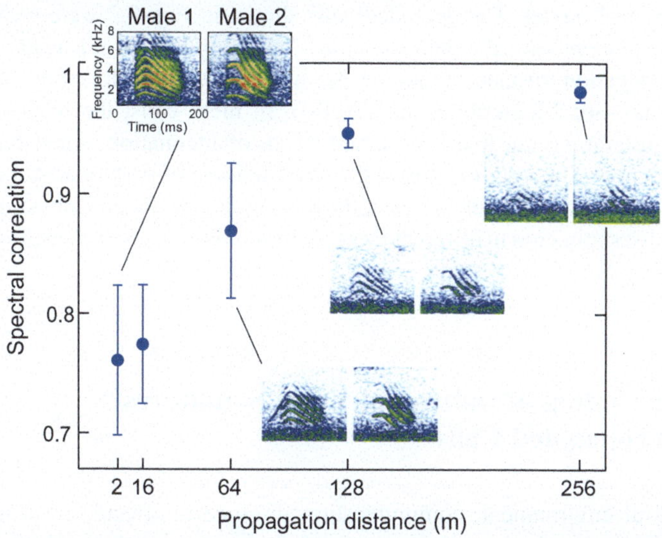

Fig. 8.2 Spectral correlations (i.e., similarity) between the distance calls of 17 pairs of male zebra finches, as a function of propagation distance. Mean correlations are represented for each distance, and error bars correspond to the standard deviation. Correlations were calculated between the mean frequency spectra of each male. As an example, the spectrograms of the same calls from two different males are shown for various distances (reproduced with permission from Mouterde et al. 2014a)

(Schnitzler and Flieger 1983; Aubin and Jouventin 2002; von der Emde 2004; Appeltants et al. 2005), understanding the underlying biological mechanisms involved in auditory scene analysis has proved particularly challenging for scientists (Shamma et al. 2011), especially on the neurophysiological level. In order to discriminate individual identity from vocalizations propagated at different distances, neurons must show a certain degree of perceptual invariance, that is, some stability in their responses to stimuli showing large variations in their acoustic parameters (Bregman 1993). A number of studies dealing with auditory scene analysis and individual discrimination or recognition in birds have revealed important insights on specific aspects of this question, such as invariance to intensity or background noise (Billimoria et al. 2008; Moore et al. 2013; Schneider and Woolley 2013), or individual recognition of undegraded signals (Chew et al. 1996; Gentner 2004). However, concern has been recently raised that studies on scene analysis do not address the complexity of the problems that need to be solved in natural settings; researchers were thus encouraged to focus more on the real-life issues faced by animals in their natural environment (Lewicki et al. 2014). Furthermore, since the auditory system of animals has evolved to process behaviorally relevant natural sounds (Mizrahi et al. 2014), using natural stimuli to investigate auditory scene analysis would be the best way to understand the neural computations at play (Theunissen and Elie 2014).

In the study presented as a common theme in this chapter, naturally degraded vocalizations were used in order to reflect ecologically relevant stimuli and behaviors for our songbird model. The degraded calls shown in Fig. 8.2 were recorded in a natural environment, on a flat field and in low wind conditions, at a height of 1.30 m and at propagation distances ranging from 2 to 256 m (for more details on the recordings, see Mouterde et al. (2014b)). In these calls, the reduction of the signal-to-noise ratio is due to the combined effects of attenuation, sound degradation, and ambient noise. In order to explore the real-life issue of partner acoustic recognition in our zebra finch model, we used these naturally degraded calls to investigate individual discrimination in different facets of the communication chain, from emitter to receiver.

8.3 The Coding of Individual Vocal Signatures in Propagated Calls

In the natural environment, communication distance is affected by a number of factors pertaining to the signal's emitter. An obvious one is the loudness of vocalizations, which may have evolved to serve a territorial function in the case of songs for example (Brenowitz 1982), and which can in any case vary due to proximal causes such as noise (Cynx et al. 1998; Brumm and Todt 2002). Active space can be enhanced (or conserved in adverse environmental conditions) by repeating the signal and thus increasing the redundancy of encoded information, or emitting the signal at a certain localization or time (Brumm and Naguib 2009). While these findings shed light on the emitters' capacity to adapt to various transmission conditions, investigating how the information is encoded in the spectro-temporal features of the vocalization itself, and how it is affected by long-distance propagation, is critical to understand how the receivers solve the issue of interpreting degraded signals at the other side of the communication chain. The first studies investigating the actual information content of propagated calls mostly dealt with species-specific information, providing significant insights into the link between birds' vocalizations and their habitat. Indeed, the coding of information in acoustic signals is generally based on features that are resistant to propagation, depending on the environmental acoustics that are characteristic of the habitat (Morton 1975; Mathevon and Aubin 1997). Generalities can be drawn: the most important acoustic features for species discrimination are usually found in a frequency bandwidth that is less degraded through propagation (Brenowitz 1982), and within this bandwidth, frequency modulation can mediate the discriminability of the information (Brémond and Aubin 1990; Mathevon et al. 1997). These observations form the basis for further investigations into the long-range degradation of finer grained information such as individual identity. In this respect, research has been rather scarce. Two previous studies using birds have been published on the subject; in the white-browed warbler, mentioned earlier in this chapter, the individual signature is encoded in songs in a succession of pure tones decreasing in frequency (Aubin et al. 2004; Mathevon et al. 2008). Conversely, in the male corncrake *Crex crex*, individuality is coded in propagated calls in the inter-pulse duration (Rek and Osiejuk 2011). In these examples, individual signatures are encoded in either the spectral or the temporal domain, with pure tones or pulses.

Let us now consider the propagation-induced information loss in complex sounds, i.e., in vocalizations displaying wide frequency bandwidths together with amplitude and frequency modulations. The conundrum in this case is to calculate the appropriate parameters for the acoustic analysis: in calls propagated at long distance, it is not possible to describe precisely the spectro-temporal features such as parameters following closely the fundamental frequency and its variation in time. These parameters, frequently used in studies investigating individual signatures (Guyomarc'h et al. 1998; Naguib et al. 2001; Vignal et al. 2008), are impossible to calculate here with reasonable precision, with the fundamental frequency progressively disappearing within the background noise, and the low resolution of the temporal characteristics hindering the calculation of even basic measurements such as the calls' duration. In the first step of our focal study, this challenge was tackled by performing a sound transmission experiment using zebra finch calls, with the aim of characterizing and quantifying the degradation of the calls' individual signature during propagation. To this end, 16 exemplars of distance calls from each of 16 female and 16 male individuals (that is, 512 calls) were recorded at five propagation distances, from 2 to 256 m. The calls were then characterized using two distinct sets of acoustic parameters. The first set, which will be called "envelope parameters", described separately the spectral envelope (amplitude in the spectral domain) and the temporal envelope (amplitude in the temporal domain) of each call. The second set used the complete spectrogram, in an attempt to circumvent the use of subjective assumptions on the nature of information-bearing acoustical features in the calls. A principal components analysis was used to reduce the high dimensionality to a manageable set of parameters describing the variability of the calls, subsequently called the "spectrogram principal component" or "SPC" parameters. To test for the presence of an individual acoustic signature in the propagated calls, discriminant function analyses were performed separately for each sex and each propagation distance (Mouterde et al. 2014b). These analyses yielded percentages of correct classification of calls (PCC), which quantified the discriminability of different individuals within each sex and for each distance; a percentage value over the chance level (1 over 16 individuals) shows the existence of individualized acoustic features in the calls.

We found that the individual vocal signature encoded in the distance calls of zebra finches is remarkably resistant to propagation-induced sound degradation. Figure 8.3 shows the percentages of correct classification, that is, the individual discriminability, as a function of propagation distance, for each set of parameters and each sex.

While the individual discriminability of calls logically decreases with distance, it is still well over the chance level at 128 m for both sets of parameters, and up to 256 m for the SPC parameters. SPC parameters yield higher discriminability values than envelope parameters, especially at longer distances, which can be explained by the fact that while the envelope parameters take spectral and temporal features into

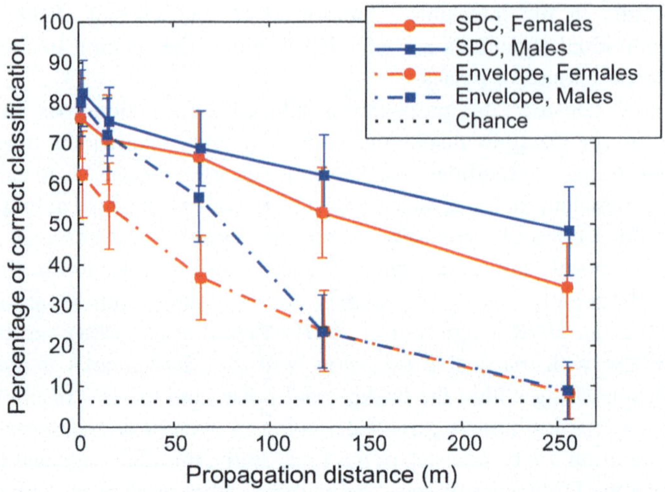

Fig. 8.3 Mean percentages of correct classification (i.e., individual discriminability) obtained for each propagation distance and both sexes. SPC parameters are represented as solid lines and envelope parameters as dash-dot lines. Standard deviations are indicated. The chance level, corresponding to 6.25% of correct classification, is shown as a horizontal dotted line (reproduced from Mouterde et al. 2014b)

account separately, SPC parameters are extracted from the full spectrogram, thus also describing the joint spectro-temporal variations in the calls such as up- or down-sweeps (Mouterde et al. 2014b). We also found that at short distances, both sets of parameters yield similar discrimination performances for males (the performance of the envelope parameters decreasing rapidly at 64 and 128 m), while for females the envelope parameters lead to noticeably lower discriminability values. Hence, the added information available in the SPC parameters is not indispensable for discriminating between males at short distances, the information extracted separately from the spectral and temporal domains (envelope parameters) being sufficient to yield equally high discrimination performances. This difference in coding properties is well illustrated in Fig. 8.4, which shows the first three discriminant functions (DFs) calculated in the analysis for the SPC parameters, represented in the spectrographic space at short, medium, and long distance. In other words, Fig. 8.4 is a description of the most important features in the spectrogram that can be used to discriminate between individuals at various distances: red bands show features whose presence is important for individual discrimination, while blue bands show features whose absence is important. One can observe that while the available frequency bandwidth shrinks with distance, fine spectral cues (as seen on the stacks of red and blue narrow bands) are useful at all distances, and for both sexes. Interestingly, at short and medium distance, the DFs pertaining to male calls do not look at all like the spectrographic representation of an average male call, the faster and more predominant down-sweep component in particular (seen on Fig. 8.1) being absent for short and medium distances and only appearing at 128 m (not shown on Fig. 8.4) and 256 m (DF2 and DF3, Fig. 8.4).

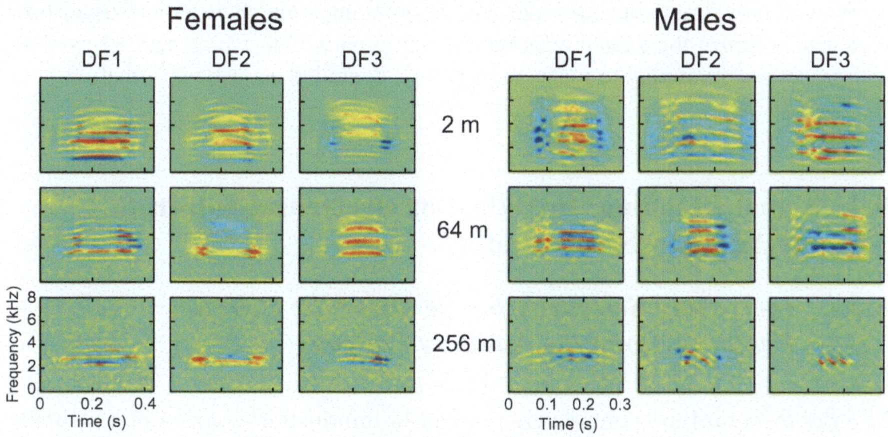

Fig. 8.4 Representation of the discriminant functions (DFs) projected into the spectrographic space for the SPC parameters. The first three DFs, obtained from the SPC parameters, are shown for females (left) and males (right). Each row indicates the propagation distance used to perform the discriminant function analysis. Positive frequency bands are shown in red and negative bands in blue. This representation describes the most important features in the spectrogram that can be used to discriminate between individuals at various distances (adapted with permission from Mouterde et al. 2014b)

This corroborates our previous conclusions, in the sense that while information about frequency modulation is not necessary to discriminate between males at close range, it becomes critical at long distance in order to maintain individual discriminability above the chance level. Thus, while it was already known that the downsweep component of male calls is highly individualized (Vignal et al. 2008), this study shows that the same level of individualization can be found at close range in male calls using parameters describing only its energy spectrum and its temporal variation in amplitude. This redundancy in the coding of the individual signature, clearly shown here for the male calls, is also found for the female calls of zebra finches, albeit to a lesser extent (Mouterde et al. 2014b); indeed, in females, the information about spectro-temporal modulations is important at all distances to obtain higher discrimination performances.

In a real-life situation, redundant coding of information within a single short call can make all the difference in terms of communicating efficiently in ever-changing conditions. Let me be clear: the furthest propagation distance (256 m) advertised here as still enabling the transmission of individualized information in the distance calls of zebra finches is the result of a transmission experiment performed on a specific day, with specific weather conditions and background noise. While testing at longer distances might have pushed the boundary of theoretical active space a little further, tougher recording conditions (wind, biotic noise) could have drastically reduced it (Brumm and Naguib 2009). In this respect, redundantly coded information in vocalizations may help maintain efficient communication at shorter distances despite the varying conditions in the environment in which communication takes place.

We will now turn to the other side of the communication chain to investigate how receivers optimize the gained information and in particular, which neurophysiological processes are at play to make sense of this degraded acoustic information.

8.4 Signal Decoding: Investigating the Neural Substrate for Auditory Scene Analysis

8.4.1 Increasing Selectivity for Behaviorally Relevant Features Along the Auditory Pathway

In order for a bird to appropriately respond to information encoded in a naturally propagated vocalization, it must have the neural ability to make sense of this auditory input. This process involves discriminating the relevant features in perceived sounds, that is, showing a higher selectivity for the type of information that is relevant to the animal. Indeed, as auditory information from the sensory system ascends the avian auditory pathway through the midbrain, thalamus and into the auditory cortex, itself divided into a primary (field L complex, a region that is analogous to the primary auditory cortex of mammals) and a secondary region (caudal mesopallium or CM; caudomedial nidopallium or NCM), neurons show an increasing selectivity in their responses for natural sounds (Theunissen et al. 2004; Woolley 2006). In particular, high selectivity for conspecific songs has been found in both primary and secondary regions of the auditory cortex (Grace et al. 2003; Hsu et al. 2004; Theunissen and Shaevitz 2006; Hauber et al. 2007). These findings support a hierarchical view of auditory processing, in which neurons in the lower pathway tune to simple low-level acoustic features and send their combined inputs to the forebrain neurons, which in turn detect more complex natural sound features (Woolley et al. 2009; Meliza et al. 2010; Theunissen and Elie 2014). This hierarchical sensory processing enables the representation of sounds as auditory objects (i.e., stable perceptual units), reflecting the system's transformation of an acoustic waveform from low-level sensory representation into perceptual representations that are behaviorally relevant to the organism (Christison-Lagay et al. 2015). A number of studies suggest that further hierarchical processing occurs within the auditory cortex (Meliza and Margoliash 2012), with secondary auditory areas encoding stimulus surprise (Gill et al. 2008) or playing a major role in learned auditory discriminations (Gentner 2004; Pinaud and Terleph 2008; Jeanne et al. 2011).

In the context of auditory scene analysis, signal decoding also implies extracting the biologically relevant information content from the noise in the perceived sound, whether the decrease in signal-to-noise ratio (SNR) stems from the noise level itself or the reduction of signal intensity due to propagation. This process starts in the peripheral auditory system, in which a selective frequency tuning process breaks complex sounds into their frequency components (Fay and Popper 2000). This tonotopic organization can improve the SNR for the representation and detection of acoustic signals by reducing the masking interference from background noise when signal and noise are in different frequency bandwidths (Klump 1996; Fay and Popper 2000). Further along the ascending pathway, neurons in Field L have been shown to adjust their spectral tuning properties to the characteristics of background noise, enhancing their frequency selectivity as a result (Nieder and Klump 1999). This dynamic sharpening of auditory filters in the auditory forebrain of birds echoes the idea, also investigated in mammals, that higher-level neural processing enables an increased tolerance for noise (Rabinowitz et al. 2013). Indeed, two electrophysiology-based studies on zebra finches found noise-invariant coding in neurons in NCM responding to familiar and unfamiliar songs from conspecifics (Moore et al. 2013; Schneider and Woolley 2013). Another study described a population of field L neurons able to discriminate between bird songs while being invariant to intensity, showing a link between invariance and discrimination performance (Billimoria et al. 2008). This increased tolerance in the auditory forebrain for variations in the signal's characteristics would indeed be an important factor of success in the task of discriminating between individuals at a distance in a natural environment.

8.4.2 Discrimination of Naturally Degraded Individual Vocal Signatures in Single Cortical Neurons

In these studies above, the overall acoustical quality of the vocalizations used as stimuli was preserved, and thus the question of how the songbird brain deals with the impact of propagation-induced degradations as naturally experienced by the animals when communicating at long range was yet to be investigated. Using the previously described database of propagated distance calls, we tested zebra finches' brains with some rather tough questions: are there single auditory neurons that are able to discriminate such fine-detailed information as individual identity in short vocalizations like distance calls, using not simply individual calls but multiple call exemplars from each individual, after they have been naturally degraded through propagation? Is such a discrimination capacity possible spontaneously, without prior experience of long-range propagation or prior knowledge of the individuals used as stimuli, and with little help from higher order neural processes such as attentional mechanisms? To answer these questions, extracellular electrophysiological recordings were performed on anaesthetized zebra finches, spanning the whole auditory forebrain, to test whether neurons from these aviary-reared subjects could discriminate individual identity in unfamiliar degraded distance calls of males or females (stimuli from each sex being tested separately, on different recording sites). From this multiunit data we identified single auditory units and analyzed their responses (Mouterde et al. 2017). We found neurons showing high discrimination of the identity of unfamiliar vocalizing birds, at short as well as long distances: these neurons are able to code distinct individuals differently, and maintain some aspects of this code across all tested distances, from 2 to 256 m (Fig. 8.5). As an example, a

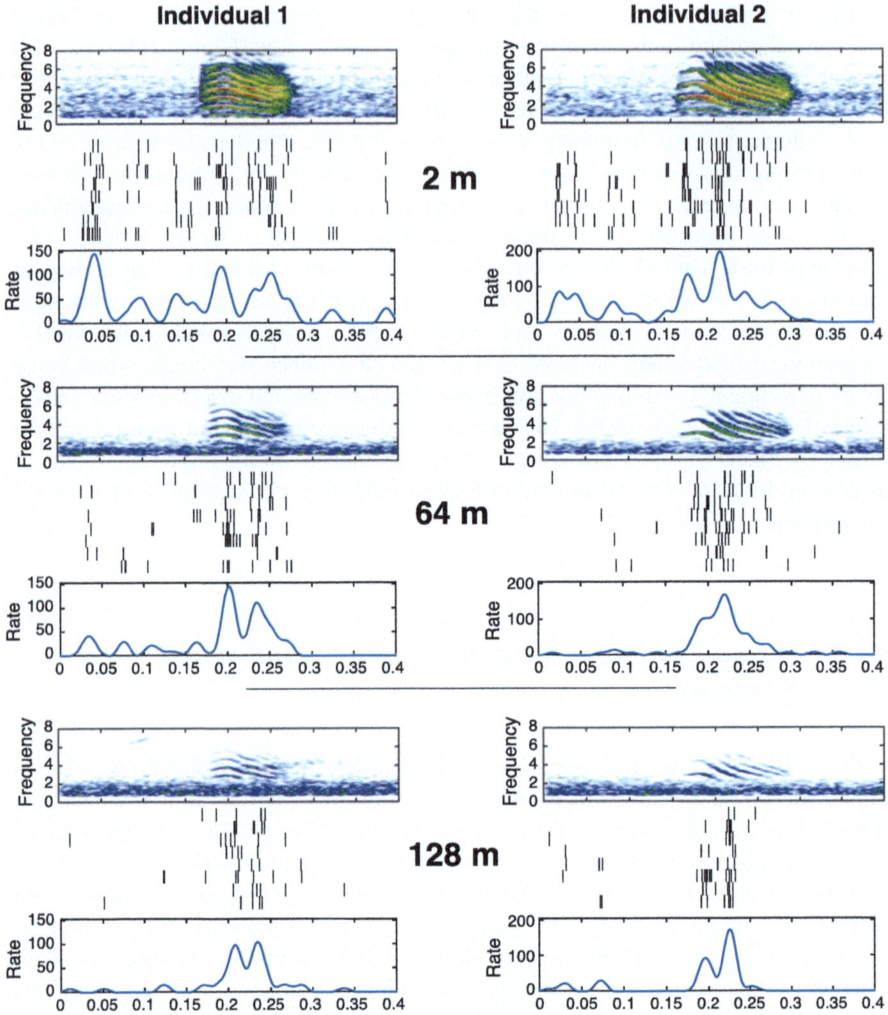

Fig. 8.5 Comparison of the responses of a unit highly discriminative for the signature of individual vocalizers to the call of two different males, at three tested distances. For each propagation distance, the spectrogram of the stimulus call is shown on top, followed by the spike trains for the eight recorded presentations, and by the peristimulus time histogram averaging these eight presentations. Sound frequency is given in kilohertz and rate in spikes per second (adapted with permission from Mouterde et al. 2017)

high-performance neuron discriminating between multiple calls from four different males showed a percentage of correct classification (PCC) over the chance level of 66.8% at 2 m, and 34.4% at 256 m (the chance level being 25% in this case). This impressive ability would require a high degree of invariance to decreasing signal-to-noise ratio and/or sound intensity, as well as the ability to extract parameters encoding individual identity in calls from previously unknown vocalizers.

In order to disentangle the overall effect of sound propagation through natural environment from the mere effect of intensity decrease, we compared the units' responses to naturally degraded calls to their responses to synthetic calls that mimicked the intensity decrease of the naturally degraded calls, but with the same high SNR as the one observed for the calls recorded at 2 m. We found that these units are remarkably invariant to the signal degradation and to large decreases in SNR induced by long-range propagation, the decline in discrimination performance being mostly a result of the intensity decrease (Mouterde et al. 2017). Interestingly, we did not find that the neural substrate for the discrimination of individuals in degraded calls was linked to specific traditional auditory areas (field L, NCM etc.), but we did find a distinctive spatial distribution of these neurons, the units most discriminative of individual identity being mostly found in superficial as well as deep regions of the auditory cortex.

Looking further into the coding properties of these discriminant neurons, we examined the spikes (or action potentials) which fired in response to the propagated calls, and calculated for each unit the length of the encoding time window yielding the highest discrimination performance for individual identity; the encoding time window is defined as the duration of the spike train assumed to correspond to a single symbol in the neural code (Theunissen and Miller 1995). This analysis also enabled us to investigate different temporal resolutions for the neural code depending on the length of the time window, exploring at one end the mean firing rate over the stimulus (rate code), and on the other end the fine temporal structure of the spike train, that is, the actual patterns of spikes within the encoding time window (temporal code). We found that while the average spike rate during the whole stimulus presentation was related to the discrimination performance, the precise timing of the spike patterns within a shorter encoding time window provided more information regarding the coding of individual identity in degraded calls. This observation is in line with a number of studies showing that neurons use a spike timing strategy to encode vocalizations (Narayan et al. 2006; Huetz et al. 2011; Gaucher et al. 2013; Lopes-dos-Santos et al. 2015). Furthermore, sensory neurons have been shown to encode different stimulus attributes on different temporal scales, resulting in a temporal multiplexing of information (Panzeri et al. 2010). We investigated how the units coded for propagation distance as well as individual identity, and indeed found units coding for both types of information, using respectively longer (~30 ms) and shorter (~10 ms) encoding time windows (Mouterde et al. 2017). This multiplexing ability enhances the encoding capacity of neural responses, enabling disambiguation of stimuli that cannot be discriminated at a single timescale and making sensory representations more stable in regard to variability. Temporal multiplexing is thus acknowledged as a strategy used by the brain to create information-rich and stable representations of noisy and variable environments (Panzeri et al. 2010), and could be a key factor in efficiently decoding the individual signature in propagated vocalizations.

It is interesting to compare the performance level of the most discriminant neurons to the performance obtained in the acoustic analysis previously described. For males for example (and using SPC parameters, see Sect. 8.3), the mean PCC over the chance level is 76.4% at 2 m and 42.1% at 256 m following the acoustic analysis, compared to respectively 66.8% and 34.4% for the most discriminant neuron tested with male calls. Although these values correspond to two different assessments of discriminability (statistical discrimination from the calls' acoustical structure versus between-calls discrimination ability by one particular neuron), this comparison implies that some neurons are able to take the best advantage of the information still remaining after long-distance propagation. Indeed, these observations highlight the fact that neurons may be able to retrieve a comparable, if somewhat lower amount of information at long distance than what was theoretically calculated using acoustical features optimized for that distance. These results show that complex processing of higher order acoustic features occurs at the level of the single neuron.

8.4.3 Ensemble Coding: The Auditory System as a Neural Network

We will now take a step back and consider the brain, processing information through a network of interconnected neurons, each of which carrying complementary information about the auditory stimuli. While single neurons can encode information on different timescales, ensembles of neurons can also process information on different spatial scales, their synchronization properties potentially generating an additional layer of sensory processing (Quian Quiroga and Panzeri 2009; Panzeri et al. 2010). Investigating the population code, defined as a neural representation in which information is conveyed by relative amounts of activity across multiple elements of an array (Covey 2000), has thus become the focus of a number of neurophysiological studies. Feng and Ratnam (2000) stated that an important and unresolved issue was to determine at which level of complexity certain features are represented in single neurons while others are represented across groups of synchronized neurons. It has been found since that information processing by single neurons can reach a level of complexity that had not been shown before in the context of auditory scene analysis (Mouterde et al. 2017). Of course, since each individual neuron receives excitatory and inhibitory inputs from a population of neurons that are lower in the ascending pathway, the analysis of single neurons in the auditory cortex is in itself a "readout" of a population code, each neuron's response being a computation of all these convergent inputs (Covey 2000). Taking into account the coordinated activity of a population of cortical neurons, however, could provide access to higher-order levels of sensory information and enhance the information-carrying capability of the auditory cortex (Woolley 2006; Kayser et al. 2009). Depending on the tuning curves of the neural population involved, a population code would either increase the range of stimulus features encoded by that population, or lead to a better discrimination of stimulus features (Panzeri et al. 2015). Previous studies on mammals suggested that ensemble coding could be critical to various auditory processes, such as auditory stream formation (defined as a series of sounds that is perceived by the listener as a coherent entity; Fishman et al. 2001; Shamma et al. 2011), sound source localization (Fitzpatrick et al. 1997; Miller and Recanzone 2009) or sound level encoding (Dean et al. 2005). Another interesting observation is that the brain encodes sensory information using a small number of active neurons at any given point in time, this strategy being referred to as "sparse coding" (Olshausen and Field 2004). Regarding auditory scene analysis, a spatially distributed population code, much like temporal multiplexing, can stabilize the sensory representation to external or internal noise (Kayser et al. 2009).

In birds, the representation of complex sounds by synchronized activation of groups of neurons has been observed in the song system (Margoliash et al. 1994) and the encoding of unexpected auditory events (Beckers and Gahr 2012). Further research is needed to analyze ensemble codes in the field of sensory processing; in particular, it remains to be determined how further gains in the neural discrimination of identity in naturally propagated calls could be achieved by examining the population code. I hypothesize that we could estimate the stimulus intensity from the ensemble responses of cortical neurons; as explained earlier, the most discriminant units for individual identity in degraded calls that we found were mostly affected by intensity decrease while being highly invariant to large decreases in SNR and degradations of the signals' spectro-temporal features (Mouterde et al. 2017). Investigating the neural population code could improve the sound level coding accuracy (Dean et al. 2005). In recent years, a number of new analysis techniques for information processing in neural networks have been developed, which would be of interest to researchers aiming to study these questions further (Mahmud and Vassanelli 2016). In addition to spatial population coding (including accounting for the various encoding timescales of the single neurons encompassed in the recorded population), imaging techniques such as functional magnetic resonance imaging have proved useful (Boumans et al. 2008), as well as the analysis of mass signals such as local field potentials (LFPs). LFPs are extracellularly recorded potentials with frequencies of up to ~500 Hz that reflect the slow fluctuations of neural activity of local cortical networks, thus providing an aggregate measure of local sub-threshold integrative processes and network state changes that are not reflected in spike trains (Kayser et al. 2009; Einevoll et al. 2013). It has been found that combining the analyses of simultaneously recorded LFPs and spikes provided more information about cortical activity (Quian Quiroga and Panzeri 2009; Einevoll et al. 2013). In particular, the nested combination of spike patterns with the phase of low-frequency network rhythms was highlighted as an especially promising code, as it was found to be most informative and provided robustness to noise (Kayser et al. 2009). Slow cortical rhythms thus help stabilize neural codes to the detrimental effects of sensory noise, and taking into account such nested combinations of codes in the study of auditory scene analyses might majorly increase our understanding of the processes involved in the discrimination of individual signatures in naturally degraded vocalizations.

As tempting as it may be, however, to think that we are now close to finding the key to how the brain as a neural network decrypts information encoded in sounds, and to predicting reliably the subsequent behavioral responses, some important issues need to be addressed. First, given the sheer size of the auditory system's network of interconnected neurons, most decoding algorithms may fail to decode stimuli owing to a high-dimensional response space or the use of incorrect assumptions about the neuronal code actually used (Quian Quiroga and Panzeri 2009). While recent studies have focused on methods for dimensionality reduction (Cunningham and Yu 2014), interesting findings suggest that a small-dimensional subspace of the experimentally measured activity may be sufficient to explain the population dynamics underlying sensory processing (Bathellier et al. 2012; Panzeri et al. 2015). Second, neural responses are sensitive to context, which constitutes a substantial source of variability. This may relate to sounds that have occurred at specific times prior to the stimulus under consideration (Covey 2000; Mizrahi et al. 2014); an illustration of this sensitivity to context is the fact that we are better at discriminating frequency relationships (e.g., recognizing a melody regardless of the key in which it is played) than we are at making absolute frequency judgments, (e.g., having absolute pitch). Context also relates to the internal state of the brain, which may or may not be mediated by the animal's environment. Cortical states can vary along a continuum of synchronized and desynchronized states. A synchronized state (commonly observed during slow-wave sleep and anesthesia) is characterized by slow fluctuations between intrinsically generated up and down states, while in a desynchronized state (commonly observed during sensory processing in awake animals), activity is strongly modulated by sensory inputs (Pachitariu et al. 2015). It has been found in mammals that variations in brain state accounted for a significant amount of the variability in population activity, the cortical state modulating the selectivity and reliability of spike patterns and impacting the fidelity of the population code as a result (Curto et al. 2009; Pachitariu et al. 2015). This cortical state can vary with behavioral and cognitive states, and may play a key role in shaping the responses of the awake cortex to sensory stimuli (Curto et al. 2009). In anaesthetized animals, the internal synchronous activity of neural networks has been found to be involved in the neural processing of context-dependent auditory information in the secondary auditory areas of zebra finches (Beckers and Gahr 2012). Since the behavioral state has a major impact on the neural interpretation of auditory stimuli, and since a neuronal code yielding high information values might not be biologically relevant in the sense that neural systems might not be capable of exploiting all of this information (Quian Quiroga and Panzeri 2009; Nagel et al. 2011), it is critical to understand how neural activity in sensory cortices relates to perception, so as to bridge the gap between neural processing and behavior. This is how we will come closer to understanding the neural interface between animals and their environment.

8.5 In Search of the Neural Interface Between Brain and Behavior

8.5.1 The Role of Conscious Perception on Neural Processing

We will now turn toward the role of attentional and learning mechanisms in the neural encoding of auditory information. In the zebra finch study described earlier (see Sect. 8.4.2), the subjects were anaesthetized with urethane, which, as we saw above, would induce a different brain state than that of awake animals. What, then, would be the potential benefits of the attentional mechanisms found in awake animals in terms of discrimination of auditory objects? A number of studies have compared neural activity in both awake and anaesthetized songbirds in response to vocalizations, and produced controversial results. Studies examining the impact of urethane anesthesia on the midbrain (Schumacher et al. 2011), primary (Narayan et al. 2006) and secondary auditory cortex (Meliza et al. 2010) of zebra finches have found that while some parameters (such as neural excitability, spike precision or inter-trial correlation) differed between anaesthetized and awake states, anesthesia did not affect the discrimination of songs or selectivity for song motifs. In contrast, other studies using songbirds found important changes in neuronal preferences between both states (Capsius and Leppelsack 1996; Schmidt and Konishi 1998; Karino et al. 2016). In an effort to address this apparent discrepancy, Karino et al. (2016) argued that anesthesia might modify the balance of neuronal preference between behaviorally significant auditory signals and nonsignificant sounds; vigilance and attention may thus be required in order to produce the appropriate behavioral responses toward particularly meaningful vocalizations. Interestingly, in a study investigating neurons in the secondary auditory cortex of anaesthetized zebra finches that responded preferentially to unexpected sounds (on the basis of recent stimulus history), Beckers and Gahr (2012) found an internally synchronized neural network that may reflect an early-stage process involved in the involuntary capturing of attention. In any case, attentional processes might indeed be useful in the task of extracting individual signatures from degraded and potentially noisy signals, which is faced by animals in their natural environment. In humans, Christison-Lagay et al. (2015) found that selective attention acts by enhancing the cortical representations of the attended sound streams, the population-level activity in the presence of competing sound streams reflecting primarily the activity in response to the attended stream in isolation compared to that of the ignored stream.

These observations spark the debate on the relative involvement of ascending vs. centrifugal pathways in the perception of behaviorally relevant auditory objects. On one hand, the neural representation of sounds is transformed systematically along the ascending auditory pathway, leading to integration in time, frequency, and space domains and allowing the extraction of behaviorally relevant features (Feng and Ratnam 2000). This "bottom-up" process is driven predictably by patterns of acoustic variations in the stimulus (Gentner 2004). On the other hand, this information flow is subject to dynamic modulation by a descending auditory pathway that runs from the cortex to the cochlea and which is largely independent of any particular acoustic feature (Feng and Ratnam 2000; Gentner 2004). These "top-down" mechanisms enhance the response of some neurons while decreasing the response of others (Jen et al. 1998), and may thus form the basis for attention, motivation, and/or reward mechanisms, enabling animals to acquire information about one or two auditory objects while ignoring others. The role of this centrifugal pathway is particularly useful for the extraction of signals in difficult listening conditions (Feng and Ratnam 2000). In this light, the interpretation of an auditory scene consists of a selection of particular auditory objects through a combination of bottom-up object salience and top-down attention, filtered by experience and expectation (Middlebrooks and Simon 2017).

While we have found that bottom-up processes can discriminate very fine-detailed information such as information about individual identity and propagation distance in highly degraded calls (Mouterde et al. 2017), the next step is to investigate how top-down processes, in the form of conscious perception, can further improve sensory processing. Can a bird learn to better discriminate degraded information in these signals over time?

8.5.2 Perceptual Learning: Pushing the Limits of Discrimination

Perceptual learning, a manifestation of experience-dependent plasticity in the sensory systems, occurs at the developmental stage as well as throughout adult life (Seitz and Dinse 2007; Dahmen and King 2007). It involves improved sensitivity independent of cognitive, motor or other, non-perceptual factors (Gold and Watanabe 2010). Interestingly, while attention and behavioral reinforcement mediate the selection and learning of only the information that is deemed to be of importance, perceptual learning has been found to occur not only under training conditions but also in situations that lack attention and reinforcement (Seitz and Dinse 2007). As an example, unreinforced learning of song discrimination has been shown to occur rapidly in zebra finches, after 3 hours of passive song exposure (Stripling et al. 2003). Seitz and Dinse (2007) have thus suggested that in order for learning to occur, sensory stimulation needs to be sufficient to drive the neural system past the point of a learning threshold. The added “layer” of attention and conscious perception however seems necessary for memory consolidation (Gilbert et al. 2001). The study of perceptual plasticity at the receiver’s level has been generally overlooked in auditory scene analysis, comparatively to the extensive research that has been conducted on the vocal plasticity of emitters in the context of difficult communication conditions (Pohl et al. 2012; Slabbekoorn 2013). While perceptual learning may increase the discrimination of specific acoustic features, how efficient can it be when approaching the limit of sensory perception? In other words, at which point will the inherent limits of the auditory system hinder any potential improvements due to perceptual learning?

As a last step in our focal study on the discrimination of degraded vocal signatures in zebra finches’ calls, we investigated the role of experience in the discrimination performance of highly degraded calls. Adult female zebra finches were used in forced-choice conditioning experiments, in which the results of two different protocols were compared. In both protocols, the females were asked to discriminate between the propagated calls of two unfamiliar males. In the first protocol, the subjects learned to discriminate between a number of calls of two males recorded at short range before being systematically challenged with the calls of the same males recorded at longer distances. In the second protocol, the subjects were challenged daily with a different pair of males and with a randomly selected propagation distance, and thus did not have the possibility to learn from their previous experience in the task. We found that the subjects were able to discriminate between the degraded calls of male zebra finches at up to 128 m without training, and up to 256 m with training. This augmented performance in the training condition suggests that the subjects enhanced their discrimination abilities through perceptual learning, even in a very difficult listening situation (Mouterde et al. 2014a).

On a physiological level, while it is a challenge to identify the changes in neuronal response properties that are causally related to the perceptual improvements, evidence for learning-related changes in the auditory cortex have accumulated, with the learning process resulting in an expanded representation of the trained features (Recanzone et al. 1993). Indeed, devoting more neurons to processing those aspects of a sensory stimulus that have particular behavioral relevance may seem like a plausible way of improving perceptual acuity (Dahmen and King 2007). In songbirds, the modification of neurons' encoding properties following experience-dependent perceptual plasticity was found in the secondary auditory regions, CMM (Gentner and Margoliash 2003), NCM (Pinaud and Terleph 2008), and in a lesser extent, CLM (Jeanne et al. 2011)—the latter authors suggesting that CLM and CMM are a part of a functional hierarchical neural circuit. These regions contribute to the recognition of familiar songs, and the learning of novel song discriminations (Gentner et al. 2004). Another important issue is to disentangle the effects of bottom-up and top-down processes in perceptual learning (Gentner 2004). While some studies supported the idea of a top-down control of perceptual learning, showing its strong interaction with attention (Ahissar and Hochstein 1993; Gilbert et al. 2001), other studies showed that perceptual learning can occur not only under training conditions but also in situations of unattended and passive sensory stimulation (Stripling et al. 2003; Seitz and Dinse 2007). Overall, it is probable that the plasticity in adult cortical networks that accompanies perceptual learning is shaped by an interaction between bottom-up sensory inputs, neuromodulator release, and task-specific top-down inputs (Gentner 2004; Polley et al. 2006; Dahmen and King 2007). Training facilitates these processes by strengthening both bottom-up sensory encoding and top-down modulation of the auditory cortex (Caras and Sanes 2017).

8.5.3 A Neural Substrate Readily Available for the Discrimination of Fine-Detailed Information

While we have found that zebra finch females could not discriminate between the naturally degraded vocal identities of males at 256 m without training, some single neurons in naive birds had the capacity to perform this task. Keeping in mind that both studies involved different subjects (but the same auditory stimuli), it seems that for a comparable level of training, this discrimination ability found at the neural level is not shown at the behavioral level. This finding is interesting in regards to previous studies that have shown in a number of sensory systems that behavioral discrimination performance tends to mirror the neuronal discrimination of the most selective single cells (Britten et al. 1992; Romo and Salinas 2003; Wang et al. 2007). In the auditory system, behavioral performance for song discrimination matched the neural performance of the best neurons using a spike-timing code, the activity of which might potentially reflect the computations performed by the entire network (Wang et al. 2007; Narayan et al. 2007). In our zebra finch study, single neurons outperformed behavioral tests in the sense that recognition at the longest propagation distance in the behavioral test was only achieved after repeated training. Narayan et al. (2007) also found several cases in which neural performance was higher than behavioral performance, and they provided two potential explanations for this observation: (1) these results stem from intersubject variability, both experiments having been made on different animals; (2) the information available in the auditory cortex might be reduced in the course of the downstream readout of information, either because of additional source(s) of noise at the readout stage, or because the readout is able to access the best neurons only after repeated exposures via a learning mechanism.

I support the latter explanation, and propose that this neural substrate for individual discrimination of degraded vocalizations that we found in zebra finches may constitute a neural basis for vocal recognition in auditory scenes, that is, a “pool of discriminability” which could then develop in the secondary auditory areas following the birds’ experience for the learned recognition of specific individuals in natural settings. In other words, this small population of highly discriminant neurons could take part in a bottom-up process creating a potential for individual discrimination, while top-down processes such as attentional or reinforcement mechanisms would develop this potential through perceptual plasticity. The more numerous, less selective neurons could serve as a pool of cells that can acquire selectivity as the bird learns to discriminate new auditory objects (Meliza et al. 2010). Indeed, these authors reported that during the acquisition of new songs by songbirds in perceptual learning tasks, CMM neurons rapidly changed their selectivity toward the reinforced novel stimuli and then lost this selectivity as the animal learned new stimuli. This “recruitment” of neurons in the wake of experience-dependent needs converges on the concept of learning threshold developed by Seitz and Dinse (2007), in which standard neural responses must be boosted by plasticity-inducing factors such as attention or signal reinforcement, or optimized sensory inputs such as multisensory stimulation, to drive the system past the point of a learning threshold and induce augmented neural responses, and potentially lead to a change in behavior.

8.6 Conclusion and Perspectives

In this chapter, I have endeavored to take a transversal view on the transmission of fine-detailed auditory information at long distance, from sender to receiver, focusing on a songbird model and emphasizing the use of stimuli and paradigms that reflected the behavioral relevance of the challenges that these animals face in their natural environment. While the individual vocal signature encoded in the distance calls of zebra finches is remarkably resistant to propagation-induced sound degradation, a number of neurons in the auditory cortex have the spontaneous ability to encode individual identity of previously unknown vocalizers, as well as propagation distance, in extremely degraded calls. This latter finding is particularly compelling since the subjects had no prior experience in hearing propagated vocalizations. Gentner and Margoliash (2003) suggested that "for vocal recognition, the predictability imparted by species-specific characteristics of vocalizations, and the constraints imposed by evolutionary history and experience, probably yield a population of neurons predisposed to represent those vocalizations." I suspect that the same process, led by evolution, can explain the proficiency with which some neurons deal with propagation-induced degradation.

Regarding the neural encoding of auditory information, the field of neurophysiology is turning toward the investigation of neural networks, following the idea that population coding drives behavior (Curto et al. 2009; Panzeri et al. 2015). One may argue that, seeing the link that was found between single neuron performance and behavior, the known sparseness of neural coding representing the natural world, and the proposed idea of a neural substrate with a potential for high-level discrimination, analyzing single neuron encoding properties is still critical to understanding the link between neural and behavioral levels (Narayan et al. 2006). There is however a general consensus for the idea that although single neurons might drive percepts and elicit behavioral responses, everyday perception is probably the result of coordinated activity by neural populations (Panzeri et al. 2010). Recently, Panzeri et al. (2017) made the case that the two processes underlying perceptual decisions (that is, how neural responses encode stimuli and how they inform behavioral choices) have mainly been studied separately, the latter one having received much less attention. They proposed to redefine the neural code as the neural features that carry sensory information used by the animal to drive appropriate behavior, since only the features that lie in the intersection between sensory coding and information readout can be used to convert sensory perception into appropriate behavioral actions. This interesting approach would help scientists target more efficiently the processes that are actually used by animals in their reactions to their sensory environment. In the case of our focal study, this investigation method, coupled with the computation of the neurons' spectro-temporal receptive fields (STRFs; Theunissen et al. 2001; Nagel et al. 2011; Meyer et al. 2015), could help pinpoint which acoustic parameters in the degraded calls are actually used in the neural code driving behavioral responses. It would be interesting to be able to compare the acoustic parameters that were found to be important for individual discrimination in the acoustic analysis with the acoustic features that birds use while performing this task. This would imply for example the recording of neural responses in a freely moving animal performing an auditory discrimination task in an operant conditioning setup, so as to link auditory input to behavioral output.

In this chapter I have focused on the transmission of individual signatures that were degraded by propagation, combining factors such as sound attenuation, spectro-temporal degradations, and ambient noise in the analysis. In their daily lives however, animals encounter other aspects of auditory scene analysis as well, such as conspecific noise, directionality and auditory source separation (Lewicki et al. 2014). In our case, the use of a loudspeaker to broadcast sounds during electrophysiological recordings effectively annihilated any spatial information that our subjects would have had access to in a natural environment. This probably lead us to underestimate the natural discriminability rather than the opposite, since animals are able to use spatially separated sound sources to enhance their discrimination performance (Dent et al. 2009; Maddox et al. 2012). In any case, combining more of these factors in future research, while investigating the interaction between neurophysiological processes at the brain level and the behavioral output of the animal, will improve our understanding of how animals process real-world scenes such as mate recognition in the natural environment, and bring further insight into how the brain processes this seemingly easy and yet computationally challenging task that is auditory scene analysis.

References

Ahissar M, Hochstein S (1993) Attentional control of early perceptual learning. Proc Natl Acad Sci 90:5718–5722

Appeltants D, Gentner TQ, Hulse SH et al (2005) The effect of auditory distractors on song discrimination in male canaries (*Serinus canaria*). Behav Process 69:331–341. https://doi.org/10.1016/j.beproc.2005.01.010

Aubin T, Jouventin P (2002) How to vocally identify kin in a crowd: the penguin model. In: Advances in the study of behavior. Elsevier, Amsterdam, pp 243–277

Aubin T, Mathevon N, Silva ML et al (2004) How a simple and stereotyped acoustic signal transmits individual information: the song of the white-browed warbler *Basileuterus leucoblepharus*. An Acad Bras Cienc 76:335–344

Bathellier B, Ushakova L, Rumpel S (2012) Discrete neocortical dynamics predict behavioral categorization of sounds. Neuron 76:435–449. https://doi.org/10.1016/j.neuron.2012.07.008

Beckers GJL, Gahr M (2012) Large-scale synchronized activity during vocal deviance detection in the zebra finch auditory forebrain. J Neurosci 32:10594–10608. https://doi.org/10.1523/JNEUROSCI.6045-11.2012

Billimoria CP, Kraus BJ, Narayan R et al (2008) Invariance and sensitivity to intensity in neural discrimination of natural sounds. J Neurosci 28:6304–6308. https://doi.org/10.1523/JNEUROSCI.0961-08.2008

Boumans T, Vignal C, Smolders A et al (2008) Functional magnetic resonance imaging in zebra finch discerns the neural substrate involved in segregation of conspecific song from background noise. J Neurophysiol 99:931–938. https://doi.org/10.1152/jn.00483.2007

Bregman AS (1993) Auditory scene analysis: hearing in complex environments. In: McAdams SE, Bigand EE (eds) Thinking in sound: the cognitive psychology of human audition. Clarendon Press/Oxford University Press, Oxford

Brémond J-C, Aubin T (1990) Responses to distress calls by black-headed gulls, *Larus ridibundus*: the role of non-degraded features. Anim Behav 39:503–511

Brenowitz EA (1982) The active space of red-winged blackbird song. J Comp Physiol A Neuroethol Sens Neural Behav Physiol 147:511–522

Britten KH, Shadlen MN, Newsome WT, Movshon JA (1992) The analysis of visual motion: a comparison of neuronal and psychophysical performance. J Neurosci 12:4745–4765

Brumm H, Naguib M (2009) Environmental acoustics and the evolution of bird song. In: Naguib M, Janik VM, Zuberbühler K, Clayton NS (eds) Advances in the study of behavior. Academic, New York, pp 1–33

Brumm H, Todt D (2002) Noise-dependent song amplitude regulation in a territorial songbird. Anim Behav 63:891–897. https://doi.org/10.1006/anbe.2001.1968

Capsius B, Leppelsack H-J (1996) Influence of urethane anesthesia on neural processing in the auditory cortex analogue of a songbird. Hear Res 96:59–70. https://doi.org/10.1016/0378-5955(96)00038-X

Caras ML, Sanes DH (2017) Top-down modulation of sensory cortex gates perceptual learning. Proc Natl Acad Sci 114:9972–9977. https://doi.org/10.1073/pnas.1712305114

Chew SJ, Vicario DS, Nottebohm F (1996) A large-capacity memory system that recognizes the calls and songs of individual birds. Proc Natl Acad Sci 93:1950–1955

Christison-Lagay KL, Gifford AM, Cohen YE (2015) Neural correlates of auditory scene analysis and perception. Int J Psychophysiol 95:238–245. https://doi.org/10.1016/j.ijpsycho.2014.03.004

Covey E (2000) Neural population coding and auditory temporal pattern analysis. Physiol Behav 69:211–220. https://doi.org/10.1016/S0031-9384(00)00203-1

Cunningham JP, Yu BM (2014) Dimensionality reduction for large-scale neural recordings. Nat Neurosci 17:1500–1509. https://doi.org/10.1038/nn.3776

Curto C, Sakata S, Marguet S et al (2009) A simple model of cortical dynamics explains variability and state dependence of sensory responses in urethane-anesthetized auditory cortex. J Neurosci 29:10600–10612. https://doi.org/10.1523/JNEUROSCI.2053-09.2009

Cynx J, Lewis R, Tavel B, Tse H (1998) Amplitude regulation of vocalizations in noise by a songbird, *Taeniopygia guttata*. Anim Behav 56:107–113. https://doi.org/10.1006/anbe.1998.0746

Dahmen JC, King AJ (2007) Learning to hear: plasticity of auditory cortical processing. Curr Opin Neurobiol 17:456–464. https://doi.org/10.1016/j.conb.2007.07.004

Dean I, Harper NS, McAlpine D (2005) Neural population coding of sound level adapts to stimulus statistics. Nat Neurosci 8:1684–1689. https://doi.org/10.1038/nn1541

Dent ML, McClaine EM, Best V et al (2009) Spatial unmasking of birdsong in zebra finches (*Taeniopygia guttata*) and budgerigars (*Melopsittacus undulatus*). J Comp Psychol 123:357–367. https://doi.org/10.1037/a0016898

Einevoll GT, Kayser C, Logothetis NK, Panzeri S (2013) Modelling and analysis of local field potentials for studying the function of cortical circuits. Nat Rev Neurosci 14:770–785. https://doi.org/10.1038/nrn3599

Fay RR, Popper AN (2000) Evolution of hearing in vertebrates: the inner ears and processing. Hear Res 149:1–10. https://doi.org/10.1016/S0378-5955(00)00168-4

Feng AS, Ratnam R (2000) Neural basis of hearing in real-world situations. Annu Rev Psychol 51:699–725. https://doi.org/10.1146/annurev.psych.51.1.699

Fishman YI, Reser DH, Arezzo JC, Steinschneider M (2001) Neural correlates of auditory stream segregation in primary auditory cortex of the awake monkey. Hear Res 151:167–187. https://doi.org/10.1016/S0378-5955(00)00224-0

Fitzpatrick DC, Batra R, Stanford TR, Kuwada S (1997) A neuronal population code for sound localization. Nature 388:871–874

Forrest TG (1994) From sender to receiver: propagation and environmental effects on acoustic signals. Am Zool 34:644–654

Fotheringham J (1997) Song transmission and auditory perception of distance in wood warblers (Parulinae). Anim Behav 53:1271–1285. https://doi.org/10.1006/anbe.1996.0361

Gaucher Q, Huetz C, Gourévitch B et al (2013) How do auditory cortex neurons represent communication sounds? Hear Res 305:102–112. https://doi.org/10.1016/j.heares.2013.03.011

Gentner TQ (2004) Neural systems for individual song recognition in adult birds. Ann N Y Acad Sci 1016:282–302. https://doi.org/10.1196/annals.1298.008

Gentner TQ, Margoliash D (2003) Neuronal populations and single cells representing learned auditory objects. Nature 424:669–674. https://doi.org/10.1038/nature01731

Gentner TQ, Hulse SH, Ball GF (2004) Functional differences in forebrain auditory regions during learned vocal recognition in songbirds. J Comp Physiol A 190:1001–1010. https://doi.org/10.1007/s00359-004-0556-x
Gilbert CD, Sigman M, Crist RE (2001) The neural basis of perceptual learning. Neuron 31:681–697
Gill P, Woolley SMN, Fremouw T, Theunissen FE (2008) What's that sound? auditory area CLM encodes stimulus surprise, not intensity or intensity changes. J Neurophysiol 99:2809–2820. https://doi.org/10.1152/jn.01270.2007
Gold JI, Watanabe T (2010) Perceptual learning. Curr Biol 20:R46–R48. https://doi.org/10.1016/j.cub.2009.10.066
Grace JA, Amin N, Singh NC, Theunissen FE (2003) Selectivity for conspecific song in the zebra finch auditory forebrain. J Neurophysiol 89:472–487
Guyomarc'h J-C, Aupiais A, Guyomarc'h C (1998) Individual differences in the long-distance vocalizations used during pair bonding in European quail (*Coturnix coturnix*). Ethol Ecol Evol 10:333–346. https://doi.org/10.1080/08927014.1998.9522847
Hauber ME, Cassey P, Woolley SMN, Theunissen FE (2007) Neurophysiological response selectivity for conspecific songs over synthetic sounds in the auditory forebrain of non-singing female songbirds. J Comp Physiol A 193:765–774. https://doi.org/10.1007/s00359-007-0231-0
Holland J (1998) Degradation of wren *Troglodytes troglodytes* song: implications for information transfer and ranging. J Acoust Soc Am 103:2154. https://doi.org/10.1121/1.421361
Hsu A, Woolley SMN, Fremouw TE, Theunissen FE (2004) Modulation power and phase spectrum of natural sounds enhance neural encoding performed by single auditory neurons. J Neurosci 24:9201–9211. https://doi.org/10.1523/JNEUROSCI.2449-04.2004
Huetz C, Gourévitch B, Edeline J-M (2011) Neural codes in the thalamocortical auditory system: from artificial stimuli to communication sounds. Hear Res 271:147–158. https://doi.org/10.1016/j.heares.2010.01.010
Jeanne JM, Thompson JV, Sharpee TO, Gentner TQ (2011) Emergence of learned categorical representations within an auditory forebrain circuit. J Neurosci 31:2595–2606. https://doi.org/10.1523/JNEUROSCI.3930-10.2011
Jen PH-S, Chen QC, Sun XD (1998) Corticofugal regulation of auditory sensitivity in the bat inferior colliculus. J Comp Physiol A 183:683–697. https://doi.org/10.1007/s003590050291
Karino G, George I, Loison L et al (2016) Anesthesia and brain sensory processing: impact on neuronal responses in a female songbird. Sci Rep 6:39143. https://doi.org/10.1038/srep39143
Kayser C, Montemurro MA, Logothetis NK, Panzeri S (2009) Spike-phase coding boosts and stabilizes information carried by spatial and temporal spike patterns. Neuron 61:597–608. https://doi.org/10.1016/j.neuron.2009.01.008
Klump GM (1996) Bird communication in the noisy world. In: Kroodsma DE, Miller EH (eds) Ecology and evolution of acoustic communication in birds. Cornell University Press, Ithaca, pp 321–338
Lewicki MS, Olshausen BA, Surlykke A, Moss CF (2014) Scene analysis in the natural environment. Front Psychol 5:199. https://doi.org/10.3389/fpsyg.2014.00199
Lohr B, Wright TF, Dooling RJ (2003) Detection and discrimination of natural calls in masking noise by birds: estimating the active space of a signal. Anim Behav 65:763–777. https://doi.org/10.1006/anbe.2003.2093
Lopes-dos-Santos V, Panzeri S, Kayser C et al (2015) Extracting information in spike time patterns with wavelets and information theory. J Neurophysiol 113:1015–1033. https://doi.org/10.1152/jn.00380.2014
Maddox RK, Billimoria CP, Perrone BP et al (2012) Competing sound sources reveal spatial effects in cortical processing. PLoS Biol 10:e1001319. https://doi.org/10.1371/journal.pbio.1001319
Mahmud M, Vassanelli S (2016) Processing and analysis of multichannel extracellular neuronal signals: state-of-the-art and challenges. Front Neurosci 10:62. https://doi.org/10.3389/fnins.2016.00248
Margoliash D, Fortune ES, Sutter M et al (1994) Distributed representation in the song system of Oscines: evolutionary implications and functional consequences. Brain Behav Evol 44:247–264
Marler P, Slabbekoorn H (2004) Nature's music, the science of birdsong. Elsevier, London

Mathevon N, Aubin T (1997) Reaction to conspecific degraded song by the wren *Troglodytes troglodytes*: territorial response and choice of song post. Behav Process 39:77–84. https://doi.org/10.1016/S0376-6357(96)00046-0

Mathevon N, Aubin T, Brémond J-C (1997) Propagation of bird acoustic signals: comparative study of starling and blackbird distress calls. C R Acad Sci III Sci Vie 320:869–876. https://doi.org/10.1016/S0764-4469(97)80871-2

Mathevon N, Aubin T, Vielliard J et al (2008) Singing in the rain forest: how a tropical bird song transfers information. PLoS One 3:e1580. https://doi.org/10.1371/journal.pone.0001580

Meliza CD, Margoliash D (2012) Emergence of selectivity and tolerance in the avian auditory cortex. J Neurosci 32:15158–15168. https://doi.org/10.1523/JNEUROSCI.0845-12.2012

Meliza CD, Chi Z, Margoliash D (2010) Representations of conspecific song by starling secondary forebrain auditory neurons: toward a hierarchical framework. J Neurophysiol 103:1195–1208. https://doi.org/10.1152/jn.00464.2009

Meyer AF, Diepenbrock J-P, Ohl FW, Anemüller J (2015) Fast and robust estimation of spectro-temporal receptive fields using stochastic approximations. J Neurosci Methods 246:119–133. https://doi.org/10.1016/j.jneumeth.2015.02.009

Middlebrooks JC, Simon JZ (2017) Ear and brain mechanisms for parsing the auditory scene. In: Middlebrooks JC, Simon JZ, Popper AN, Fay RR (eds) The auditory system at the cocktail party. Springer, Cham, pp 1–6

Miller LM, Recanzone GH (2009) Populations of auditory cortical neurons can accurately encode acoustic space across stimulus intensity. Proc Natl Acad Sci 106:5931–5935. https://doi.org/10.1073/pnas.0901023106

Mizrahi A, Shalev A, Nelken I (2014) Single neuron and population coding of natural sounds in auditory cortex. Curr Opin Neurobiol 24:103–110. https://doi.org/10.1016/j.conb.2013.09.007

Moore RC, Lee T, Theunissen FE (2013) Noise-invariant neurons in the avian auditory cortex: hearing the song in noise. PLoS Comput Biol 9:e1002942. https://doi.org/10.1371/journal.pcbi.1002942

Morton ES (1975) Ecological sources of selection on avian sounds. Am Nat 109:17–34

Mouterde SC, Elie JE, Theunissen FE, Mathevon N (2014a) Learning to cope with degraded sounds: female zebra finches can improve their expertise in discriminating between male voices at long distances. J Exp Biol 217:3169–3177. https://doi.org/10.1242/jeb.104463

Mouterde SC, Theunissen FE, Elie JE et al (2014b) Acoustic communication and sound degradation: how do the individual signatures of male and female zebra finch calls transmit over distance? PLoS One 9:e102842. https://doi.org/10.1371/journal.pone.0102842

Mouterde SC, Elie JE, Mathevon N, Theunissen FE (2017) Single neurons in the avian auditory cortex encode individual identity and propagation distance in naturally degraded communication calls. J Neurosci 37:3491–3510. https://doi.org/10.1523/JNEUROSCI.2220-16.2017

Nagel K, Kim G, McLendon H, Doupe A (2011) A bird brain's view of auditory processing and perception. Hear Res 273:123–133. https://doi.org/10.1016/j.heares.2010.08.008

Naguib M (1995) Auditory distance assessment of singing conspecifies in Carolina wrens: the role of reverberation and frequency-dependent attenuation. Anim Behav 50:1297–1307

Naguib M, Hammerschmidt K, Wirth J (2001) Microgeographic variation, habitat effects and individual signature cues in calls of chiffchaffs *Phylloscopus collybita canarensis*. Ethology 107:341–355

Narayan R, Graña G, Sen K (2006) Distinct time scales in cortical discrimination of natural sounds in songbirds. J Neurophysiol 96:252–258. https://doi.org/10.1152/jn.01257.2005

Narayan R, Best V, Ozmeral E et al (2007) Cortical interference effects in the cocktail party problem. Nat Neurosci 10:1601–1607. https://doi.org/10.1038/nn2009

Nieder A, Klump GM (1999) Adjustable frequency selectivity of auditory forebrain neurons recorded in a freely moving songbird via radiotelemetry. Hear Res 127:41–54. https://doi.org/10.1016/S0378-5955(98)00179-8

Olshausen B, Field D (2004) Sparse coding of sensory inputs. Curr Opin Neurobiol 14:481–487. https://doi.org/10.1016/j.conb.2004.07.007

Pachitariu M, Lyamzin DR, Sahani M, Lesica NA (2015) State-dependent population coding in primary auditory cortex. J Neurosci 35:2058–2073. https://doi.org/10.1523/JNEUROSCI.3318-14.2015
Panzeri S, Brunel N, Logothetis NK, Kayser C (2010) Sensory neural codes using multiplexed temporal scales. Trends Neurosci 33:111–120. https://doi.org/10.1016/j.tins.2009.12.001
Panzeri S, Macke JH, Gross J, Kayser C (2015) Neural population coding: combining insights from microscopic and mass signals. Trends Cogn Sci 19:162–172. https://doi.org/10.1016/j.tics.2015.01.002
Panzeri S, Harvey CD, Piasini E et al (2017) Cracking the neural code for sensory perception by combining statistics, intervention, and behavior. Neuron 93:491–507. https://doi.org/10.1016/j.neuron.2016.12.036
Pinaud R, Terleph TA (2008) A songbird forebrain area potentially involved in auditory discrimination and memory formation. J Biosci 33:145–155. https://doi.org/10.1007/s12038-008-0030-y
Pohl NU, Leadbeater E, Slabbekoorn H et al (2012) Great tits in urban noise benefit from high frequencies in song detection and discrimination. Anim Behav 83:711–721. https://doi.org/10.1016/j.anbehav.2011.12.019
Polley DB, Steinberg EE, Merzenich MM (2006) Perceptual learning directs auditory cortical map reorganization through top-down influences. J Neurosci 26:4970–4982. https://doi.org/10.1523/JNEUROSCI.3771-05.2006
Quian Quiroga R, Panzeri S (2009) Extracting information from neuronal populations: information theory and decoding approaches. Nat Rev Neurosci 10:173–185. https://doi.org/10.1038/nrn2578
Rabinowitz NC, Willmore BDB, King AJ, Schnupp JWH (2013) Constructing noise-invariant representations of sound in the auditory pathway. PLoS Biol 11:e1001710. https://doi.org/10.1371/journal.pbio.1001710
Recanzone GH, Schreiner CE, Merzenich MM (1993) Plasticity in the frequency representation of primary auditory cortex following discrimination training in adult owl monkeys. J Neurosci 13:87–103
Rek P, Osiejuk TS (2011) No male identity information loss during call propagation through dense vegetation: the case of the corncrake. Behav Process 86:323–328
Romo R, Salinas E (2003) Cognitive neuroscience: flutter discrimination: neural codes, perception, memory and decision making. Nat Rev Neurosci 4:203–218. https://doi.org/10.1038/nrn1058
Schmidt MF, Konishi M (1998) Gating of auditory responses in the vocal control system of awake songbirds. Nat Neurosci 1:513–518
Schneider DM, Woolley SMN (2013) Sparse and background-invariant coding of vocalizations in auditory scenes. Neuron 79:141–152. https://doi.org/10.1016/j.neuron.2013.04.038
Schnitzler H-U, Flieger E (1983) Detection of oscillating target movements by echolocation in the Greater Horseshoe bat. J Comp Physiol A 153:385–391. https://doi.org/10.1007/BF00612592
Schumacher JW, Schneider DM, Woolley SMN (2011) Anesthetic state modulates excitability but not spectral tuning or neural discrimination in single auditory midbrain neurons. J Neurophysiol 106:500–514. https://doi.org/10.1152/jn.01072.2010
Seitz AR, Dinse HR (2007) A common framework for perceptual learning. Curr Opin Neurobiol 17:148–153. https://doi.org/10.1016/j.conb.2007.02.004
Shamma SA, Elhilali M, Micheyl C (2011) Temporal coherence and attention in auditory scene analysis. Trends Neurosci 34:114–123. https://doi.org/10.1016/j.tins.2010.11.002
Slabbekoorn H (2013) Songs of the city: noise-dependent spectral plasticity in the acoustic phenotype of urban birds. Anim Behav 85:1089–1099. https://doi.org/10.1016/j.anbehav.2013.01.021
Stripling R, Milewski L, Kruse AA, Clayton DF (2003) Rapidly learned song-discrimination without behavioral reinforcement in adult male zebra finches (*Taeniopygia guttata*). Neurobiol Learn Mem 79:41–50
Theunissen FE, Elie JE (2014) Neural processing of natural sounds. Nat Rev Neurosci 15:355–366. https://doi.org/10.1038/nrn3731

Theunissen F, Miller JP (1995) Temporal encoding in nervous systems: a rigorous definition. J Comput Neurosci 2:149–162. https://doi.org/10.1007/BF00961885

Theunissen FE, Shaevitz SS (2006) Auditory processing of vocal sounds in birds. Curr Opin Neurobiol 16:400–407. https://doi.org/10.1016/j.conb.2006.07.003

Theunissen FE, David SV, Singh NC et al (2001) Estimating spatio-temporal receptive fields of auditory and visual neurons from their responses to natural stimuli. Netw Comput Neural Syst 12:289–316

Theunissen FE, Amin N, Shaevitz SS et al (2004) Song selectivity in the song system and in the auditory forebrain. Ann N Y Acad Sci 1016:222–245. https://doi.org/10.1196/annals.1298.023

Vicario DS, Naqvi NH, Raksin JN (2001) Sex differences in discrimination of vocal communication signals in a songbird. Anim Behav 61:805–817. https://doi.org/10.1006/anbe.2000.1651

Vignal C, Mathevon N, Mottin S (2004) Audience drives male songbird response to partner's voice. Nature 430:448–451. https://doi.org/10.1038/nature02645

Vignal C, Mathevon N, Mottin S (2008) Mate recognition by female zebra finch: analysis of individuality in male call and first investigations on female decoding process. Behav Process 77:191–198. https://doi.org/10.1016/j.beproc.2007.09.003

von der Emde G (2004) Distance and shape: perception of the 3-dimensional world by weakly electric fish. J Phys Paris 98:67–80. https://doi.org/10.1016/j.jphysparis.2004.03.013

Wang L, Narayan R, Grana G et al (2007) Cortical discrimination of complex natural stimuli: can single neurons match behavior? J Neurosci 27:582–589. https://doi.org/10.1523/JNEUROSCI.3699-06.2007

Wiley RH, Richards DG (1982) Adaptations for acoustic communication in birds: sound transmission and signal detection. In: Kroodsma DE, Miller EH (eds) Acoustic communication in birds. Academic, New York, pp 131–181

Woolley SMN (2006) Stimulus-dependent auditory tuning results in synchronous population coding of vocalizations in the songbird midbrain. J Neurosci 26:2499–2512. https://doi.org/10.1523/JNEUROSCI.3731-05.2006

Woolley SMN, Gill PR, Fremouw T, Theunissen FE (2009) Functional groups in the avian auditory system. J Neurosci 29:2780–2793. https://doi.org/10.1523/JNEUROSCI.2042-08.2009

Zann R (1984) Structural variation in the zebra finch distance call. Z Tierpsychol 66:328–345. https://doi.org/10.1111/j.1439-0310.1984.tb01372.x

Zann RA (1996) The zebra finch: a synthesis of field and laboratory studies. Oxford University Press, Oxford

Chapter 9
Mother–Offspring Vocal Recognition and Social System in Pinnipeds

Isabelle Charrier

Abstract In many species, parents and offspring have developed the ability to vocally identify each other. In avian species, a strong relationship between individual recognition system and social structure has been shown, with recognition systems being more elaborated in species exposed to strong selective pressures (e.g., colonial species vs. solitary species). Pinnipeds (seals, fur seals, sea lions, and walrus) are an excellent mammalian clade model for comparative studies of individual vocal recognition as they present a high diversity in both their social structures and breeding systems, and they use vocal signals in all their social interactions. The investigation of mother–pup vocal recognition systems demonstrates some clear evidence that pinniped species with the highest selective pressures for mother–pup recognition have developed the most complex recognition system. Indeed, such species show a high index of vocal stereotypy (IVS), a rapid onset of vocal recognition, a multi-parametric vocal signature mainly based on temporal analysis whereas species living in less constraining environments show a moderate to low IVS, a delayed onset of vocal recognition, a multi-parametric signature mainly based on a frequency analysis. Our understanding on how ecological and social constraints drive communication systems in vertebrates is essential. Our current knowledge on different taxa show that species encountering similar constraints for individual recognition has developed similar communication systems, suggesting common communication strategies in vertebrates.

9.1 Introduction

In many colony-living species of birds and mammals, parents and offspring have developed the ability to identify each other, leading to mutual benefits (Halliday 1983). Indeed, parents avoid misdirected care and thus ensure their reproductive

I. Charrier (✉)
Paris Saclay Institute of Neurosciences, UMR 9197 CNRS, Université Paris Sud, Orsay, France
e-mail: isabelle.charrier@u-psud.fr

T. Aubin, N. Mathevon (eds.), *Coding Strategies in Vertebrate Acoustic Communication*, Animal Signals and Communication 7,
https://doi.org/10.1007/978-3-030-39200-0_9

Table 9.1 Biological and social characteristics of the three pinnipeds families

	Social structure	Breeding system	Lactation duration	Allosuckling	Selective pressures for mother–young recognition
Otariids (16 species)	Colonial (high to moderate)	Polygyny (extreme to moderate)	4 month to 3 years	Rare	High to moderate
Odobenids (1 species)	Colonial	Polygyny	2–3 years	Rare	High
Phocids (18 species)	Colonial to solitary	Polygyny to serial monogamy	4 days to 2.5 months	Rare to common	High to low

success while for young, such recognition is essential for their survival since most parents/mothers only feed their own offspring. The degree of recognition (mutual or unilateral) varies in regards to the social structure and environmental constraints of the species (Beecher 1989; Aubin and Jouventin 2002; Insley et al. 2003). In mammals, care of the young is mainly provided by the mother. Except in rare cases of cooperative mammals, only the mother provides care to her offspring and rejects sometimes aggressively any non-filial young (Le Boeuf and Briggs 1977; Harcourt 1992a; Maestripieri 1992).

Pinnipeds (seals, fur seals, sea lions, and walrus) are an excellent mammalian clade model for comparative studies of individual vocal recognition. First, pinnipeds use vocal signals, in air and/or under water, in most of their social interactions: territorial defense, mate selection, mother–young care, predator avoidance (Insley et al. 2003). Second, they show a high diversity in their social structures (from solitary to highly colonial species), breeding systems (from serial monogamy to highly polygynous species), and maternal attendance (short to long lactation, high or low level of allonursing) (Table 9.1). Phocids (i.e., true seals) live solitary or in small groups, with the exception of colonial phocids such as elephant and gray seals that form large aggregations of individuals during the breeding season. In general, phocid females stay in permanence with their young that they suckle for several weeks (4 days to 2.5 month) (Riedman 1990). Allonursing and fostering can be observed but it is not a common trait. Otariids (i.e., fur seals and sea lions) show different characteristics as they form large colonies during the breeding season, but the density of animals varying among species. Females exclusively nurse their young for several months (4–36 month) (Riedman 1990) and they can be highly aggressive toward non-filial pups (Harcourt 1992b). Throughout lactation, females alternate foraging trips at sea with ashore suckling periods, and the first separation occurs as soon as 10–15 days after birth. Finally, odobenids (walrus) show similar characteristics than otariids since they live in groups, more and less large but females stay always densely packed together. Walrus females stay in permanence with their calf even while foraging at sea since the young is able to swim few hours after birth (Kovacs and Lavigne 1992; Stewart and Fay 2001). Separations between the mother and her calf can however be frequent due to their instable habitat (fast and pack ice),

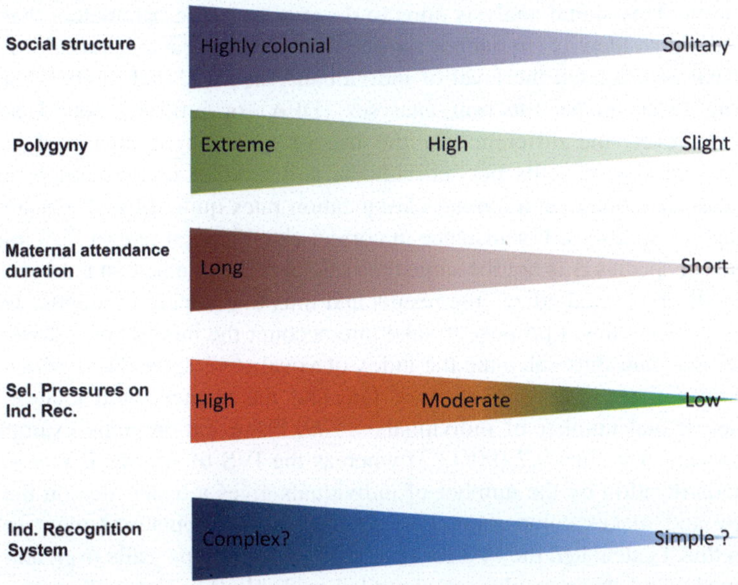

Fig. 9.1 Gradient of biological and social traits of pinnipeds, the resulted selective pressures on mother–pup recognition, and their hypothetic influence on the individual recognition system

but also due to the approach of predators or other disturbance (aircraft, human activities) around the colony that can induce stampede. Walrus females nurse their young for up to 3 years, and the social bond between the mother and her calf is among the strongest in mammals (Knudtson 1998).

Such gradient in the social and breeding systems of pinnipeds results in different selective pressures for mother–pup vocal recognition. A comparative approach will allow a better understanding on how social organization can shape individual vocal recognition systems (Fig. 9.1). From the different quantitative and experimental works carried out on several species of pinnipeds, I aim to demonstrate that the gradient found in selective pressures for individual recognition is also found at the complexity level of their recognition systems. I define the complexity of a vocal recognition system by its characteristics or "dimensions": occurrence (i.e., presence/absence, mutual or unilateral recognition), ontogeny (rapid or slow development of individual recognition), individual vocal stereotypy (i.e., low or high level of individuality in calls), and complexity (number and characteristics of acoustic parameters involved in identification, and their resistance to degradations during propagation in the natural environment). From the current knowledge on mother–young vocal recognition in pinnipeds, I will describe the different dimensions of their recognition systems and thus discuss the link between social and recognition systems in pinnipeds.

A prerequisite for individual vocal recognition is the use of vocalizations showing an individual stereotypy. Thus, a first step is to analyze the signals to determine if they are sufficiently individualized and thus may allow a reliable individual identification. This signal analysis aims to describe acoustic parameters that could encode individual identity (frequency modulation, spectral features, duration). Many studies have investigated the level of individuality in pups' and mothers' calls by performing discriminant function analyses (DFA) or artificial neural network (ANN). However, the difference in the use of numbers of measured acoustic parameters, number of calls per individuals, and number of individuals in each study makes the comparison among classification rates quite difficult (Insley et al. 2003; Khan et al. 2006). For instance, a correct classification rate of 90% on three individuals of species A is not the same than one of 90% obtained on ten individuals on species B. To "standardize" the results and thus to compare in a better way the results of these studies, I propose to take into account the number of individuals in these analyses, and thus calculate the index of vocal stereotypy (IVS) which is the ratio between the correct classification rate and the chance, with chance being defined as: 1/total number of individuals $\times 100$. From the previous example, the IVS of species A is thus 2.7 (90/33.3) whereas the IVS of species B is 9 (90/10). Such standardization by the number of individuals gives a better idea on the vocal stereotypy and makes comparisons among species or populations more reliable. Based on this, I calculated the IVS for both mothers' and pups' calls from all studies on pinnipeds, and these results are compiled in Table 9.2. For both mothers and pups, IVS varies with the selective pressures for individual vocal recognition. Indeed, the individual vocal stereotypy decreases when ecological constraints for mother–pup recognition decrease (Fig. 9.2).

These results are consistent with those found on penguins with species showing differences in their ecological constraints (colony density, presence of territories and nests, background noise). Acoustic analyses on contact calls on four species of penguins revealed that penguins without nest (King and Emperor penguins), so with the highest constraints for individual recognition, show a higher individual vocal stereotypy compared to penguins species with nest facing less constraints for individual recognition (Aubin and Jouventin 2002). Even if individual vocal characteristics are highly linked to physical traits of the emitter (vocal tract length, resonance cavities) (Riede and Fitch 1999), and thus can explain a certain level of individual stereotypy in their vocalizations, ecological constraints also greatly modulate this level.

9.2 Vocal Recognition and Ontogeny

9.2.1 Evidence for Vocal Recognition

An individual vocal signature, revealed by analysis, will not necessarily mean that a given species uses the vocal signal to identify individuals. For instance, in gray seals (*Halichoreus grypus*), one colony in Canada was studied and mother–pup vocal recognition occurs (McCulloch et al. 1999) whereas another colony studied in Scotland did not show such vocal recognition (McCulloch and Boness 2000).

Table 9.2 Indice of Vocal Stereotypy (IVS) of the different studied pinnipeds species in regards to their ecological constraints and selective pressures for individual recognition

Family species	Pop density	Mating system	Allosuckling	Maternal absence	Sel. pres. for IR	Cl. rate mothers	Nb ind	IVS	Cl. rate pups	Nb ind	IVS	References for classification rates
OTARIIDS (/16)												
Subantarctic FS	H	Extreme P	Rare	Y	H	84	10	8.4	83	10	8.3	Page et al. (2002)
Northern FS	H	Extreme P	Rare	Y	H	83	8	6.64	79	8	6.32	Insley (1992)
Antarctic FS	H	Extreme P	Rare	Y	H	74	10	7.4	52	9	4.68	Page et al. (2002)
South am FS	H	P		Y	H	70	15	10.5	60	13	7.8	Phillips and Stirling (2000)
Australian FS	H	Extreme P		Y	H	76	13	9.88	75	9	6.75	Tripovich et al. (2006)
New Zealand FS	H	P		Y	H	88	6	5.28	67	12	8.04	Page et al. (2002), Dowell (2005)
Steller SL	H	Extreme P	Rare	Y	H	71	26	18.46	–	–	–	Campbell et al. (2002)
South am SL	H	P		Y	H	95	17	16.15	89	19	16.91	Fernandez-juricic et al. (1999)
						63	13	9.1	53	11	5.83	Trimble and Charrier (2011)
Australian SL	M-L	Moderate P		Y	M	65	14	9.1	77	15	11.55	Charrier and Harcourt (2006)
ODOBENIDS (/1)												
Atlantic walrus	H	P	Rare	Short	H	64	8	5.12	67	16	10.72	Charrier et al. (2010) (DFA)
						74	8	5.92	71	16	11.36	Charrier et al. (2010) (ANN)

(continued)

Table 9.2 (continued)

Family species	Pop density	Mating system	Allosuckling	Maternal absence	Sel. pres. for IR	Cl. rate mothers	Nb ind	IVS	Cl. rate pups	Nb ind	IVS	References for classification rates
PHOCIDS (/18)												
Elephant seal	H	High P	Common	N	M	54	8	4.32	64	8	5.12	Insley (1992)
Gray seal	H-M	Moderate P	Common	N/short	M	–	–	–	31	20	6.2	McCulloch et al. (1999)
Weddell seal	M-L	Moderate P	Rare	Short	M	56	9	5.04	44	10	4.4	Collins et al. (2005, 2006)
Harp seal	M-L	Weak P		Short	L	–	–	–	8–55	12–47	9.21[a]	Van Opzeeland et al. (2004, 2009)
Harbor seal	L	Weak P	Rare	Short	L	–	–	–	42.6	15	6.39	Sauvé et al. (2015a)
Hawaiian monk seal	L	Weak P	Common	N	L	–	–	–	14	9	1.26	Job et al. (1995)

Population density: high H, moderate M, low L; mating system: polygyny P; maternal absence: frequent and for several days Y, no absence N; Selective Pressures for Individual Recognition: high H, moderate M, low L; Classification Rate in %

[a]Average values

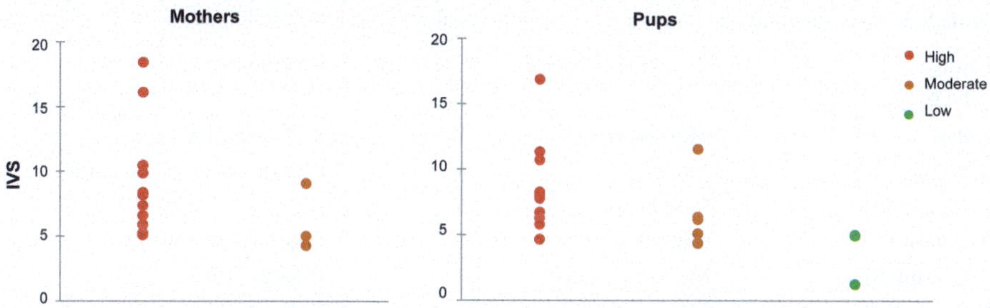

Fig. 9.2 Indice of Vocal Stereotypy (IVS) in mothers' and pups' calls of pinniped species showing high, moderate, and low selective pressures for individual vocal recognition

Several hypotheses have been suggested to explain such difference within a given species. The density of animals is very high in Canada and low in Scotland. A difference of habitat also occurs between these two studied colonies: in Canada the colony is located on a sandy beach, so an open area without any landmark, whereas in Scotland, the colony is established on a rocky area. Strong differences were found for maternal attendance. Indeed, females in Scotland often go foraging during lactation, so frequently separated from their young, whereas in Canada, females fast and stay in permanence with their pup. The level of allonursing is quite high in Scotland but rare in Canada. So, based on these biological and environmental traits, both populations show significant differences; however, both show needs for a mother–pup vocal recognition, and thus, they should have developed it (frequent separations between mothers and pups in Scotland, and high dense colony in Canada with no landmark), and the colony in Scotland exhibits even greater needs than those in Canada. So these findings are quite paradoxical as individual recognition is only developed in the Canadian colony. It has been suggested that the individual recognition found in the Canadian colony could be a residual behavior from an ice-breeding ancestry. Living on an unstable environment such as ice increases the chance of separations between mothers and pups, and thus this would have led to the development of vocal recognition between mothers and pups, and this has been maintained over time, even if selective pressures for individual recognition have decreased with time.

Such example reinforces the idea that it is essential to experimentally test the animals in order to assess the occurrence of such individual discrimination. In the last 30 years, mother–pup vocal recognition has been tested by playback experiments in 13 species, including seven otariids, four phocids, and one walrus subspecies (Table 9.3). Depending on the species, only the mother or the pup was tested, but for some species, both sides were investigated, with mutual vocal recognition demonstrated in northern fur seals (*Callorhinus ursinus*), subantarctic fur seals (*Arctocephalus tropicalis*), and Australian sea lions (*Neophoca cinerea*). Further investigations are still needed especially in phocids to draw firm conclusions, but currently Otarioidea (otarids and odobenids) show a well-developed vocal recognition system as well as some colonial phocids (elephant, gray, and harbor seals)

Table 9.3 Experimental tests on the mother–pup vocal recognition in pinnipeds

Family	Recognition tested	Findings	References
OTARIDS			
Galapagos FS	Mother by pup	Y	Trillmich (1981)
Subantarctic FS	Mutual	Y	Charrier et al. (2001, 2002)
Antarctic FS	Mother by pup	Y	Aubin et al. (2015)
Northern FS	Mutual	Y	Insley (2000, 2001)
California SL	Mother by pup[a]	Y	Hanggi (1992)
Galapagos SL	Mother by pup	Y	Trillmich (1981)
Australian SL	Mutual	Y	Charrier et al. (2009), Pitcher et al. (2012)
PHOCIDS			
Gray seal	Pup by mother	Y	McCulloch and Boness (2000)
	Pup by mother	N	McCulloch et al. (1999)
Northern elephant seal	Pup by mother	Y	Petrinovich (1974)
Harbor seal	Pup by mother	Y	Sauvé et al. (2015b)
Weddell seal	Pup by mother	N	Van Opzeeland et al. (2012)
ODOBENIDS			
Atlantic walrus	Pup by mother	Y	Charrier et al. (2010)
Pacific walrus	Mutual[a]	Y	Charrier pers.com

Updated from Insley et al. (2003)
[a]Indicates experiments on captive animals ($n = 1$); Y: occurrence of recognition; N: lack of recognition

whereas most non-colonial phocids do not exhibit mother–pup recognition (Table 9.3).

9.2.2 *Onset of Vocal Identification*

If vocal recognition exists, it seems essential to investigate when this identification is established, but also to examine the potential factors affecting the development of this cognitive process. Studies on the development of mother–pup vocal recognition are quite rare, and only four otariids species have been studied so far. Pups identify their mother's calls 10–30 days after birth in Galapagos sea lion (*Zalophus wollebaeki*, $n = 8$ (Trillmich 1981)), 10 days after birth in Galapagos fur seal (*Arctocephalus galapagoensis*, $n = 4$; Trillmich 1981), between 2 and 5 days after birth in subantarctic fur seal (*A. tropicalis*, $n = 9$ (Charrier et al. 2001)), and between 10 days and 2 month in Australian sea lion (*N. cinerea*, $n = 10$ (Pitcher et al. 2009)). The study species showing the highest ecological constraints (*A. tropicalis*) is the one in which the vocal recognition has been established more rapidly and especially before the first separation between the mother and her pup. For the other three species, the colony densities are lower, and thus the risk of confusion among individuals is weaker, and thus the development of recognition of the mother's voice by pups requires more time. Such late vocal discrimination in pups might be compensated by an early vocal recognition of the pup by the mother. Indeed, observations on different otariids species suggest that females can discriminate their pup's voice few hours after birth (Trillmich 1981, Charrier obs. pers.); however, this has only been experimentally shown in Australian sea lion (Pitcher et al. 2010) ($n = 17$ females). Indeed, 48 h after parturition, Australian sea lion females can discriminate between calls of a given pup and those from their own pup. For species showing mutual vocal recognition, it is likely that recognition is established first in females and later in pups. The time difference between females and pups may vary with the strength of ecological constraints. Further investigations are still needed in phocids, for which both the occurrence of vocal recognition and the selective pressures for individual recognition vary greatly.

9.3 Individual Vocal Signature

Then, an essential step that highly interested me for years is to decipher the individual vocal signatures involved in such individual identification processes. By performing playback experiments, using modified signals and/or synthetic signals in which a particular acoustic parameter has been modified or removed, we can determine the different parameters involved in this identification process. Finally, propagation tests in the natural environment of the study models are used to determine the active/efficiency space of their communication signals. This allows determining a theoretical maximal distance at which a vocalization can be reliably detected by the young or the mother in the colony.

9.3.1 Cracking the Code of Individual Recognition

How individual vocal characteristics are coded was experimentally tested in only three otariids species: the subantarctic and Antarctic fur seal (*A. tropicalis* and *A. gazella* respectively), and the Australian sea lion (*N. cinerea*). In these three species, both mothers and pups use a multi-parametric vocal signature to decode the identity of the receiver. The main acoustic features involved in this identification process are the frequency modulation (FM), the amplitude modulation (AM), and the energy spectrum (ES, i.e., repartition of energy among the frequency bandwidth, or the timbre) of the call (Charrier et al. 2002, 2003, 2009; Pitcher et al. 2012; Aubin et al. 2015). However, we can detect some differences among these three studied species showing high to moderate selective pressures for individual recognition (Table 9.4). Indeed, the two species with high selective pressures for individual recognition (i.e., subantarctic and Antarctic fur seals) perform a temporal analysis of the calls using FM and/or AM as well as a timbre analysis to discriminate among

Table 9.4 Individual vocal signatures and acoustic features involved in the recognition process

Species	Temporal analysis		Spectral analysis		Sel. pressures for ind. rec	References
	AM	FM	Exact frequencies (pitch)	Distribution of energy (timbre)		
Subantarctic fur seal (pup)	×	✓	×	✓	High	Charrier et al. (2003)
Subantarctic fur seal (m)	×	✓	nt	✓	High	Charrier et al. (2002)
Antarctic fur seal (pup)	✓	✓	×	✓	High	Aubin et al. (2015)
Pacific walrus (m)	✓	✓	×	✓	High	$N = 1$, Charrier pers. obs.
Australian sea lion (pup)	✓	✓	✓	×	Moderate	Charrier et al. (2009)
Australian sea lion (m)	✓	×	✓	×	Moderate	Pitcher et al. (2012)

×: not used; ✓: used; nt: not tested

individual voices, whereas the species with moderate selective pressures (i.e., Australian sea lion) performs also a temporal analysis using both AM and FM, a pitch analysis by paying attention to the exact frequency values of the calls but does not use the timbre (i.e., distribution of energy within frequencies). Before drawing firm conclusions, we need to further investigate species showing low selective pressures for individual recognition, such as non-colonial phocids. These future studies may reveal an individual vocal signature mostly relying on a pitch analysis (simple code as the risk of confusion among individuals is quite low). Indeed, the comparative study on penguins (Aubin and Jouventin 2002) showed that species without nest use a temporal analysis whereas species with nest use a spectral analysis (using both pitch and timbre), a signature thus considered as less complex (Aubin and Jouventin 2002).

From a production point of view, AM and FM are considered more difficult to produce (Greenwalt 1968; Brackenbury 1982; Aubin and Jouventin 2002) than spectral cues, as the emitter needs to control perfectly two sound dimensions at the same time (AM: amplitude and time or FM: frequency and time) and this has to be maintained over time (either over several years in the case of mate recognition, or over a breeding season for parent–offspring recognition). In contrast, spectral features (either timbre or pitch) do not require a fine motor control as temporal features, and thus seem less complex to produce. Finally, in terms of coding possibilities, an identity coding based on temporal features such as AM and FM offers a larger set of individual signatures as it combines two dimensions (Greenwalt 1968; Aubin and Jouventin 2002), and thus limits the risk of confusion among individual voices. In contrast, a code based on one dimension, such as spectral cues (pitch and/or timbre), offers a lower diversity of vocal signatures (with pitch cues offering less diversity than timbre cues), and thus it can potentially lead to confusion among individuals.

In summary, vocal signatures based on a temporal analysis are considered as more complex (stronger motor control, and high diversity of possible vocal signatures) than those relying on a spectral analysis. However, independently of the type of coding (temporal or spectral), the individual vocal signature used by a species is efficient and thus well adapted to the ecological constraints faced by the species. In other words, a species facing weak ecological constraints, and thus low selective pressures for individual recognition, does not need to develop a complex vocal signature as a "simple" signature offers sufficient vocal signatures to avoid confusion among individuals.

9.3.2 Propagation of the Vocal Signature

The use of multiple parameters is a way to secure the code and thus to optimize the chance of detection and identification especially for species living in constraining environment such as noisy and confusing colonial environment. Propagations tests performed on different pinniped species have shown different efficiency of propagations in their natural environment; however, it is important to consider these results in the context of mother–young reunions and their distance ranges. Indeed, for instance, in the case of the Australian sea lion, mothers' and pups' calls can be reliably identified up to 32 and 64 m, respectively, as the energy spectrum that codes for individual identity is still reliably detected at these distances (Charrier et al. 2009; Pitcher et al. 2012). In the wild, mothers and pup exchange vocalizations over distances up to 50 m, and this can start even further as mothers start calling when still in the water. Moreover, pups usually stay around the last suckling spot to increase their chance to detect their calling mother returning to the colony. Such midrange propagation efficiency of the individual vocal signature in Australian sea lions seems thus sufficient in the natural range of mother–pup reunion. For Atlantic walrus (*Odobenus rosmarus rosmarus*), there is a strong difference in distance efficiency between propagation on ice and above water for both mothers and calves' calls. All studied acoustic features reliably propagate up to 16 and 32 m on ice, and up to 128 m above water (Charrier et al. 2010). Most of the time, mothers and calves stay quite close, and if they got separated during group movements or panic in the presence of a disturbance or a predator, the distances to reunite again are within 10–20 m. These means that either on ice or above water, the acoustic features showing a great individual stereotypy, such as FM and energy spectrum, can be assessed at a natural communication range, and thus could be reliably used in the context of individual identification. Similarly, the propagation tests of harbor seal pups vocalizations above water show an efficiency of individualized acoustic features up to 512 m, a distance range that is way above the observed distance at which mother–pup reunions occurs (Sauvé et al. 2015b). These propagation studies have demonstrated that the active space of the acoustic features used or that could be used in mother–pup vocal recognition show propagation properties adapted to the environmental conditions in which mother–pup reunions occurs. Further investigations could assess if animals actually perceive and identify each other at these propagations distances, or if they can perform even better, as shown in king penguins that are able to extract the vocal signal when intensity is 6 dB below that of the background noise (i.e., cocktail party effect (Aubin and Jouventin 1998)). Indeed, a recent study on Antarctic fur seal (*A. gazella*) showed that pups identify their mother's calls using AM, FM, and energy spectrum. Propagation experiments revealed that if FM propagates reliably up to 64 m, both AM and energy spectrum are degraded for distances over 8 m (Aubin et al. 2015). Playback experiments performed on groups of pups with female's calls (groups include the pup whom mother's calls were used) at different distances (8, 32 and 64 m) showed that when distances decreased, the number of responding pups were also decreasing, with 1 or 2 pups responding at 8 m. Such behavioral experiments clearly show that at long range, the identification of the mother can lead to some errors of identification (several pups responded), whereas at short range, when all acoustic features of the individual signature are not degraded and thus reliable, the identification is more robust. Redundancy of information secures the identification process especially for species living in a constraining environment such as noisy colonies.

9.4 Interactions of Acoustic with Non-acoustic Cues and Individual Recognition

It has been suggested that sensory cues besides vocalizations, such as olfactory and visual and/or spatial cues could potentially be involved in mother–pup individual recognition (Kaufman et al. 1975; Terhune et al. 1979). Anecdotal descriptions of reunions suggested the involvement of acoustic and olfactory cues with spatial and visual cues helping the localization of the individuals (e.g., area in the colony where the pup was left for spatial cues; body size and color fur pattern for the visual cues). However, vocal signals remain the primary signals allowing for efficient individual identification at both short and long range (some females calling their pups while still in the water). Visual and olfactory cues could be involved in a second step of the identification process, when mothers and pups are at close range, and they can be used as an intermediate or final check during reunions. Australian sea lion mothers have been experimentally tested with olfactory cues alone, and they are able to discriminate the smell of their pups from those of a non-filial pup (Pitcher et al. 2010). Australian sea lions mothers have been also shown to use body size and color pattern of their pup pelage to discriminate among pups of different age-classes (Wierucka et al. 2017). This age-class visual discrimination of pups by females likely plays an important role to facilitate mother–pup reunion. Indeed, this species exhibits an extended pupping period (up to 7 month (Marlow 1975)), and thus pups of different age classes (i.e., different body size, and different contrast/brightness color fur pattern) occur in the same time in the colony. When a female comes back from a foraging trip at sea, she has to find her pups among others (pups often form crèches while their mothers are foraging at sea), and thus distinguishing the appropriate age-class of her own pup will facilitate their reunion, but also reduce potential injury to pups by non-mother females (Wierucka et al. 2017). It is likely that females can individually discriminate their pups using visual cues as other mammals do (Parr and de Waal 1999; Kendrick et al. 2001), but this has not yet being tested.

Most studies on animal communication and individual recognition focus on a single sensory modality, and thus there is a lack of investigation on the synergy of sensory cues in identification process. A recent work performed on Australian sea lions involving several sensory cues highlighted the predominant role of acoustic cues in a multimodal context for both mothers and pups. Indeed, the addition of visual cues to acoustic cues did not enhance the pups' responsiveness (Wierucka et al. 2018a). In females, the addition of olfactory and visual cues to acoustic cues enhanced the investigation behavior of females (i.e., sniffing) but did not enhance their vocal responsiveness (Wierucka et al. 2018b). Finally, when examining the relative importance of acoustic and olfactory cues in the recognition of pups by mothers, acoustic cues dominate olfactory cues (Wierucka et al. 2018b). Indeed, the vocal response of females relied only on acoustic cues and was not influenced by the identity or presence of olfactory cues (i.e., females' responses were similar whatever the olfactory cue was filial or non-filial). Such findings highlight the importance of understanding the relative role of sensory cues in communication and recognition processes. In a multimodal context, there are environmental and biological factors influencing the use of cues such as their active space, as well as the costs and benefits to assess and integrate them (Hebets and Papaj 2005). In a mother–young recognition context, the costs and benefits of obtaining sensory cues is quite different between mothers and offspring, especially in otariids where females can be highly aggressive toward non-filial pups (Riedman 1990; Harcourt 1992b). Thus, assessing reliable olfactory and visual cues for pups would require a close approach of the calling female, and thus presents a high risk of injury. In contrast, females do not risk anything in approaching calling pups to assess additional cues. Even if vocal cues alone are sufficient to identify the pups, other cues may serve as a final check before accepting to suckle the pup.

9.5 Conclusions

Based on the current knowledge of mother–pup vocal recognition in pinnipeds, we can draw some general conclusions on the link between social and recognition systems. The wide ranges of both social structures and breeding strategies have resulted in differences in selective pressures for mother–pup individual recognition (Fig. 9.1). Even if further investigations are still needed in non-colonial phocids, showing low selective pressures for individual recognition, the gradient found for the selective pressures is also found on the complexity of the individual recognition system. Indeed, at the different levels or dimensions characterizing a recognition system such as vocal stereotypy, ontogeny, individual signature, there is a clear evidence that species with the highest selective pressures for mother–pup recognition have developed a more complex recognition system (i.e., high index of vocal stereotypy (IVS), rapid onset of vocal recognition, multi-parametric vocal signature, temporal analysis involved in recognition process) compared to species showing lower selective pressures (i.e., moderate to low IVS, delayed onset of vocal recognition, multi-parametric signature but spectral analysis involved in identification process).

Such findings on this mammalian clade, the pinnipeds, are consistent with those found on colonial birds such as penguins. These studied vertebrate species showing similar communication network and ecological constraints (e.g., group-living mammals, colonial birds) have developed similar communication systems. This suggests the occurrence of general communication strategies in vertebrates.

References

Aubin T, Jouventin P (1998) Cocktail-party effect in king penguin colonies. Proc R Soc B 265:1665–1673

Aubin T, Jouventin P (2002) How to vocally identify kin in a crowd: the penguin model. Adv Study Behav 31:243–277

Aubin T, Jouventin P, Charrier I (2015) Mother vocal recognition in Antarctic fur seal *Arctocephalus gazella* pups: a two-step process. PLoS One 10:1–13. https://doi.org/10.1371/journal.pone.0134513

Beecher MD (1989) Signalling systems for individual recognition: an information theory approach. Anim Behav 38:248–261

Brackenbury JH (1982) The structural basis of voice production and its relationship to sound characteristics. In: Kroodsma D, Miller EH (eds) Acoustic communication in birds. Academic, New York, pp 53–73

Campbell GS, Gisiner RC, Helweg DA, Milette LL (2002) Acoustic identification of female Steller sea lions (*Eumetopias jubatus*). J Acoust Soc Am 111:2920. https://doi.org/10.1121/1.1474443

Charrier I, Harcourt RG (2006) Individual vocal identity in mother and pup Australian sea lions (*Neophoca cinerea*). J Mammal 87:929–938. https://doi.org/10.1644/05-MAMM-A-344R3.1

Charrier I, Mathevon N, Jouventin P (2001) Mother's voice recognition by seal pups. Nature 412:873–874. https://doi.org/10.1038/35091139

Charrier I, Mathevon N, Jouventin P (2002) How does a fur seal mother recognize the voice of her pup? An experimental study of *Arctocephalus tropicalis*. J Exp Biol 205:603–612

Charrier I, Mathevon N, Jouventin P (2003) Vocal signature recognition of mothers by fur seal pups. Anim Behav 19:543–550. https://doi.org/10.1006/anbe.2003.2073

Charrier I, Pitcher BJ, Harcourt RG (2009) Vocal recognition of mothers by Australian sea lion pups: individual signature and environmental constraints. Anim Behav 78:1127–1134. https://doi.org/10.1016/j.anbehav.2009.07.032

Charrier I, Aubin T, Mathevon N (2010) Mother–calf vocal communication in Atlantic walrus: a first field experimental study. Anim Cogn 13:471–482

Collins KT, McGreevy PD, Wheatley KE, Harcourt RG (2005) Individual variation of in-air female "pup contact" calls in Weddell seals, *Leptonychotes weddellii*. Behaviour 142:167–189. https://doi.org/10.1163/1568539053627668

Collins KT, Terhune JM, Rogers TL et al (2006) Vocal individuality of in-air Weddell seal (*Leptonychotes weddellii*) pup "primary" calls. Mar Mamm Sci 22:933–951. https://doi.org/10.1111/j.1748-7692.2006.00074.x

Dowell SA (2005) Mother-pup recognition behaviour, pup vocal signatures and allosuckling in the New Zealand fur seal, *Arctocephalus forsteri*. Master Thesis, University of Canterbury. 156pp

Fernandez-Juricic EF, Enriquez V, Campagna C, Ortiz CL (1999) Vocal communication and individual variation in breeding South American sea lions. Behaviour 136:495–517

Greenwalt CH (1968) Bird song: acoustics and physiology. Smithsonian Institution Press, Washington, DC

Halliday TR (1983) Information and communication. In: Halliday TR, Slater PJB (eds) Animal behaviour: communication. Blackwell, Oxford, pp 43–81

Hanggi E (1992) The importance of vocal cues in mother–pup recognition in a California sea lion. Mar Mamm Sci 8(4):430–432

Harcourt R (1992a) Factors affecting early mortality in the South American fur seal (*Arctocephalus australis*) in Peru: density-related effects and predation. J Zool 226:259–270. https://doi.org/10.1111/j.1469-7998.1992.tb03838.x

Harcourt R (1992b) Maternal aggression in the South American fur seal in Peru. Can J Zool 70:320–325. https://doi.org/10.1139/z92-048

Hebets EA, Papaj DR (2005) Complex signal function: developing a framework of testable hypotheses. Behav Ecol Sociobiol 57:197–214. https://doi.org/10.1007/s00265-004-0865-7

Insley SJ (1992) Mother-offspring separation and acoustic stereotypy: a comparison of call morphology in two species of pinnipeds. Behaviour 120:103–122. https://doi.org/10.1007/s13398-014-0173-7.2

Insley SJ (2000) Long-term vocal recognition in the northern fur seal. Nature 406:404–405. https://doi.org/10.1038/35019064

Insley SJ (2001) Mother-offspring vocal recognition in northern fur seals is mutual but asymmetrical. Anim Behav 61:129–137. https://doi.org/10.1006/anbe.2000.1569

Insley S, Phillips A, Charrier I (2003) A review of social recognition in pinnipeds. Aquat Mamm 29:181–201

Job D, Boness D, Francis J (1995) Individual variation in nursing vocalizations of Hawaiian monk seal pups, *Monachus schauinslandi* (Phocidae, Pinnipedia), and lack of maternal recognition. Can J Zool 73:975–983

Kaufman G, Siniff D, Reichle R (1975) Colony behavior of weddell seals, *Leptonychotes weddelli*, at Hutton Cliffs, Antarctica

Kendrick KM, da Costa AP, Leigh AE et al (2001) Sheep don't forget a face. Nature 414:165–166. https://doi.org/10.1038/35102669

Khan CBC, Markowitz H, McCowan B (2006) Vocal development in captive harbor seal pups, *Phoca vitulina richardii*: age, sex, and individual differences. J Acoust Soc Am 120:1684. https://doi.org/10.1121/1.2226530

Knudtson P (1998) The world of the walrus. Sierra Club Books, San Francisco

Kovacs K, Lavigne D (1992) Maternal investment in otariid seals and walruses. Can J Zool 70 (10):1953–1964

Le Boeuf BJ, Briggs KT (1977) The cost of living in a seal harem. Mammalia 41:167–195

Maestripieri D (1992) Functional aspects of maternal aggression in mammals. Can J Zool 70:1069–1077. https://doi.org/10.1139/z92-150

Marlow B (1975) The comparative behaviour of the Australasian sea lions *Neophoca cinerea* and *Phocarctos hookeri* (Pinnipedia: Otariidae). Mammalia 39:159–230

McCulloch S, Boness D (2000) Mother–pup vocal recognition in the grey seal (*Halichoerus grypus*) of Sable Island, Nova Scotia, Canada. J Zool 251:449–455

McCulloch S, Pomeroy P, Slater P (1999) Individually distinctive pup vocalizations fail to prevent Allo-suckling in grey seals. Can J Zool 77:716–723

Page B, Goldsworthy SD, Hindell MA (2002) Individual vocal traits of mother and pup fur seals. Bioacoustics 13:121–143

Parr LA, de Waal FB (1999) Visual kin recognition in chimpanzees. Nature 399:647–648. https://doi.org/10.1038/21345

Petrinovich L (1974) Individual recognition of pup vocalization by northern elephant seal mothers. Z Tierpsychol 34:308–312. https://doi.org/10.1111/j.1439-0310.1974.tb01803.x

Phillips AV, Stirling I (2000) Vocal individuality in mother and pup South American fur seals, *Arctocephalus australis*. Mar Mamm Sci 16:592–616. https://doi.org/10.1111/j.17487692.2000.tb00954.x

Pitcher BJ, Ahonen H, Harcourt RG, Charrier I (2009) Delayed onset of vocal recognition in Australian sea lion pups (*Neophoca cinerea*). Naturwissenschaften 96:901–909. https://doi.org/10.1007/s00114-009-0546-5

Pitcher BJ, Harcourt RG, Charrier I (2010) Rapid onset of maternal vocal recognition in a colonially breeding mammal, the Australian sea lion. PLoS One 5:771–776. https://doi.org/10.1371/journal.pone.0012195

Pitcher BJ, Harcourt RG, Charrier I (2012) Individual identity encoding and environmental constraints in vocal recognition of pups by Australian sea lion mothers. Anim Behav 83:681–690. https://doi.org/10.1016/j.anbehav.2011.12.012

Riede T, Fitch T (1999) Vocal tract length and acoustics of vocalization in the domestic dog (*Canis familiaris*). J Exp Biol 202(20):2859–2867

Riedman M (1990) The pinnipeds: seals, sea lions, and walruses. University of California Press, Berkeley, CA

Sauvé CC, Beauplet G, Hammill MO, Charrier I (2015a) Acoustic analysis of airborne, underwater, and amphibious mother attraction calls by wild harbour seal pups (*Phoca vitulina*). J Mammal 96:591–602

Sauvé CC, Beauplet G, Hammill MO, Charrier I (2015b) Mother-pup vocal recognition in harbour seals: influence of maternal behaviour, pup voice and habitat sound properties. Anim Behav 105:109–120. https://doi.org/10.1016/j.anbehav.2015.04.011

Stewart R, Fay F (2001) Walrus. In: The new encyclopedia of mammals. Oxford University Press, Oxford, pp 174–179

Terhune J, Terhune M, Ronald K (1979) Location and recognition of pups by adult female harp seals. Appl Anim Ethol 5:375–380

Trillmich F (1981) Mutual mother-pup recognition in Galapagos fur seals and sea lions: cues used and functional significance. Behaviour 78:21–42

Trimble M, Charrier I (2011) Individuality in South American sea lion (*Otaria flavescens*) mother-pup vocalizations: implications of ecological constraints and geographical variations? Mamm Biol 76:208–216. https://doi.org/10.1016/j.mambio.2010.10.009

Tripovich JS, Rogers TL, Canfield R, Arnould JPY (2006) Individual variation in the pup attraction call produced by female Australian fur seals during early lactation. J Acoust Soc Am 120:502. https://doi.org/10.1121/1.2202864

Van Opzeeland IC, Van Parijs SM, Van Opzeeland I, Van Parijs S (2004) Individuality in harp seal, *Phoca groenlandica*, pup vocalizations. Anim Behav 68:1115–1123. https://doi.org/10.1016/j.anbehav.2004.07.005

Van Opzeeland IC, Corkeron PJ, Risch D et al (2009) Geographic variation in vocalizations of pups and mother-pup behavior of harp seals *Pagophilus groenlandicus*. Aquat Biol 6:109–120. https://doi.org/10.3354/ab00170

Van Opzeeland IIC, Van Parijs S, Van Parijs SM et al (2012) Individual variation in pup vocalizations and absence of behavioral signs of maternal vocal recognition in Weddell seals (*Leptonychotes weddellii*). Mar Mamm Sci 28:158–172. https://doi.org/10.1111/j.1748-7692.2011.00505.x

Wierucka K, Pitcher BJ, Harcourt R, Charrier I (2017) The role of visual cues in mother–pup reunions in a colonially breeding mammal. Biol Lett 13:20170444. https://doi.org/10.1098/rsbl.2017.0444

Wierucka K, Charrier I, Harcourt R, Pitcher BJ (2018a) Visual cues do not enhance sea lion pups' response to multimodal maternal cues. Sci Rep 8(1):9845. https://doi.org/10.1038/s41598-018-28171-w

Wierucka K, Pitcher BJ, Harcourt RG, Charrier I (2018b) Multimodal mother–offspring recognition: the relative importance of sensory cues in a colonial mammal. Anim Behav 146:135–142. https://doi.org/10.1016/j.anbehav.2018.10.019

Chapter 10
Acoustic Coding of Information in a Complex Social Network: Identity Signaling in Northern Elephant Seals

Caroline Casey

Abstract The process of learning plays a pivotal role when an animal must correctly identify individuals within a dynamic social group. In male–male competition for access to reproductive success, a male's ability to learn the calls that are associated with one's rival can help to maintain structured dominance relationships and reduce the costs associated with fighting. Male northern elephant seals operate in one of the most competitive breeding systems among mammals, and selection pressures for accurate rival assessment are extreme. Through a long-term field effort that has tracked individual males over multiple breeding seasons, we have found that male elephant seals are operating in a large, spatially dynamic social network, which is conducive to the emergence of individual acoustic recognition and associative learning. Given their accessibility during the breeding season, the northern elephant seal provides an excellent comparative framework for studies of the relationship between signal function and an individual's social environment.

10.1 Introduction

10.1.1 Learning and Communication

Communication is a fundamental feature of all animal social systems, and serves to support significant life history functions such as navigation, breeding, foraging, and parental care. While there are many ways to define communication, most agree that it involves the general process of information exchange between signalers and receivers, during which the receiver can extract biologically important information and potentially exhibit some kind of behavioral response. The potential information contained in animal vocalizations may range from motivational state to individual

C. Casey (✉)
Department of Ecology and Evolutionary Biology, University of California Santa Cruz, Santa Cruz, CA, USA
e-mail: cbcasey@ucsc.edu

T. Aubin, N. Mathevon (eds.), *Coding Strategies in Vertebrate Acoustic Communication*, Animal Signals and Communication 7,
https://doi.org/10.1007/978-3-030-39200-0_10

identity to competitive ability. While some responses to signal characteristics seem to be reflexive, learning often plays an important role in how animals interpret and respond to signal features associated with a caller. Individuals possess flexibility in their decision-making, and acquire valuable information as a consequence of their own behavior and the behavior of others. Imagine for a moment a male great tit defending his territory during the breeding season. While perched atop a branch, he hears the song of a nearby rival. The song may be produced by a male with whom he has already interacted, during which he successfully defended his territory, or alternatively, was defeated. How the listening male chooses to respond to his competitor (either by attacking or retreating from the invader) may be influenced by the male's previous experience with this individual. In this way, combined with inherent tendencies and direct assessments about size or strength, previous experiences and their consequences can alter the choices that animals make within communicative contexts.

While some behaviors are innate, such as an individual's reflexive fear of predators, many others are influenced by the process of learning. Learning can be defined as a behavioral change that arises from experience that is not attributed to maturation or development (Domjan 1998). Within the world of an animal, a particular action may be followed by an event that is positive, negative, or neutral. The type and magnitude of the consequence alters the probability of whether an emitted response is likely to be repeated in the future. The laws of associative learning have been well established through a long history of laboratory studies that carefully measured the conditions under which behavior changes over time for many species. These behavioral laws also apply in the natural world, where animals must make decisions within their environment based on what they have learned through prior experience. In the case of the great tit, the call of his neighbor is associated with an experience (being beaten by or beating the bird who is producing the call), and he must recall this event, and decide whether to attack or retreat in response.

10.1.2 The Role of Associative Learning During Individual Recognition

While individuals may be genetically predispositioned to respond to various species-specific signals (e.g., through phenotype or allele matching), learning can play a pivotal role when a receiver must correctly identify individuals within a dynamic social group based on familiarity (Bradbury and Vehrencamp 1998). This process is referred to as individual recognition, and is commonly established through the process of associative learning. Individual recognition is accomplished when a receiver is able to discriminate between familiar conspecifics based on previous experience and memory of distinct features associated with that individual (Bradbury and Vehrencamp 1998; Hauser 1996). The information value of a signal is dependent on the extent to which it is associated with, and therefore accurately represents, the individual (Seyfarth and Cheney 2003).

Communication strategies based on recognition are likely to evolve in social systems where the risks associated with misclassifying individuals are high. Under these conditions, recognition is dependent on both signal design and a receiver's perceptual tuning to key elements of the signal. In order to be meaningful for individual recognition, the identity signal, or individual signature, must be detectable amongst other features (both biotic and abiotic) within the environment, and must be memorable to the receiver. To be memorable, the signature should be stable across motivational contexts and time, and should exhibit very little within-individual variation. When being different from other individuals is advantageous, individually distinctive signals will spread via negative frequency-dependent selection, such that individuals who look, sound, or smell unique within social groups will be favored (Tibbets and Dale 2007; Sheehan and Tibbets 2009; Dale et al. 2001).

10.1.3 Selection Pressures that Drive the Emergence of Individual Recognition

The complexity of a signal's design is directly related to the number of classes that an individual must distinguish between. The more individuals within a social system, the more difficult the perceptual task becomes, and the higher the degree of "uniqueness" that an effective signal must exhibit (Bradbury and Vehrencamp 1998; Krams et al. 2012; Freeberg et al. 2012). This becomes even more imperative if individuals are highly mobile within the environment, thereby making the task of recognition more difficult due to the lack of reliable spatial cues. The relationship between signal complexity and an animal's social and ecological conditions has been explored across a variety of different colonial systems (Charrier et al. 2001; Aubin and Jouventin 2002; Mathevon et al. 2003), including species of cliff swallows. Medvin et al. (1993) evaluated two different species of swallow (the cliff swallow, *Hirundo pyrrhonota*, and the barn swallow, *H. rustica*) which are both colonial nesters. However, cliff swallows build nests that are more dispersed where there are fewer chicks for a mother to distinguish between. Barn swallows live in tightly packed, larger groups in which a mother has to distinguish her young amongst several hundred other chicks. Medvin et al. measured a number of call features from chicks in each environmental condition, and found that calls produced by chicks living in denser nesting environments exhibit far more variation in both the temporal and spectral domains (Medvin et al. 1993) than do chicks raised in more dispersed groups where the risk of misidentification is low.

The design of signature signals is also dependent on the ecological constraints of the environment itself. With regard to acoustic signaling in noisy environments, the risk of misidentification of target individuals is high. The range over which signature information is still perceptible to the receiver, or the active communication space of the signal depends (among other things) on the specific features within the call that are responsible for encoding individual identity. Among the various kinds of signaling strategies, signature information is most often exhibited in the acoustic domain (e.g., Aubin and Jouventin 2002; Charrier et al. 2002; Searby and Jouventin 2003; Mathevon et al. 2003). For example, the territorial call of the male corncrake (*Crex crex*) encodes individual identity through temporal features of their vocalization, and this information can be transmitted up to 100 m through dense vegetation without degradation (Rek and Osiejuk 2011). Alternatively, if identity information is encoded in frequency-related features of the call (specifically higher frequencies), then signature information may disappear faster due to the susceptibility of high frequencies to be lost with increasing distance (Mathevon et al. 2008). Detectability may be improved by possessing a loud signal (Brumm 2004), having a signal that exhibits a frequency range well matched to penetrate the environment, or having a signal that is highly repetitive (Brumm and Slater 2006).

10.1.4 Intra-sexual, Mate, and Kin Individual Recognition

Interactions between individuals range from antagonistic (e.g., exchange of information between individuals competing for resources) to affiliative (e.g., exchange of information between parents and offspring), and the potential costs of misidentification have a large effect on signal design. The degree to which individual recognition facilitates coordination between individuals varies depending on the nature of the interaction. However, any system that relies on individual recognition likely involves frequent interactions between group members under social conditions where a receiver has to distinguish between multiple individuals. In these circumstances, the receiver has the opportunity to learn the unique qualities of an individual of interest, and can associate these cues with information that assists the receiver in navigating his/her social environment (e.g., finding offspring, mates, or identifying rivals).

The ability to accurately assess and respond to other individuals is paramount when the consequences associated with misidentification have a detrimental effect on fitness. This can be demonstrated clearly during conflict resolution between individuals, during which the optimal outcome for one individual is different than that of the other. In some cases, the suboptimal outcome can be energy expenditure, injury, or even death. Under these circumstances where the cost of misidentification is high, it becomes increasingly important that individuals correctly identify their rivals. As females are typically the resource of interest, this effect is most pronounced during male–male competition, during which males compete for access to breeding opportunities. Males may produce signals that contain information about the relative fighting ability or threat level of the sender, and receivers may have the opportunity to extract information from those signals and make appropriate decisions about whether to attack or flee (Searcy and Nowicki 2005). Under these consequences, misidentification of rivals can have detrimental social effects, as males may accidently challenge more dominant individuals and lose, or alternatively, allow a subordinate male to displace his position within the social hierarchy.

10.1.5 Honest Signaling Versus Individual Signatures

Depending on the social system within which information exchange occurs, males competing for resources may rely on acoustic signals that communicate phenotypically linked acoustic cues that indicate the size or strength of the sender, or alternatively, that convey individual identity. In the former system, information about an individual's size, age, or level of arousal is encoded within specific features in the call, thereby giving the receiver an honest indication of the signaler's fighting ability. Variation among callers allows receivers to distinguish between males that possess varying levels of resource holding potential (e.g., Kitchen et al. 2003). This kind of signaling approach is likely more common in unstable social systems in which repeated social interactions between individuals are infrequent, and animals must acquire information about their competitor's level of threat quickly. In this scenario, receivers can gain important information about a signaler even without prior experience with that individual. For example, male American bison (*Bison bison*) utilize call amplitude to assess the physical condition and motivation of potential competitors (Wyman et al. 2012). Conversely, in more stable social groups in which males have the opportunity for repeated interactions, individuals competing for resources may be more likely to rely on acoustic signals that convey individual identity (Tibbetts and Dale 2007). These signals are often consistent within a given individual but variable between group members, and require association with competitors and consequences become meaningful. In this case, learning the identity and relative fighting ability of one's rivals allows a male to assess the strength and/or stamina of his familiar competitors (Tibbetts and Dale 2007). Such associative learning can lead to improved rival assessment over time.

Although both types of signaling strategies are useful in assessing one's opponents, they can be distinguished by observing the responses of individuals to the calls of unfamiliar rivals. Signals conveying individual identity are only useful in informing listeners about the fighting ability of familiar rivals with whom they have had previous experience. Alternatively, honest signals can inform listeners about the impending threat of both familiar and unfamiliar senders. Recognition of one's previous opponent can be the most secure strategy for rival assessment when both the competition level and the cost of physical fights are extremely high. In a system where great size and strength are traits of any male who survives to adulthood, signals conveying honest information about male quality may not be that informative, and an alternative means for rival assessment (such as individual recognition) may be necessary.

10.1.6 Male–Male Competition Among Northern Elephant Seals

Owing to extreme selection pressures for rival assessment, the northern elephant seal (*Mirounga angustirostris*) provides an optimal social model to explore the role that learning plays in the communication system among breeding males. Because

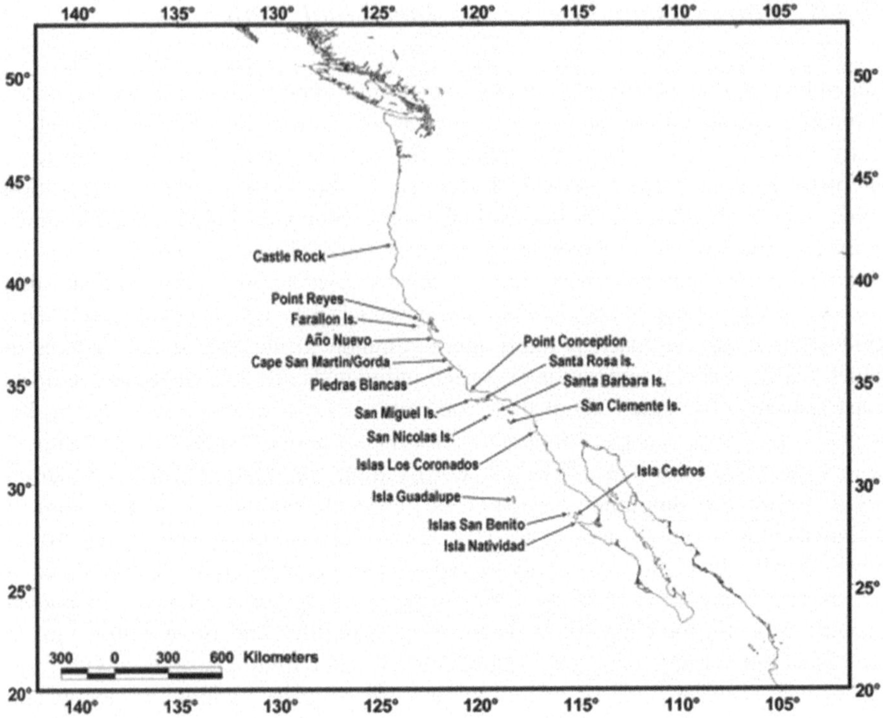

Fig. 10.1 Present locations of known northern elephant seal breeding colonies along the coasts of California and Mexico. Primary rookeries are located at San Miguel Island, San Nicolas Island, Año Nuevo, Point Reyes, the Farallon Islands, and Piedras Blancas (in California), and Isla Guadalipe and Islas San Benito (in Mexico) (from Lowry et al. 2014)

elephant seals congregate in predictable breeding locations each year, they enable the direct observation and close-range recording of acoustic displays emitted during competitive interactions, a detailed assessment of male dominance relationships, as well as the ability to experimentally test the significance of these calls to listening individuals. Additionally, the behavior of northern elephant seals has been studied extensively over the past several decades, providing the opportunity to ask both proximate and ultimate questions about communication and learning in this species.

Historic populations of northern elephant seals were rapidly and systematically extirpated along the coastline and offshore islands of North America beginning in the early 1800s to satisfy demand for rendered oil. More than 200,000 seals had been killed by the turn of the century, and the species was considered extinct by 1884. A small remnant population, rediscovered in 1911 on Isla de Guadalupe (150 miles from mainland Mexico), likely consisted of less than 20 individuals. Following the collapse of the sealing industry and subsequent protection by the Mexican government, these survivors enabled the population to recover to more than 210,000 animals, which can be observed each winter (December–March) at eight main island or mainland rookeries along the coasts of California and Mexico during the annual breeding season (Lowry et al. 2014; see Fig. 10.1).

The northern elephant seal is the largest seal in the northern hemisphere, with males weighing between 1500 and 2300 kg, and females reaching approximately one-third the size of males at 400–900 kg (Le Boeuf and Laws 1994). The species is also extremely sexually dimorphic. Aside from males being much larger than females, they also develop a long fleshy proboscis and a broad, thick, calloused chest shield that begins to form during puberty. Reproduction in this species is annually synchronous, and mature female seals aggregate by the thousands on breeding beaches each winter. Mature males arrive at these annual breeding sites early in the season to establish their dominance status prior to the arrival of adult females. The status of males within this social network has previously been divided into descriptive ranks based on each individual's proximity to female harems throughout the breeding season: alpha males hold stable positions within female harems, beta males hold flanking positions relative to harems, and younger, peripheral males are totally excluded from access to females (Le Boeuf 1974; Casey et al. 2015). It is worth noting that there is no evidence that females exhibit selection with regards to competing males. Their behavior is driven more by suitable haul-out space than by proximity to dominant individuals (Le Boeuf and Laws 1994).

Subadult and adult males remain ashore at their breeding colonies until the females have weaned their pups, completed their estrous cycles, and returned to sea—a tenure that may span 100 days without access to food or water (White and Odell 1971; Le Boeuf 1974; Deutsch et al. 1990). Compared to females, males live much shorter lives (Le Boeuf and Laws 1994). Only 5% survive to physical maturity (Condit et al. 2014), with less than 1% ever gaining reproductive access to females (Le Boeuf and Laws 1994). This asymmetry in life history and reproductive success underpins one of the most competitive breeding systems known among mammals.

10.1.7 Acoustic Signaling in Male Northern Elephant Seals

While the position of males within the dominance hierarchy is initially established through physical confrontations on breeding rookeries (Le Boeuf 1974; Haley 1994), the majority of agonistic encounters between males are resolved through the use of ritualized threat displays (Le Boeuf 1974). These displays are composed of multimodal components including visual posturing that emphasizes body size; incredibly loud, pulsed vocalizations directed at opponents; and occasional seismic cues that are produced by slamming their heavy chests against the ground (Bartholomew and Collias 1962; Sandegren 1976). These vocalizations are among some of the loudest recorded among any mammal in air, and can reach up to 131 dB re: 20 μPa (peak) (Southall et al. 2019). The vocalizations produced by males during their displays, traditionally called "clap threats," contain 3–20 broadband units emitted at high levels with repetition rates of a few pulses per second (Le Boeuf and Peterson 1969). Individuals show reliable substructure within the repeatable units comprising the rhythmic portion of each call, which enables even a human listener to easily discriminate between individuals. These acoustic displays appear to

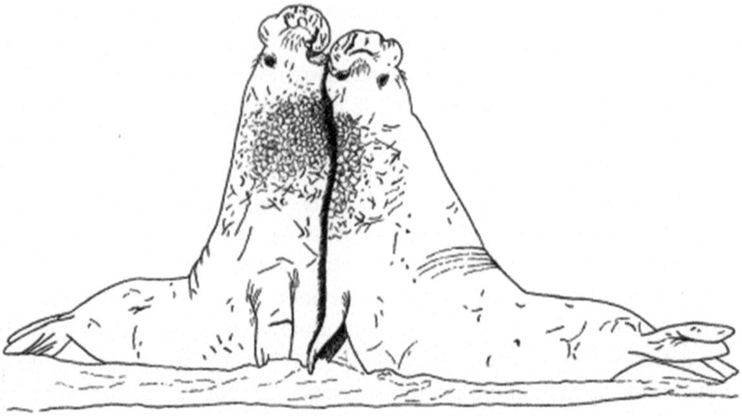

Fig. 10.2 Two equally matched adult male northern elephant seals engaged in battle at the beginning of the breeding season. Drawing by E. Levy

be of particular importance to male northern elephant seal, as calling males are often able to control the movement of rivals over large distances (Bartholomew and Collias 1962; Sandegren 1976), even under circumstances where visual cues are limited (Shipley and Strecker 1986). In over 8000 observations of competing males during multiple breeding seasons, over 80% of interactions involved vocalizations on the part of at least one of the males (Casey et al. 2015).

Throughout the breeding period, dominant individuals closely monitor the behavior of subordinate males who retreat from the vocal displays of their dominant rivals. Fighting is rare during most of the season (Cox 1981), presumably because of the effectiveness of communicative exchanges, the extreme energetic cost of engaging in battle during a prolonged fasting period, and the risk of serious injury or death (Le Boeuf and Laws 1994). Less than 6% of competitive interactions that occur between males involve physical contact (Casey et al. 2015). Battles only occur when repeated vocal exchanges are insufficient to end an escalating conflict, and neither individual retreats—typically between males who have not fought previously (Casey et al. 2015; see Fig. 10.2) and whose relative dominance status is uncertain (Bartholomew and Collias 1962).

Several investigators have noted individual differences in call characteristics between males within specific breeding sites (Bartholomew and Collias 1962; Shipley et al. 1981). This variation has been attributed to phenotypic traits and/or social dominance rank (Bartholomew and Collias 1962), as well as maturational effects (Schusterman 1978). Alternatively, it has been suggested that these signals function to convey individual identity and that males relate distinctive features of a threat call to a specific male through learned associations (Sandegren 1976; Shipley et al. 1981). Regardless of the potential function of male threat displays, several independent researchers have commented on the apparent stability of a male's vocalization both within seasons and across multiple years (Le Boeuf and Petrinovich 1974). Until recently, it was still unclear whether the acoustic displays of male northern elephant seals function as honest signals that opponents can decode without prior experience, or whether they are individual identifiers which males must learn in order to economize their effort during the energetically demanding breeding season.

10.2 The Case for Identity Signaling in Male Northern Elephant Seals

Our research team has spent nearly a decade attempting to decode the signals of these giants at our field site at the Año Nuevo breeding colony in central California. During this time, we carefully monitored adult male elephant seals while ashore during the annual breeding season. Information about individual phenotype, fine-scale spatial use patterns, and pairwise competitive interactions was collected, cataloged, and combined with detailed analyses of their specialized acoustic displays. In essence, we determined who these individuals were, where they went during the breeding season, what they sounded like, and how successful they were at operating in their dynamic social network.

We took a comprehensive approach to this work, and individually identified subadult and adult males (using small plastic flipper tags and black hair dye) reliably seen at the Año Nuevo breeding colony study site (for details see Casey et al. 2015). We closely monitored their movements throughout the season, and observed male–male interactions to characterize where each male fell within the overall dominance hierarchy. With regard to vocal displays, we were interested in first determining the distinctiveness of each male's call and which features of the vocalization were most stable within an individual. From our own experience, we found that a practiced human observer could accurately identify a male based solely on the temporal characteristics of his call. At the beginning of each breeding season, we were delighted when we heard the vocalizations of our favorite males (even before reading their identification tags) when they emerged from the water, as it meant they had survived another year at sea. For a subset of individuals that we were able to track both within and across seasons, we also assessed the reliability of these vocalizations across years and social contexts. We wanted to examine, for example, whether a male's call would change in response to varying levels of perceived threat. Data pertaining to the size and dominance status of callers was then compared to the vocalizations produced by these males, to determine whether specific call parameters may be correlated with a male's resource holding potential.

After an analysis of the calls recorded from multiple adult males within and between years, we found that the vocal displays produced by adult male elephant seals were individually unique (Fig. 10.3), with one temporal and one spectral call component sufficient to allow for accurate differentiation of individuals (Fig. 10.4). This analysis confirmed previous observations that there are notable and consistent differences in the calls of adult males, and these signals remain the same, even when emitted during different behavioral states (e.g., calling alone versus during a fight).

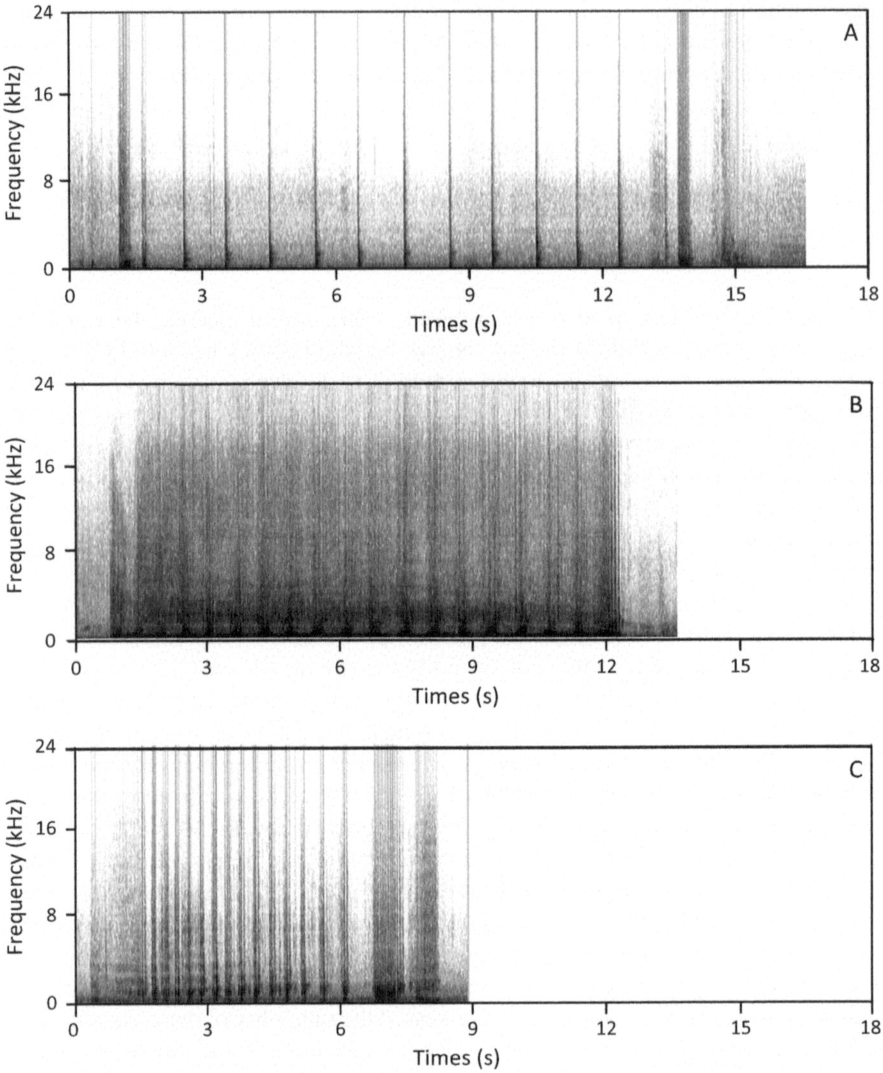

Fig. 10.3 Comparison of representative vocalizations produced by three adult male northern elephant seals of equal dominance status within the hierarchy at the Año Nuevo breeding colony. All three calls were recorded at 0-degree orientation and within 10 m of the vocalizing animal. Spectrograms using a Hamming window and 90% overlap are provided for Male 3C (Panel A) Male 10A (Panel B) and Male X579 (Panel C); sampling rate 24 kHz, FFT size 256

Each call appears to be composed of a series of repeating units, which is perceived by human listeners as the tempo of the vocalization. Within these repeating units, some males exhibited substructural components that help to acoustically differentiate individuals. In agreement with previous observations of this species, we also

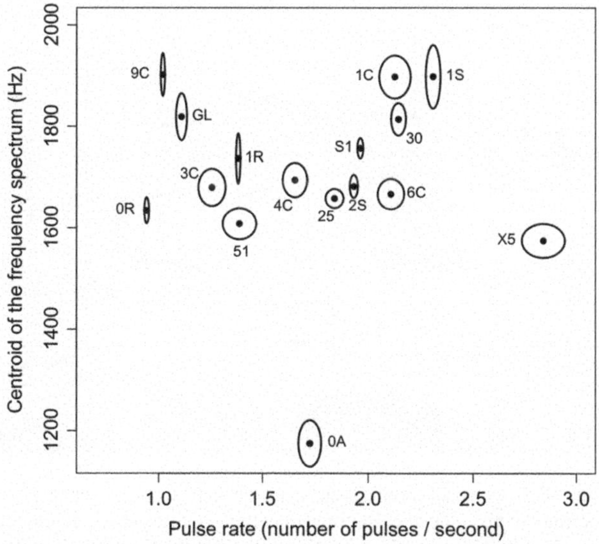

Fig. 10.4 Calls can be reliably assigned to individuals using two acoustic parameters (mean ± SE): the centroid of the frequency spectrum and the number of pulses per call (the two main factors that separate individuals on the first discriminant function of the cross-validated DFA). Adapted from Casey et al. (2015)

discovered that the calls (including call amplitude) produced by adult males are stable across multiple seasons. The repetitive nature of these high-amplitude calls may potentially aid in signal detection in high density social environments. One of our study males produced a call with a peak sound pressure level reliably measured to be 140 at 1 m, which is one of the loudest sounds ever recorded among mammals (Suer et al. 2011). Interestingly, despite his call amplitude, this individual was not particularly dominant within the hierarchy and spent the majority of his time on the outskirts of a harem held by another adult male.

While we observed a great deal of variation between the calls of adult males, and a range of social ranks among breeding individuals, we found no correlation between the kind of call a male exhibits and his dominance status. For example, the calls illustrated in Fig. 10.3 were all produced by alpha males from different harems recorded at the Año Nuevo breeding colony in 2011.

The variation in call structure between individuals and the stability of calls within males over time supported the hypothesis that these calls function as identity signals that males must learn and subsequently remember throughout the breeding season (Fig. 10.3). The presence of vocal signatures by itself, however, was not sufficient to illustrate that social recognition does indeed occur within this complex social system (Insley et al. 2003). The role that learning plays in the social lives of males can only be investigated with a clear experimental approach during which the level of an individual's experience with the caller is systematically assessed. To address this, the results of our vocal analysis were then applied to two field playback experiments that tested predictions about whether these unique signals effectively function to convey individual identity.

In the first set of playback experiments we targeted 10 beta males reliably present at the Año Nuevo study site. We first sought to determine whether the acoustic

Fig. 10.5 Playback configuration in the field testing each male's responsiveness to the calls of familiar rivals. The focal animal receiving the playback is circled, and the playback speaker is positioned ~7 m away from, and on axis with, the target animal. Researchers are positioned outside of the direct line of view of the male of interest, and are able to initiate the playback remotely

portion of this multimodal display was enough to elicit a behavioral response from the receiver, as suggested by observations of males retreating from or challenging rivals when visual cues are limited. In other words, we asked "can vocalizations alone substitute for callers themselves?" We also wished to explore whether the calls of known rivals would prompt differential responses depending on relative dominance status between the caller and receiver. For these experiments, each of the 10 target males received the playback of their most familiar dominant and subordinate rival, and their behavioral response was recorded and compared to observations of these interactions under natural conditions (Fig. 10.5). These subordinate and dominant treatments were selected based on our own observations in the field. The results of these experiments confirmed our observations that males would respond differentially to the calls of their familiar subordinate and dominant rivals. In these experiments, 75% of the animals exhibited a behavioral response to the playback condition, and of those that responded, 86% did so in the predicted direction—i.e., males exposed to their most familiar dominant rival retreated from the speaker, and those exposed to their most familiar subordinate rival attacked the speaker.

In a second set of playback experiments performed at a distant breeding location, we examined whether previous experience with the caller was necessary to elicit a behavioral response from the animal that was listening. In these experiments, we traveled to the Piedras Blancas breeding colony located 480 km south of the Año Nuevo breeding location. We observed the social interactions of competing males at this site, and selected 10 beta males to receive the same subordinate and dominant treatments given to individuals at the Año Nuevo colony. The only difference between these two conditions is that males from the Piedras Blancas breeding colony had never interacted with the individuals whose call they were being presented. In these experiments, only 15% resulted in the male responding the playback stimuli at all, and only two in the "predicted" direction. In the majority of cases the males attended to the sounds being produced by the playback speaker, but did not exhibit a retreat or approach response, but instead opted to stay put where they were with no measurable change in body position relative to the playback speaker.

The results of these experiments demonstrated that the vocal component of male multimodal displays is enough to elicit a behavioral response from the receiver, suggesting that the calls themselves can functionally substitute for the individual during agonistic interactions. Furthermore, when presented with the playbacks of familiar dominant and subordinate associates, males were able to discriminate between the vocalizations of known individuals and respond appropriately based on the relative dominance status of the caller. In order for the playback to encourage a significant response from the animal being tested, however, it appeared that each male must have had prior experience with the individual whose call he was receiving. Males with no prior experience with a caller exhibited little response to playbacks, confirming that males do not respond to phenotypically linked information within the construct of the call, but rather associate variation in call structure with the consequences of previous challenges with familiar rivals. This does not mean that the males exposed to unfamiliar calls were indifferent; rather, their responses could represent the safest strategy when assessing the calls of unfamiliar challengers. Mid-ranking males may have a good deal to gain by asserting dominance over new contestants, but a substantial amount to lose if their new opponent is far larger or more motivated to attack. This dichotomy in motivation to respond may leave mid-ranking males at a draw when first assessing the calls of unfamiliar opponents. In these cases, males may require additional information about their opponent before deciding to attack or retreat.

Clearly, there is important information contained within the calls of individuals, but are males truly capable of discriminating slight differences in two main parameters that acoustically separate these signature calls—pitch and rhythm? We explored this question through a follow-up set of playback experiments, during which we experimentally modified the most distinctive characteristics of the vocalizations of familiar dominant rivals and then tested how this influenced a male's responsiveness to playbacks of these signals (Mathevon et al. 2017). We tested 10 beta males with a range of calls that had been altered with respect to both tempo and pitch, the two acoustic features of the vocalizations that were most reliable within individuals (Casey et al. 2015, Fig. 10.4). Some of these alterations were within the normal range of intraindividual variation, while others fell outside of this range. The playbacks revealed that both the natural calls and the experimental calls from dominant individuals that were modified within the range of typical intraindividual variation elicited expected retreat behavior from tested males, indicating that they were still able to recognize these familiar contestants. Conversely, the experimental calls with alterations outside of this range (but within the range of the population) were ignored by the tested males (Mathevon et al. 2017). This finding is consistent with previous observations of the natural behavior of males in response to calls from familiar dominant versus unknown rivals. While the use of spectral features for individual discrimination has been found in a number of mammals including pinnipeds (Aubin et al. 2015; Charrier et al. 2003), the northern elephant seal is the first mammal species shown to learn, remember, and use sound metrical patterns across a wide range of tempi in a biologically representative context (Mathevon et al. 2017).

10.3 The Social Conditions that Support Individual Recognition in Northern Elephant Seals

Clearly associative learning plays an important role in this system, and males must learn the calls of their rivals in order to optimize energy conservation. What specific social conditions might encourage this kind of signaling strategy? During male–male competition over access to mating opportunities, the use of individually distinct vocalizations should occur in a system where individual recognition is more informative and less costly than a system based on vocalizations conveying only information about one's size or fighting ability. Male northern elephant seals must therefore have the opportunity to not only learn the distinctive call characteristics of their most familiar competitors, but also have sufficient experience with their rivals for these cues to become meaningful. These behavioral patterns should be influenced by ecological conditions at the breeding site, including availability of suitable haul-out space, which in turn influences the distribution of seals within the colony. A communication system based on a competitor's ability to learn and remember these identity signals should be expected to exhibit the following social and environmental conditions: (1) males returning to the same general location on the breeding beaches each year (fine-scale site fidelity), (2) a strong degree of familiarity between rivals, and (3) the use of reliable acoustic signatures during competitive interactions that have predictable outcomes.

10.4 Fine-Scale Site Fidelity of Male Northern Elephant Seals

Fine-scale site fidelity constitutes an important structural component in animal social networks, as it sets limits to the space over which social interactions can occur and provides the opportunity for repeated interactions between group members (Wolf et al. 2007). While loyalty to a particular breeding colony has been broadly considered for both sexes, details pertaining to fine-scale site fidelity within the colony (both within a season and throughout an animal's lifetime) among male elephant seals remain sparse. Male tenure within a breeding season has been broadly considered, and roughly 20% of the males within one colony remain until all the females have departed (Le Boeuf 1974). Additionally, dominant males have been reported to be successful breeders for up to three consecutive years (Le Boeuf 1974) at the Año Nuevo breeding colony, indicating some level of fidelity to a specific breeding location.

While there has historically been less of an emphasis on tagging males within the Año Nuevo population, recent improvements in our ability to track individual males have enabled us to follow focal individuals across multiple years. As a consequence, we now know that the majority of our study males return to the same breeding colonies between seasons. This development offers an incredible opportunity to explore the behavioral strategies that males exhibit across different age classes during breeding events, which may ultimately lead to their reproductive success later in life. Of the 11 alpha males we observed during the 2016 breeding season for example, 10 had been observed at the Año Nuevo breeding rookery in prior years. Additionally, all of them arrived early in the breeding season relative to other males, and stayed until every last female departed to sea. We often witnessed successful males return after their foraging trips to within 5 m of the location where they had been seen during the previous year, even before any females arrived at these breeding beaches. Once individuals have reached sexual maturity (between 4–6 years of age), males may avoid moving to new breeding sites, as they may need to fight to establish a new position within an unfamiliar dominance hierarchy.

One of our most well-studied individuals at Año Nuevo provides an example of the level of insight these kinds of detailed observations can provide. Timmy (tag number U46) was born during the 2008 breeding season and was immediately abandoned by his mother. He was subsequently nursed by another female, and was successfully weaned at the end of the season although he was below the average weaning weight. Since 2008, Timmy has been sighted at the breeding colony each season. As a subadult he was observed drifting between locations around the colony and was always seen on the perimeter of groups of females. During the 2015–2016 season (at the age of 7), he began to compete more directly with older, more dominant individuals, and was consistently observed near a single harem interacting with an established group of adult males. Although he lost nearly every fight he instigated in the winter of 2016, he was noted as being a very motivated and energetic individual, often getting chased out of the harem by a more dominant male only to turn around and try his luck again. During the 2017 breeding season, Timmy returned to the same area he was in the previous year, and quickly asserted his dominance and controlled access to a small group of females for approximately one month. He was later displaced by a larger individual, and yielded to holding a beta position within the hierarchy for the remainder of the season. We hope that Timmy will return to Año Nuevo next year, and perhaps finally attain his alpha status on the rookery. Ultimately his success in becoming dominant may be related to prior knowledge and ability to navigate the complex social network of male northern elephant seals. Although anecdotal, Timmy's story provides some insight into the different behavioral strategies that enable a male to one day achieve breeding opportunities as an adult, and has motived our research team to continue to track focal individuals throughout their lifetimes. These kinds of observations suggest that males are operating in a stable, complex social environment that is conducive to the emergence of individual recognition and associative learning.

10.5 The Social Network of Male Northern Elephant Seals

The stability of male seals in time and space will influence the number of social partners that an individual interacts with within and across breeding seasons, and the strength of association and familiarity between rival males. Paired with associative learning capabilities that allow individuals to link acoustic signatures to predictable social consequences, the use of reputation signals may produce an evolutionary stable strategy within the context of dominance interactions in familiar social groups. Additionally, signal design is likely influenced by the number of individuals a male must discriminate between. In our study, individual male elephant seals were observed to engage with as many as 43 opponents over the course of a single breeding season. While the relationship between signal and social complexity still remains to be investigated in this species, small differences between individuals in call substructure likely aid in this process. Through close observation of males within and between breeding seasons, we have been able to characterize important details of this dynamic social network.

A male's success within this social system depends on managing knowledge of his relative dominance status compared to others within the hierarchy. While the Año Nuevo colony in California has approximately 400 breeding males that may be present across an ~3 km^2 breeding area, the site is divided into 11 evenly distributed stable harems, each with anywhere between 50 and 150 adult females and their pups. One alpha adult male (age 8–11 years old) controls access to a single harem throughout the season, while beta males (most of whom are 7 and older) sit on the outskirts of these groups awaiting the perfect time to attempt to breed with one of the females while the alpha is not looking (Fig. 10.6). Over the course of one breeding season, we observed alpha males maintaining stable relationships with an average of 38 other individuals (Casey et al. 2015—Fig. 10.7). Alpha males must learn to balance their time between vigorously defending their harems and also breeding with females, while minimizing energy and water loss during periods of extended fasting. The defensibility of the alpha male's harem varies with the density of females, as larger harems become increasingly more vulnerable to infiltration by other individuals (Le Boeuf and Laws 1994).

The costs of overexertion are evident in observations of males ascending to the peak of the social hierarchy early in the season, only to be overthrown by their close competitors after engaging in excessive chasing and fighting. However, this is not to say that these individuals are incapable of returning in subsequent seasons to resume their position as top males within the hierarchy; we observed one male who was ousted half-way through the breeding season by two beta males, only to come back the following year to reign again as alpha at the same harem. It is important to note

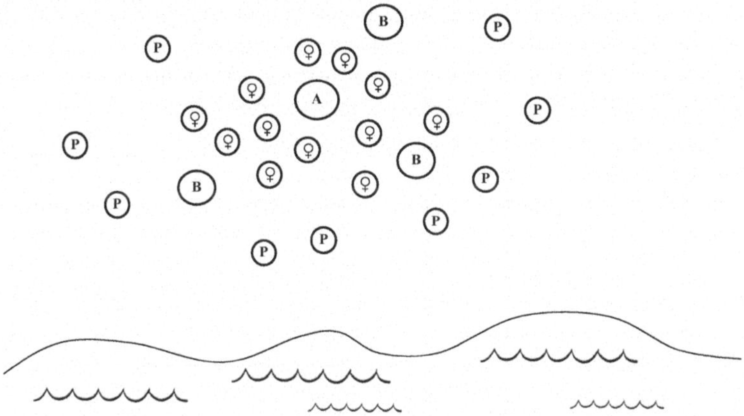

Fig. 10.6 Aerial schematic of a typical harem. The "A" dot represents the alpha male, who is positioned in the center of a group of females, and is surrounded by several "B" dots, representing beta males. Peripheral males (P dots) remain on the outskirts of the harem without access to females

Fig. 10.7 An example of the social network of northern elephant seal males. Bottom row of larger dots = alpha males and corresponding names that controlled female harems over the season; second from bottom dots = beta males that reliably held harem flanking positions with opportunistic access to females; top two rows of lighter dots = peripheral males that typically lacked access to harems. Each arrow represents an interaction, drawn from a winner to a loser. The thickness of the arrows is proportional to the number of observed competitive events between two males. Adapted from Casey et al. (2015)

that in this case we observed no change in the call characteristics emitted by this individual—further indicating that the calls of adults do not change relative to their fighting ability. Substituting fighting and chasing with calling may represent the most efficient strategy for avoiding harm and conserving energy in a system where the costs associated with excessive activity may have unfavorable consequences for individual reproductive success.

Beta males display a number of different strategies in pursuit of breeding opportunities throughout the season. Some individuals attempt to tire the alpha through continual harassment, while others tuck their noses under and slink across the sand and into the harem to mate with females while the alpha is asleep or distracted. We have even seen beta males sleep through the first half of the season in a nearby sand dune, and then attempt to overthrow the nearest alpha who had already exhausted much of his energy reserves. An alpha's harem is most susceptible to beta males toward the end of the season when females begin to depart to sea. As females are leaving the harem and approaching the shoreline, they must somehow evade dozens of subordinate males that are eagerly waiting for their opportunity to finally mate. This process is incredibly dangerous for females, as hormonally crazed beta and peripheral males will fight relentlessly around her as she attempts to run this dangerous male gauntlet. Indeed, we have witnessed females who are crushed to death as multiple males attempt to mate with her during this process. Some alpha males will escort females to the waterline to prevent subordinate males from copulating with her during her departure, which leaves his harem vulnerable to lurking betas. This is often the period where we witness beta males attain the most breeding success during the season, as alpha males are already exhausted and cannot escort females and guard their harem at the same time. There is obvious variation in the behavioral tactics of males, and this topic warrants further investigation in relationship to reproductive success.

What is clear from these observations is the level of familiarity between alpha and beta males, which provides individuals with the opportunity to learn the distinctive features of their closest competitor's calls during agonistic interactions. These relationships are stable both in association and the space over which they occur. In addition to constantly jockeying with one another for prime positions relative to female groups, beta males also spend a great deal of time chasing away peripheral males (ages 4–7 years old) from the harem. A smart alpha male may have a few familiar beta rivals he tolerates near his harem, and in turn those beta males indirectly assist in keeping away peripheral individuals (Fig. 10.6). Over a single season, we observed beta males maintain stable relationships with an average of 26 other individuals (Casey et al. 2015—Fig. 10.7).

Peripheral males rarely ever infiltrate these female groups, and show far less spatial stability within the colony compared to both alphas and betas. While higher ranking individuals can be consistently seen in association with one harem throughout the season, these peripheral subadults tend to float around the colony, loosely associated with different social groups. Interestingly, studies of male development have noted that younger animals rarely vocalize on the rookery during the height of the breeding season, as their activities are closely monitored by older individuals. Additionally, the calls of juveniles appear to be highly variable, and show particular instability with regard to call structure and pulse rate (Shipley et al. 1986). While it would be challenging for listeners to associate these unstable calls with known individuals, this may actually be advantageous for subadult males. If you are a loser, you may not want to be remembered! This could be advantageous to subadult males, who rarely win competitive contests and so might wish to remain acoustically inconspicuous until they have a fair chance at competing for reproductive access.

Notably, more than 97% of relationships observed between males over the course of a single breeding season were fully asymmetric (i.e., the interactions within a given dyad of males were always won by the same individual). These dyads are thus characterized by well-established "dominant–subordinate" relationships, illustrating the stability of the hierarchy between males (Casey et al. 2015). These observations support the existence of a communication system based on learning, in that the consequence associated with an individual's call is predictable within a given season. As individual vocal signatures remain stable over successive years and the majority of males appear to return to the same breeding-beaches annually, it is possible that males may recognize familiar rivals across breeding seasons. Prior to each season, the relative dominance status of individuals may thus be influenced by the long-term memory of past competitors. Alternatively, previous dominance status may have no bearing on these social relationships at the beginning of a new season, and one's competitive ability may be based on a number of factors such as foraging success and/or individual motivation. Given the strong association between males within and across seasons, multiyear recognition of rivals warrants further investigation in this species.

10.6 Vocal Ontogeny of Male Acoustic Displays

Despite many years of study, little is known about the ontogeny of northern elephant seal male vocal displays within an individual, and whether they are genetically based or shaped by auditory experience. Early work noted that the calls of juvenile males are more variable than those of adults (Shipley et al. 1986), suggesting that males go through a transitional period during development in which their calls lack the characteristic structure of adult males. This draws certain parallels to the ontogeny of passerine bird songs, in which the vocalizations produced by juveniles are a highly variable version of the adult repertoire, and then undergo a crystallization period and become relatively fixed beyond a certain age (Marler and Peters 1982). Studies of song development in passerine birds have shown that a common feature of vocal development includes a stage of babbling—characterized by overproduction and variability in structure—that is regarded as a "training" and a motor phase during which individuals have the opportunity to develop their own calls in relationship to the acoustic signals of adults (Snowdon and Hausberger 1997). We often observe subadult male northern elephant seals producing calls in social isolation from the harem and in a nondirected context, and that are highly variable in frequency, duration, and amplitude. The urge to produce these calling bouts appears to be hormonally driven, and can last for hours if uninterrupted. While it is difficult to tease apart the environmental and genetic influences shaping sound production, the northern elephant seal presents a particularly intriguing model. These seals exhibit a deficit in genetic variability due to human-induced population decline and subsequent inbreeding (Hoelzel 1999), which could allow for potential isolation of genetic factors contributing to vocal ontogeny. Further studies are required to determine the extent to which learning and the acoustic environment shape the development of these specialized acoustic displays, and the exact timing of call crystallization in this species.

10.7 Conclusions

The behavior of animals is best observed under natural social conditions; however, unraveling the information that is encoded within acoustic signals, and understanding how individuals use this information to adjust their behavior during social encounters, presents several challenges. Recent studies have correlated acoustic features with phenotypic traits and/or the ability to convey individual identity (Sanvito et al. 2007); however, few studies in mammals have confirmed the biological function of these signals through systematic experiments that evaluate the response of animals to the calls of individuals within their social network (e.g., Reby et al. 2005; Charlton et al. 2010). A complete understanding of signal form and function requires both analysis of call variables as well as targeted experiments that test the capacity of an individual to discriminate between the calls of conspecifics (Sayigh et al. 1999). This type of work is often prohibitively difficult, as it requires the ability to follow target individuals over time, extensive knowledge of the social relationships between individuals within a breeding group, detailed analysis of acoustic displays, and the opportunity to conduct playback experiments with focal animals. Despite the logistical challenges, these combined approaches are critical to understanding the role that recognition and learning play in mediating competition between individuals, and may provide insight into the social conditions driving the development of communication strategies among different animal groups. Given the growing interest in status signaling and selective pressures influencing the structure of animal social networks (Bergman et al. 2003; Charlton et al. 2011; Taylor et al. 2010; Ilany et al. 2013; Arnott and Elwood 2009), future studies should combine both descriptive and experimental methods to gain a true understanding of the information gained by listeners during social exchanges.

From a comparative perspective, our most comprehensive understanding of vocal communication comes from studies of passerine birdsong. Detailed field observations have been paired with careful laboratory experiments to address questions of mechanism, ontogeny, function, and phylogeny of vocal signaling in several songbird species (Konishi 1989). This holistic approach has resulted in birdsong being regarded as a model system for understanding the evolution of acoustic communication across several animal taxa, including those distantly related from birds. However, as these species operate in different social environments and display a wide diversity of acoustic signaling, it seems evident that we should continue to search for new models for cross-species comparisons, so that we may understand the relative importance of acoustic signaling across varying evolutionary landscapes.

The northern elephant seal has presented us with a remarkable opportunity to understand the role that sound plays in the reproductive behavior of mammals that rely on sound during important life history events. Calls can be easily recorded and attributed to known individuals and linked with specific behaviors; individuals can be followed throughout their vocal (and physical) development; and the function of different sound types can be experimentally tested and manipulated using playback experiments. Through this work, we have been able to decode the information contained within the specialized calls produced by males, and evaluate the role that learning plays during this process. What is notable about this system is the complexity of the social relationships between males, the ability of individuals to use this information, and the stability of these social relationships in time and space. As is usually the case, our findings have opened the door to several new questions pertaining to memory, development, and how these signals may have been influenced by this species' dramatic population decline and recovery from near-extinction. Future research focused on this species will continue to afford a detailed comparative framework for studies of acoustic communication in other mammalian species.

References

Arnott G, Elwood RW (2009) Assessment of fighting ability in animal contests. Anim Behav 77:991–1004

Aubin T, Jouventin P (2002) Acoustic systems are adapted to breeding ecologies: individual recognition in nesting penguins. Anim Behav 64(5):747–757

Aubin T, Jouventin P, Charrier I (2015) Mother vocal recognition in Antarctic fur seal *Arctocephalus gazella* pups: a two-step process. PLoS One 10:1–13

Bartholomew GA, Collias NE (1962) The role of vocalization in the social behaviour of the northern elephant seal. Anim Behav 10(1):7–14

Bergman TJ, Beehner JC, Cheney DL, Seyfarth RM (2003) Hierarchical classification by rank and kinship in baboons. Science 302(5648):1234–1236

Bradbury JW, Vehrencamp SL (eds) (1998) Principles of animal communication. Sinauer Associates, Sunderland, MA

Brumm H (2004) The impact of environmental noise on song amplitude in a territorial bird. J Anim Ecol 73(3):434–440

Brumm H, Slater PJ (2006) Ambient noise, motor fatigue, and serial redundancy in chaffinch song. Behav Ecol Sociobiol 60(4):475–481

Casey C, Charrier I, Mathevon N, Reichmuth C (2015) Rival assessment among northern elephant seals: evidence of associative learning during male–male contests. R Soc Open Sci 2(8):150228

Charlton BD, Zhihe Z, Snyder RJ (2010) Giant pandas perceive and attend to formant frequency variation in male bleats. Anim Behav 79:1221–1227

Charlton BD, Ellis WA, McKinnon AJ, Brumm J, Nilsson K, Fitch WT (2011) Perception of male caller identity in koalas (*Phascolarctos cinereus*): acoustic analysis and playback experiments. PLoS One 6(5):20329

Charrier I, Mathevon N, Jouventin P, Aubin T (2001) Acoustic communication in a black-headed gull colony: how do chicks identify their parents? Ethology 107(11):961–974

Charrier I, Mathevon N, Jouventin P (2002) How does a fur seal mother recognize the voice of her pup? An experimental study of *Arctocephalus tropicalis*. J Exp Biol 205:603–612

Charrier I, Mathevon N, Jouventin P (2003) Vocal signature recognition of mothers by fur seal pups. Anim Behav 19:543–550

Condit R, Reiter J, Morris PA, Berger R, Allen SG, Boeuf BJ (2014) Lifetime survival rates and senescence in northern elephant seals. Mar Mamm Sci 30:122–138

Cox CR (1981) Agonistic encounters among male elephant seals: frequency, context, and the role of female preference. Am Zool 21(1):197–209

Dale J, Lank DB, Reeve HK (2001) Signaling individual identity versus quality: a model and case studies with ruffs, queleas, and house finches. Am Nat 158(1):75–86

Deutsch CJ, Haley MP, Le Boeuf BJ (1990) Reproductive effort of male northern elephant seals: estimates from mass loss. Can J Zool 68(12):2580–2593

Domjam M (1998) The principles of learning and behavior, 4th edn. Brooks & Cole, Pacific Grove, CA

Freeberg TM, Dunbar RI, Ord TJ (2012) Social complexity as a proximate and ultimate factor in communicative complexity. Philos Trans R Soc Lond B 367(1597):785–1801

Haley MP (1994) Fighting and dominance in male northern elephant seals. Behav Ecol Sociobiol 34:427

Hauser MD (1996) The evolution of communication. MIT press, Cambridge, MA

Hoelzel AR (1999) Impact of population bottlenecks on genetic variation and the importance of life-history; a case study of the northern elephant seal. Biol J Linn Soc 68(1–2):23–39

Ilany A, Barocas A, Kam M, Ilany T, Geffen E (2013) The energy cost of singing in wild rock hyrax males: evidence for an index signal. Anim Behav 85:995–1001

Insley SJ, Phillips AV, Charrier I (2003) A review of social recognition in pinnipeds. Aquat Mamm 29(2):181–201

Kitchen DM, Seyfarth RM, Fischer J, Cheney DL (2003) Loud calls as indicators of dominance in male baboons (*Papio cynocephalus ursinus*). Behav Ecol Sociobiol 53(6):374–384

Konishi M (1989) Birdsong for neurobiologists. Neuron 3(5):541–549

Krams I, Krama T, Freeberg TM, Kullberg C Lucas JR (2012) Linking social complexity and vocal complexity: a paired perspective. Philos Trans R Soc Lond B 367(1597):1879–1891

Le Boeuf BJ (1974) Male-male competition and reproductive success in elephant seals. Am Zool 14 (1):163–176

Le Boeuf BJ, Laws RM (eds) (1994) Elephant seals: population ecology, behavior, and physiology. University of California Press, Berkeley, CA

Le Boeuf BJ, Peterson RS (1969) Dialects in elephant seals. Science 166:1654–1656

Le Boeuf BJ, Petrinovich LF (1974) Dialects of northern elephant seals, *Mirounga angustirostris*: origin and reliability. Anim Behav 22(3):656–663

Lowry MS, Condit R, Hatfield B, Allen SG, Berger R, Morris PA, Le Boeuf BJ, Reiter J (2014) Abundance, distribution, and population growth of the northern elephant seal *(Mirounga angustirostris*) in the United States from 1991 to 2010. Aquat Mamm 40(1):20–31

Marler P, Peters S (1982) Developmental overproduction and selective attrition: new processes in the epigenesis of birdsong. Dev Psychol 15(4):369–378

Mathevon N, Charrier I, Jouventin P (2003) Potential for individual recognition in acoustic signals: a comparative study of two gulls with different nesting patterns. C R Biol 326(3):329–337

Mathevon N, Aubin T, Vielliard J, Da Silva ML, Sebe F, Boscolo D (2008) Singing in the rain forest: how a tropical bird song transfers information. PLoS One 3(2):1580

Mathevon N, Casey C, Reichmuth C, Charrier I (2017) Northern elephant seals memorize the beats and timbre of their rival's voices. Curr Biol 27(15):2352–2356

Medvin MB, Stoddard PK, Beecher MD (1993) Signals for parent-offspring recognition: a comparative analysis of the begging calls of cliff swallows and barn swallows. Anim Behav 45 (5):841–850

Reby D, McComb K, Cargnelutti B, Darwin C, Fitch WT, Clutton-Brock T (2005) Red deer stags use formants as assessment cues during intrasexual agonistic interactions. Proc R Soc B 272 (1566):941–947

Ręk P, Osiejuk TS (2011) No male identity information loss during call propagation through dense vegetation: the case of the corncrake. Behav Process 86(3):323–328

Sandegren FE (1976) Agonistic behavior in the male northern elephant seal. Behaviour 57:136–158

Sanvito S, Galimberti F, Miller EH (2007) Vocal signaling of male southern elephant seals is honest but imprecise. Anim Behav 73:287–299

Sayigh LS, Tyack PL, Wells RS, Solow AR, Scott MD, Irvine AB (1999) Individual recognition in wild bottlenose dolphins: a field test using playback experiments. Anim Behav 57(1):41–50

Schusterman RJ (1978) Vocal communication in pinnipeds. In: Markowitz H, Stevens V (eds) Studies of captive wild animals. Nelson Hall, Chicago

Searby A, Jouventin P (2003) Mother-lamb acoustic recognition in sheep: a frequency coding. Proc R Soc B 270(1526):1765–1771

Searcy WA, Nowicki S (2005) The evolution of animal communication: reliability and deception in signaling systems. Princeton University Press, Princeton, NJ

Seyfarth RM, Cheney DL (2003) Signalers and receivers in animal communication. Annu Rev Psychol 54(1):145–173

Sheehan MJ, Tibbetts EA (2009) Evolution of identity signals: frequency-dependent benefits of distinctive phenotypes used for individual recognition. Evolution 63:3106–3113

Shipley C, Strecker G (1986) Day and night patterns of vocal activity of northern elephant seal bulls. J Mammal 67(4):775–778

Shipley C, Hines M, Buchwald JS (1981) Individual differences in threat calls of northern elephant seal bulls. Anim Behav 29(1):12–19

Shipley C, Hines M, Buchwald JS (1986) Vocalizations of northern elephant seal bulls: development of adult call characteristics during puberty. J Mammal 67:526–536

Snowdon CT, Hausberger M (eds) (1997) Social influences on vocal development. Cambridge University Press, Cambridge, UK

Southall BL, Casey C, Holt M, Insley S, Reichmuth C (2019) High-amplitude vocalizations of male northern elephant seals and associated ambient noise on a breeding rookery. J Acoust Soc Am 146(6):4514–4524

Suer J, Mackie D, Windmill JFC (2011) So small, so loud: extremely high sound pressure level from a pygmy aquatic insect (*Corixidae*, *Micronectinae*). PLoS One 6(6):21089

Taylor AM, Reby D, McComb K (2010) Size communication in domestic dog, *Canis familiaris*, growls. Anim Behav 79:205–210

Tibbetts EA, Dale J (2007) Individual recognition: it is good to be different. Trends Ecol Evol 22 (10):529–537

White FN, Odell DK (1971) Thermoregulatory behavior of the northern elephant seal, *Mirounga angustirostris*. J Mammal 52(4):758–774

Wolf JB, Mawdsley D, Trillmich F, James R (2007) Social structure in a colonial mammal: unravelling hidden structural layers and their foundations by network analysis. Anim Behav 74(5):1293–1302

Wyman MT, Mooring MS, Mccowan B, Penedo MCT, Reby D, Hart LA (2012) Acoustic cues to size and quality in the vocalizations of male North American bison, *Bison bison*. Anim Behav 84 (6):1381–1391

Chapter 11
How Songbird Females Sample Male Song: Communication Networks and Mate Choice

Nina Bircher and Marc Naguib

Abstract The behavioural decisions animals take directly influence their fitness and thus have a fundamental impact on evolutionary processes. In many animals, acoustic signals play an important role in social decisions with mate choice being among the most apparent ones. Male bird song has played a key role along this line, yet the understanding of how female birds use song to prospect, assess and choose mates in their natural environment is surprisingly limited. A main reason for this limited understanding is that it is very difficult to follow a female during her prospecting and decision process and quantify her experience with different males before she makes a final decision. Here we integrate insights from communication networks, male song traits and female prospecting behaviour to stimulate a more integrative approach on the role of signalling in behavioural and reproductive decisions.

11.1 Introduction

The reproductive decisions animals take directly influence their fitness and thus have a fundamental impact also on broader evolutionary processes. Mate choice has been well studied across many taxa and social systems and often involves elaborate signalling by one of the sexes, usually the males. Since Darwin highlighted sexual selection as a driving force in the evolution of such signals (Darwin 1859), many studies have provided evidence that individuals with more elaborate signals on average are preferred by the opposite sex and have a fitness benefit (Andersson and Simmons 2006). Yet more recent studies have emphasised that females also vary in their preferences (Riebel 2009; Buchanan et al. 2013) and that males with the most elaborate signals thus might not always be the most preferred mate. Moreover, under

N. Bircher (✉) · M. Naguib
Behavioural Ecology Group, Wageningen University & Research, Wageningen, The Netherlands
e-mail: nina.bircher@wur.nl

T. Aubin, N. Mathevon (eds.), *Coding Strategies in Vertebrate Acoustic Communication*, Animal Signals and Communication 7,
https://doi.org/10.1007/978-3-030-39200-0_11

natural conditions females will never have perfect information, make errors, and also will differ individually in the information they have available. Thus, variation in decisions is expected even if females have the same decision rules. Such variation in preferences, decisions, information accuracy and availability would also explain why the high variation in sexually selected signals is maintained (Holveck and Riebel 2007; Riebel et al. 2009; Rodríguez et al. 2013; Zandberg et al. 2017).

In many animals, acoustic signals play an important role in social decisions with mate selection being among the most apparent ones. Females usually are the choosier sex and use the elaborate signals produced by males to assess potential mates. Male vocal signals can convey a variety of information such as quality, motivation, resource holding potential or personality (Bradbury and Vehrencamp 2011). Yet, because of the large transmission range of sound signals compared to the natural spacing of individuals, the actual relation between the signal and the decisions made by receivers can be rather complex. A signal (1) can affect spatial movements of prospecting individuals without that they approach closely and (2) will often be heard by many receivers at the same time allowing direct comparison from the distance. Such a network-view of animal acoustic communication has led to substantial research efforts on the consequences of signalling relations by interacting individuals on eavesdropping conspecifics, and on audience effects on vocal communication and decision-making (McGregor 2005). A network of signallers may also lead to interferences and signal detection problems (Wiley 2015). Some models and studies of mate choice assume that females sample males one at the time, either memorising expression levels of a certain trait of all visited males while moving along or simply choosing the first male expressing a trait above a certain threshold (Janetos 1980; Real 1991). However, when females are able to receive the signals of several males at the same time, this will allow them to compare potential mates simultaneously (Gibson and Langen 1996). Such a way of assessing males in a network fashion may thus reduce the time and energy females have to invest into mate sampling (Otter and Ratcliffe 2005).

To understand the broader principles of animal communication and specifically its role in mate choice, bird song has played a key role as it is important for both territory defence against other males and mate attraction (Searcy and Yaukawa 1996; Catchpole and Slater 2008; Naguib and Riebel 2014). Yet despite decades of research, our understanding of how female birds use song to assess and choose mates in their natural environment is surprisingly limited, as many studies on female preference for male song have been conducted in highly standardised laboratory situations (Riebel 2009; Leboucher et al. 2012; Honarmand et al. 2017). Studies in the wild linking female choice to male characteristics have either correlated male traits with the timing of mating or mating success, or related female extra-pair mating decisions to male traits (Catchpole and Slater 2008). Thus comparatively little is known about the process of female choice under natural conditions, as it is very difficult to follow a female during her prospecting and decision process and quantify her experience with different males before she makes a final decision (Otter and Ratcliffe 2005). Moreover, the actual choice females make is presumably the outcome of a combination between her preference function, previous experience

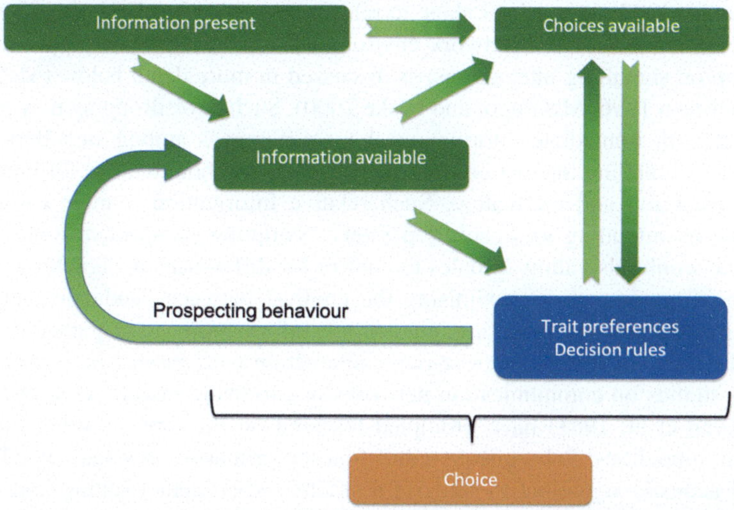

Fig. 11.1 Schematic representation of factors affecting female choice. The choice a female will make will depend on her own preferences and decision rules as well as the information and options she has available. Her sampling behaviour will affect the information she has available and also can change her decision rules

and the information she actually has available which will be affected by her sampling strategy, the assessment period, and use of information. Thus, her choice may not necessarily reflect her actual preference (Fig. 11.1).

11.2 Communication Networks

Many acoustic signals have a transmission range that is far larger than the average spacing of the animals using them (Brenowitz 1982; Calder 1990; Naguib et al. 2008). Thus, such signals will be heard not only by one receiver but by many receivers, and if several individuals vocalise at the same time, a receiver may perceive their signals simultaneously. The social environment in which most communication occurs therefore has been suggested to be a network of several signallers and receivers (McGregor and Dabelsteen 1996; McGregor 2005). The size of such a network in terms of area covered and number of individuals it encompasses depends on (1) the natural spacing of individuals and (2) the active space of a signal, and thus the area in which a conspecific is still able to perceive it (Brumm and Naguib 2009) as well as on the movement and prospecting behaviour of receivers (Amrhein et al. 2004; Roth et al. 2009). A network environment may create costs and benefits for receivers in addition to those that are normally considered for signaller–receiver dyads. For example, it becomes more difficult for receivers to distinguish information from individual signallers when multiple of them are active at the same time

(Poesel et al. 2007) or at a later stage to relocate-specific signallers after they have moved around. However, a network environment also provides the opportunity to eavesdrop on signalling interactions, as discussed in more detail below (McGregor and Dabelsteen 1996; McGregor and Peake 2000). Such eavesdropping may provide a receiver with immediate information about differences among signallers, especially when signalling interactions provide information about relative differences in quality, condition and motivation. Such relative information is more difficult to determine by attending to signals separately. Sampling prospective males at the same time would also allow females to control for differences in signalling created by the environment, thus maximising the contrast between candidate males. By attending to signals and signalling interactions, information can be gathered at little cost and before engaging in a close-range interaction with another individual.

Early studies on communication networks in chorusing insects (Otte 1974) and frogs (Ryan et al. 1981) have prompted research on the topic in other taxa and signalling modalities. Subsequent studies focused primarily on visual signalling in fish and acoustic signalling in birds in a strictly experimental setting and indeed provided convincing evidence that individuals attend to several signallers and their interactions (Naguib 2005; Peake 2005). Yet in nature, part of a communication network will involve individuals moving around to obtain information from signals while signalling also varies in space and time (Amrhein et al. 2004; Roth et al. 2009). Thus, despite communication networks having become a central part of our understanding of animal communication, little information is available on the relation between signalling behaviour and receiver movement and decisions under natural conditions.

Due to the long-range and dynamic nature of songbird interactions, following and recording several individuals at the same time is challenging. Experimental studies on avian acoustic networks have thus mostly been limited to three-party networks and focused on male individuals, also because of the often secluded female behaviour (McGregor 2005). Several of these studies provide evidence of communication network-based behaviour, such as eavesdropping and audience effects (Liu 2004; Vignal et al. 2004). These show that the audience extracts information from asymmetries in signalling interactions while such interactions at the same time are affected by the presence of bystanders in close vicinity. More recent work extended the network level showing that whole neighbourhoods of territorial males share signalling traits (Snijders et al. 2015a) and even are vocally interacting and eavesdropping on one another (Fitzsimmons et al. 2008a, b; Foote et al. 2008, 2010).

11.3 Female Mate Sampling

Despite decades of research on the topic and the well-established importance of male song in female choice, surprisingly little is known about the actual behavioural processes by which females assess and choose their mates. Research focus has primarily been directed on investigating which mates are chosen and what the potential benefits are, rather than on the processes by which prospective mates are sampled and evaluated (Gibson and Langen 1996). Existing models for mate sampling strategies can be broadly organised by two criteria: whether they consider the sampling of males to be sequential or simultaneous and whether they consider sampling to be dependent on the female alone (static models) or also on the distribution of the sampled males (dynamic models) (Rosenthal 2017). Evidence from empirical studies suggests that various strategies are used across taxa, within populations, and even within individuals (Jennions and Petrie 1997; Beckers and Wagner 2011). During such sampling, females may assess various male traits. Yet little is known how they integrate such traits, memorise the information, and then use it in their final decision-making. In sequential sampling, females may also encounter males in different contexts and unless they engage in close-range mating display interactions, they may not have standardised equal information from potential mates. In communication networks females additionally might assess potential mates not sequentially but simultaneously and thus may evaluate a large number of males without ever approaching them. Females could even position themselves in a way that maximises the number of assessable males at the same time (Otter and Ratcliffe 2005).

Most field studies that followed female movement patterns investigated how many males were sampled, interpreting a close approach as evidence for an individual being assessed (Bensch and Hasselquist 1992; Neudorf and Pitcher 1997; Roth et al. 2009). These studies on the one hand showed apparent sequential sampling of males, which in nightingales (*Luscinia megarhynchos*), for instance, also peaked at the time of male highest nocturnal singing activity (Roth et al. 2009). Such studies, despite the correlative nature, strongly suggest that male singing affects female spatial behaviour. Females sampling males sequentially may use information from individual songs or specifically attend to male–male vocal interactions at these times, integrating information extracted from these interactions in their decision-making.

Yet, close approach during male sampling does not rule out that females also assess males from the distance. A close approach might result from females either sampling male signals with a short transmission range or resources and display sites (Dale et al. 1990; Uy et al. 2001) or from females sampling long-range signals from far away and then approaching for closer inspection (Fig. 11.2). In case of long ranging signals such as most songs of birds or calls of anurans, movement patterns thus can be difficult to interpret. Female barking treefrogs (*Hyla gratiosa*) for example approach only one male in a chorus, which based on movement pattern alone, would indicate that no sampling occurs. However, instead they may evaluate several males from a distance and then approach only a selected male. In that case, an approach indicates a final choice rather than assessment (Murphy and Gerhardt 2002). Moreover, females may not only use long-range signals to decide from a distance with whom to mate, but also to select a subsample of males which is then assessed further using different signal modalities during each ‘selection round’ (Uy and Safran 2013). Such long-distance preselection would strongly increase the efficiency of female mate sampling. However, the extent to which acoustic signals are suitable for long-range selection or subsampling of potential mates depends on

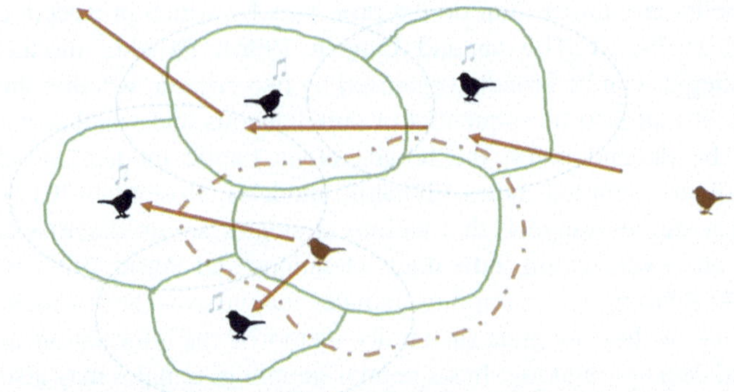

Fig. 11.2 In many animal species female (brown) assessment of male (black) signals will take place in a communication network, as the active space of male signals (grey circles) is larger than the spacing between individuals. A territorial female can listen to close-by neighbours without leaving the territory (green area) and sample males by moving around off territory, even without close-range interactions. Females not belonging to the neighbourhood can sample males by visiting the territories or by moving along territory edges so they can receive signals from several males at the time. The movement patterns of females will also determine which information they have available for decision-making

the spacing of individuals and on the information content of a distant signal. A recent study on the tree cricket (*Oecanthus henryi*) found that males are spaced too far apart for most females to assess any of them simultaneously and female thus would have to move for sampling even long-range acoustic signals (Deb and Balakrishnan 2014). Likewise in nightingales some song characteristics, such as tonal whistle-like elements, travel far and are suggested to specifically function in female long-range attraction (Kunc et al. 2007), yet other song traits, such as broadband trills, which carry relevant quality information such as age (Sprau et al. 2013), do not even fully transmit towards a territorial neighbour (Naguib et al. 2008). Future studies on female sampling of signalling males therefore need to consider the active space of the signal in question along with female spatial behaviour to obtain an accurate indication of the number of males sampled and on the information that she may have available for decisions in mate choice.

11.4 Song Traits for Female Choice in a Communication Network

Bird song is a signal containing multiple sexually selected traits (Gil and Gahr 2002). Many aspects of singing behaviour reflect male condition and thus can serve as an honest signal of the quality of a potential mate for females: Repertoire size for example can indicate male age and thus survival (Rivera-Gutierrez et al. 2010; Nemeth et al. 2012) as well as paternal effort (Bartsch et al. 2014). Song output can be related to the parasitic load (Garamszegi 2005), immune response (Saino et al. 1997) and dominance rank (Otter et al. 1997). Also the production quality can provide important information relevant in female choice (Forstmeier et al. 2002; Ballentine et al. 2004). Indeed, aspects of male singing behaviour have been shown to be important for female choice for primary mates (Radesäter et al. 1987; Alatalo et al. 1990; Buchanan and Catchpole 1997) and also correlate with a male's success in siring extra-pair offspring (Houtman 1992; Hasselquist et al. 1996; Kempenaers et al. 1997; Møller et al. 1998; Chiver et al. 2008). The role of such individual traits in female choice has been reviewed in more detail elsewhere (Gil and Gahr 2002).

In addition to assessing the individual song of males, females may also gain information on relative song traits from eavesdropping on male–male interactions. Territorial males of many songbird species frequently engage in counter-singing interactions (Todt and Naguib 2000). Interacting males may thereby choose both the timing and patterning of their response: Time specific responses such as overlapping the song of the other and pattern-specific responses such as matching the song of the other by using the same song type or frequency content might convey information about the signaller's relative quality and motivation (Naguib 2005). Song overlapping occurs when a bird starts singing while the song of another bird has not yet finished and thus is being overlapped by the song of the other. Overlapping or being overlapped and matching or being matched may have different behavioural consequences and experimentally overlapping or matching the song of another male usually results in increased vocal and/or spatial responses (Naguib and Mennill 2010). In male nightingales for example, song overlapping and their response to overlapping and alternating playbacks predict their mating success (Kunc et al. 2006; Schmidt et al. 2006). This could suggest that females directly attend to these interactions. Yet overall less is known about whether the frequency with which a male uses either of these strategies is associated with specific characteristics of the singer (such as dominance rank, age, condition or personality) and what the specific functions are during natural interactions (Searcy and Beecher 2009; Naguib and Mennill 2010). Bischoff et al. (2009) showed that males exposed to parasites in early life used less song overlapping in adult life than males raised in parasite-free nests (Bischoff et al. 2009). Thus, singing strategies during interactions might even function as a signal of a male's past growth and development. Whether they can reveal information about a male's current condition is less studied.

Currently only a few studies have investigated whether females use broader communication network information to direct their prospecting and mate choice. Most of these studies have thereby focused on female eavesdropping on dyadic male interactions. For example, in a study on captive domestic canaries (*Serinus canaria*), Leboucher and Pallot (2004) exposed females to playback of a vocal interaction between males with the songs of one male overlapping the song of the other male in two consecutive experiments. In the first experiment some males also displayed the sexually attractive A-phrase, so called "sexy syllables," in addition to overlapping or being overlapped. Female preference was subsequently assessed by counting copulation solicitation displays. Females showed a preference for the overlapping song in the experiment where overlapping or being overlapped was the main difference between males, but not when some males also displayed A-phrases, possibly because the phrases were contradicting male performance in overlapping (Leboucher and Pallot 2004). In another study on the domestic canary, females showed an overall preference for overlapping song and this preference was especially strong on the day before laying the first egg (Amy et al. 2008). Moreover, female domestic canaries hearing overlapping song laid eggs with greater egg yolk ratio than females exposed to overlapped songs, indicating that eavesdropping on male interactions might also influence maternal investment (Garcia-Fernandez et al. 2010). A study in nightingales showed that prior to mating, males that eventually became paired overlapped more songs of a simulated intruder than males that remained unmated during the breeding season (Kunc et al. 2006) which raises the question if females may mate preferably with more aggressively or dominantly singing males and thus put a selection pressure on males to interact with each other in specific ways to appear sexually attractive (Kunc et al. 2006). Likewise in an interactive playback study males who eventually became mated changed the timing of their singing less in response to alternating or overlapping playback while those who stayed unpaired, interrupted their singing significantly more when they were overlapped, i.e. when they were challenged by a presumably more dominant singer (Schmidt et al. 2006). This study may also be taken to suggest that females attend to the way males interact with others (Naguib et al. 2011). Indeed two experimental studies on use of network information by females in the wild both have shown that females appear to be directly or indirectly affected by the way their mate interacts with a rival. Mennill et al. (2002) involved high- and low-ranking black-capped chickadee (*Poecile atricapillus*) males in interactive playbacks, making them either 'loose' (overlapping or frequency matching their song) or 'win' (alternating with and avoiding frequency matching their song) the interaction. Females paired to high-ranking males losing the contest to the simulated intruder subsequently were more likely to mate with other males: the proportion of extra-pair offspring in the nests of these males was significantly higher compared to high-ranking males that received an alternating playback or that were not involved in a playback (Mennill et al. 2002, 2003). Eavesdropping on male counter-singing might thus have directly influenced female reproductive choices. Likewise the outcome of experimentally induced male interactions in great tits (*Parus major*) showed that females were highly attentive to these interactions with females being paired to males that had been overlapped subsequently foraying more often into the territory of neighbouring males than females paired to males that had won the contest. Moreover, in most cases these females visited the territory of males that was more successful in the playback contest than their mate (Otter et al. 1999). The playbacks did however not influence extra-pair paternity decisions (Otter et al. 2001). This attentiveness of females to male song contests has also been found in a more recent study on great tits: Nearby females approached the playback site more closely when the male had sung more songs and overlapped a simulated intruder more often (Snijders et al. 2017). Yet, whether or not the strength of a male's response to an intruder influences the reproductive choices of eavesdropping females remains to be investigated. Even though there is emerging evidence that females pay close attention to the outcome of male vocal interactions, it remains unclear whether they actually move around specifically to listen to other males and their singing interactions and possibly even adapt their reproductive decisions based on information gathered by such eavesdropping.

Almost all studies on female eavesdropping on male acoustic interactions found that females seem to prefer more aggressive (overlapping and/or matching) males. An exception is the study conducted by Bartsch et al. (2014) on song type matching in nightingales. Song type matching generally is interpreted as aggressive signal (Todt 1981; Falls 1985; Vehrencamp et al. 2007) and female nightingales here spent more time close to the apparent looser of an interaction (the matched male) (Bartsch et al. 2014). Female preference for non-aggressive males has also been reported for females observing visual displays between males (Ophir and Galef 2003; Bierbach et al. 2013) (but see Doutrelant and McGregor 2000), possibly due to direct costs females expect from more harassing contests winners.

Another line of evidence for females using male–male interactions as information source is the observation that females in some cases induce male contests. In the great snipe (*Gallinago media*) female calls from the boundaries of male territories within leks increase the number of competitive male–male contests (Saether 2002). In the territorial black-capped chickadee females build their nests close to the territorial boundaries, which leads to more territorial disputes between mates and their neighbours (Ramsay et al. 1999). Female European robins (*Erithacus rubecula*) produce specific calls only when they are fertile (Tobias and Seddon 2002) which can be interpreted as a way to increase female choice through sperm competition, yet these calls could also increase the number of interactions between males and thus facilitate the assessment of copulation partners (Otter and Ratcliffe 2005).

Apart from assessing male interactions, a female may also be able to evaluate several males simultaneously when the males are not interacting but singing independently within a distance that allows her to perceive and discriminate their signals. For example, a female great tit may be able to hear (Blumenrath and Dabelsteen 2004) and discriminate (Blumenrath et al. 2007) her mate and her closest neighbours even when sitting in a nest box, therefore also at dawn when information gathering from singing males may be essential. Individual differences in dawn song have been shown to reflect male condition in several bird species (Alatalo et al. 1990; Otter et al. 1997; Poesel et al. 2004) and to predict their territorial behaviour in a contest (Snijders et al. 2015b). Similarly, female field sparrows (*Spizella punella*) have been shown to respond more strongly to nocturnal intruder song during the pre-fertile and fertile phase in their mating season, which may indicate that far-reaching rare nocturnal song in a diurnal bird species serves to attract females for extra-pair mating (Celis-Murillo et al. 2016), and nightingale whistle songs which are common in nocturnal song have long propagation range allowing initial assessment of multiple males from far away (Naguib et al. 2008). Given the long transmission range of these signals the question arises as to how the decision processes are guided by potential long-range assessment of possibly multiple males at the same time, before a subsequent actual choice at close range is made.

11.5 Integrating Information from Multiple Interactions

Several playback studies have provided intriguing evidence that both male and female birds respond to their neighbours' fate in interactions with other neighbours and strangers as discussed above. Gathering information by eavesdropping, they can assess relative differences between those individuals and then direct their own behaviour based on that information. For example, the experiment by Peake et al. (2002) shows that male great tits are able to base their interactions with a simulated intruder on how this individual performed in an interaction with a neighbour of known rank with respect to themselves (Peake et al. 2002). Similarly, female great tits are capable of inferring the ranking of two males based on their respective reactions towards the same intruder and altered their behaviour using this information (Otter et al. 1999). Also in nightingales males use their neighbours response as yard stick to assess unfamiliar competitors (Naguib et al. 2004). Song birds may thus be able to evaluate the threat level of an intruder or the quality of a potential mate by listening to separate vocal interactions and this could be an important tool to assess individuals around them. The question remains however, whether songbirds are able to rank a multitude of different individuals and whether they would need to eavesdrop directly on all interactions to do so or whether they could attend to some and then interpolate information about the remaining individuals based on memory (Pepperberg 2005). One mechanism with which animals might do that is transitive inference, thus observing a series of linear relationships between objects ($A > B > C > D > E$) and then inferring the relationship between objects not directly compared (e.g. the relationship of $B > D$). The ability of animals to solve such transitive tasks could have important adaptive value in rank estimation in social animals: If dominance is not directly assessable by physical traits an animal incapable of transitive inference would have to interact with every single member of the group in potentially risky encounters. To rank six different individuals for example, all 15 combinations would have to be directly experienced (Vasconcelos 2008). Thus, if capable of transitive inference, female songbirds might be able to gain information about far more potential mates than they have listened to when directly interacting which each other. It is however unclear whether female animals actually exhibit transitive preferences when assessing mates (Bateson and Healy 2005; Kirkpatrick et al. 2006; Dechaume-Moncharmont et al. 2013; Lea and Ryan 2015). When presented with only two males, A and B, with B being more attractive regarding different call characteristics, female tungara frogs (*Physalaemus pustulosus*) preferred male B. However, when presented with A, B and an additional male C, which was less attractive than both A and B, female frogs preferred the intermediate male A (Lea and Ryan 2015). Rather than assigning each male an absolute 'score of attractiveness' and consistently preferring a potential mate with high score over one with a lower score (e.g. B would always be more attractive than A, independent of whether C is around or not), females might use comparative evaluation where the attractiveness of a male depends on the social context rather than his absolute quality (Bateson and Healy 2005).

11.6 Conclusions

In the wild, sampling of males will be limited by many factors including mate guarding, harassment by other females, predation risk, time constraints, the spatial arrangement of males and their signalling behaviour. Thus, assessing males simultaneously from the distance by acoustic signals could not only reduce time spent for searching but also search-associated risks. Indeed, some studies found female movements into the territory of other males (so called extra-territorial forays) to be correlated with aspects of male song (Chiver et al. 2008), indicating that females specifically visited males producing higher quality song. Although many studies also support female choice of an extra-pair mate based on song traits (Hasselquist et al. 1996; Byers 2007; Dalziell and Cockburn 2008), others failed to find such a correlation and a recent meta-analysis found no effect of song on extra-pair paternity occurrence (Garamszegi 2004). It is thus possible that females use song for an initial assessment from a distance to decide on whom to approach and then use additional male traits such as plumage coloration for a final assessment (Otter and Ratcliffe 2005). More studies considering both song traits and short-range signals when following female movements are needed to determine whether females use different signals in a hierarchical way based on their transmission range to assess potential mates and to assess which information females have available when making choices. Overall it remains an exciting field for future research to determine the relation between female prospecting behaviour, male individual and interactive singing behaviour as well as the underlying cognitive processes of simultaneous and subsequent assessments.

References

Alatalo RV, Glynn C, Lundberg A (1990) Singing rate and female attraction in the pied flycatcher: an experiment. Anim Behav 39:601–602

Amrhein V, Kunc HP, Naguib M (2004) Non-territorial nightingales prospect territories during the dawn chorus. Proc R Soc B 271:S167–S169

Amy M, Monbureau M, Durand C, Gomez D, Thery M, Leboucher G (2008) Female canary mate preferences: differential use of information from two types of male-male interaction. Anim Behav 76:971–982

Andersson M, Simmons LW (2006) Sexual selection and mate choice. Trends Ecol Evol 21:296–302

Ballentine B, Hyman J, Nowicki S (2004) Vocal performance influences female response to male bird song: an experimental test. Behav Ecol 15:163–168

Bartsch C, Wenchel R, Kaiser A, Kipper S (2014) Singing onstage: female and male common nightingales eavesdrop on song type matching. Behav Ecol Sociobiol 68:1163–1171

Bateson M, Healy SD (2005) Comparative evaluation and its implications for mate choice. Trends Ecol Evol 20:659–664

Beckers OM, Wagner WE (2011) Mate sampling strategy in a field cricket: evidence for a fixed threshold strategy with last chance option. Anim Behav 81:519–527

Bensch S, Hasselquist D (1992) Evidence for female choice in a polygynous warbler. Anim Behav 44:301–311
Bierbach D, Sassmannshausen V, Streit B, Arias-Rodriguez L, Plath M (2013) Females prefer males with superior fighting abilities but avoid sexually harassing winners when eavesdropping on male fights. Behav Ecol Sociobiol 67:675–683
Bischoff LL, Tschirren B, Richner H (2009) Long-term effects of early parasite exposure on song duration and singing strategy in great tits. Behav Ecol 20:265–270
Blumenrath SH, Dabelsteen T (2004) Degradation of great tit (*Parus major*) song before and after foliation: implications for vocal communication in a deciduous forest. Behaviour 141:935–958
Blumenrath SH, Dabelsteen T, Pedersen SB (2007) Vocal neighbour-mate discrimination in female great tits despite high song similarity. Anim Behav 73:789–796
Bradbury JW, Vehrencamp SL (2011) Principles of animal communication. Sinauer Associates, Sunderland, MA
Brenowitz EA (1982) The active space of red-swinged blackbird song. J Comp Physiol A 147:511–522
Brumm H, Naguib M (2009) Environmental acoustics and the evolution of bird song. Adv Study Behav 40:1–33
Buchanan KL, Catchpole CK (1997) Female choice in the sedge warbler, *Acrocephalus schoenobaenus*: multiple cues from song and territory quality. Proc R Soc B 264:521–526
Buchanan KL, Grindstaff JL, Pravosudov VV (2013) Condition dependence, developmental plasticity, and cognition: implications for ecology and evolution. Trends Ecol Evol 28:290–296
Byers BE (2007) Extrapair paternity in chestnut-sided warblers is correlated with consistent vocal performance. Behav Ecol 18:130–136
Calder WA III (1990) The scaling of sound output and territory size: are they matched? Ecology 71:1810–1816
Catchpole CK, Slater PJB (2008) Bird song: biological themes and variation. Cambridge University Press, New York
Celis-Murillo A, Benson TJ, Sosa-López JR, Ward MP (2016) Nocturnal songs in a diurnal passerine: attracting mates or repelling intruders? Anim Behav 118:105–114
Chiver I, Stutchbury BJM, Morton ES (2008) Do male plumage and song characteristics influence female off-territory forays and paternity in the hooded warbler? Behav Ecol Sociobiol 62:1981–1990
Dale S, Amundsen T, Lifjeld JT, Slagsvold T (1990) Mate sampling behaviour of female pied flycatchers: evidence for active mate choice. Behav Ecol Sociobiol 27:87–91
Dalziell AH, Cockburn A (2008) Dawn song in superb fairy-wrens: a bird that seeks extrapair copulations during the dawn chorus. Anim Behav 75:489–500
Darwin C (1859) On the origin of species by means of natural selection. Murray, London
Deb R, Balakrishnan R (2014) The opportunity for sampling: the ecological context of female mate choice. Behav Ecol 24:967–974
Dechaume-Moncharmont FX, Freychet M, Montreuil S, Cezilly F (2013) Female mate choice in convict cichlids is transitive and consistent with a self-referent directional preference. Front Zool 10:69
Doutrelant C, McGregor PK (2000) Eavesdropping and mate choice in female fighting fish. Behaviour 137:1655–1669
Falls JB (1985) Song matching in western meadowlarks. Can J Zool 63:2520–2524
Fitzsimmons LP, Foote JR, Ratcliffe LM, Mennill DJ (2008a) Eavesdropping and communication networks revealed through playback and an acoustic location system. Behav Ecol 19:824–829
Fitzsimmons LP, Foote JR, Ratcliffe LM, Mennill DJ (2008b) Frequency matching, overlapping and movement behaviour in diurnal countersinging interactions of black-capped chickadees. Anim Behav 75:1913–1920
Foote JR, Fitzsimmons LP, Mennill DJ, Ratcliffe LM (2008) Male chickadees match neighbors interactively at dawn: support for the social dynamics hypothesis. Behav Ecol 19:1192–1199
Foote JR, Fithsimmons LP, Mennill DJ, Ratcliffe LM (2010) Black-capped chickadee dawn choruses are interactive communication networks. Behaviour 147:1219–1248

Forstmeier W, Kempenaers B, Meyer A, Leisler B (2002) A novel song parameter correlates with extra-pair paternity and reflects male longevity. Proc R Soc B 269:1479–1485
Garamszegi LZ (2004) Extrapair paternity and the evolution of bird song. Behav Ecol 15:508–519
Garamszegi LZ (2005) Bird song and parasites. Behav Ecol Sociobiol 59:167–180
Garcia-Fernandez V, Amy M, Lacroix A, Malacarne G, Leboucher G (2010) Eavesdropping on male singing interactions leads to differential allocation in eggs. Ethology 116:662–670
Gibson RM, Langen TA (1996) How do animals choose their mates? Trends Ecol Evol 11:468–469
Gil D, Gahr M (2002) The honesty of bird song: multiple constraints for multiple traits. Trends Ecol Evol 17:133–141
Hasselquist D, Bensch S, vonSchantz T (1996) Correlation between male song repertoire, extra-pair paternity and offspring survival in the great reed warbler. Nature 381:229–232
Holveck MJ, Riebel K (2007) Preferred songs predict preferred males: consistency and repeatability of zebra finch females across three test contexts. Anim Behav 74:297–309
Honarmand M, Krause ET, Naguib M (2017) Implications of nutritional stress as nestling or fledgling on subsequent attractiveness and fecundity in zebra finches (*Taeniopygia guttata*). PeerJ 5:33628
Houtman AM (1992) Female zebra finches choose extra-pair copulations with genetically attractive males. Proc R Soc B 249:3–6
Janetos AC (1980) Strategies of female mate choice: a theoretical analysis. Behav Ecol Sociobiol 7:107–112
Jennions MD, Petrie M (1997) Variation in mate choice and mating preferences: a review of causes and consequences. Biol Rev 72:283–327
Kempenaers B, Verheyren GR, Dhondt AA (1997) Extrapair paternity in the blue tit (*Parus caeruleus*): female choice, male characteristics, and offspring quality. Behav Ecol 8:481–492
Kirkpatrick M, Rand AS, Ryan MJ (2006) Mate choice rules in animals. Anim Behav 71:1215–1225
Kunc HP, Amrhein V, Naguib M (2006) Vocal interactions in nightingales (*Luscinia megarhynchos*): more aggressive males have higher pairing success. Anim Behav 72:25–30
Kunc HP, Amrhein V, Naguib M (2007) Vocal interactions in common nightingales (*Luscinia megarhynchos*): males take it easy after pairing. Behav Ecol Sociobiol 61:557–563
Lea AM, Ryan MJ (2015) Irrationality in mate choice revealed by túngara frogs. Science 349:964–966
Leboucher G, Pallot K (2004) Is he all he says? Intersexual eavesdropping in the domestic canary, *Serinus canaria*. Anim Behav 68:957–963
Leboucher G, Vallet E, Nagle L, Béguin N, Bovet D, Hallé F, Draganoiu TI, Amy M, Kreutzer M (2012) Studying female reproductive activities in relation to male song: the domestic canary as a model. Adv Study Behav 44:183–223
Liu WC (2004) The effect of neighbours and females on dawn and daytime singing behaviours by male chipping sparrows. Anim Behav 68:39–44
McGregor PK (2005) Communication networks. Cambridge University Press, Cambridge
McGregor PK, Dabelsteen T (1996) Communication networks. In: Kroodsma DE, Miller EH (eds) Ecology and evolution of acoustic communication in birds. Cornell University Press, Ithaca, pp 409–425
McGregor PK, Peake TM (2000) Communication networks: social environments for receiving and signalling behaviour. Acta Etol 2:71–81
Mennill DJ, Ratcliffe LM, Boag PT (2002) Female eavesdropping on male song contests in songbirds. Science 296:873
Mennill DJ, Boag PT, Ratcliffe LM (2003) The reproductive choices of eavesdropping female black-capped chickadees, *Poecile atricapillus*. Naturwissenschaften 90:577–582
Møller AP, Saino N, Taramino G, Galeotti P, Ferrario S (1998) Paternity and multiple signaling: effects of a secondary sexual character and song on paternity in the barn swallow. Am Nat 151:236–242
Murphy CG, Gerhardt HC (2002) Mate sampling by female barking treefrogs (*Hyla gratiosa*). Behav Ecol 13:472–480

Naguib M (2005) Singing interactions in song birds: implications for social relations, territoriality and territorial settlement. In: McGregor PK (ed) Communication networks. Cambridge University Press, Cambridge, pp 300–319

Naguib M, Mennill D (2010) The signal value of bird song: empirical evidence suggests song overlapping is a signal. Anim Behav 80:e11–e15

Naguib M, Riebel K (2014) Singing in space and time: the biology of birdsong. In: Witzany G (ed) Biocommunication of animals. Springer, Dordrecht, pp 233–247

Naguib M, Amrhein V, Kunc HP (2004) Effects of territorial intrusions on eavesdropping neighbors: communication networks in nightingales. Behav Ecol 6:1011–1015

Naguib M, Schmidt R, Sprau P, Roth T, Floercke C, Amrhein V (2008) The ecology of vocal signaling: male spacing and communication distance of different song traits in nightingales. Behav Ecol 19:1034–1040

Naguib M, Kunc HP, Sprau P, Roth T, Amrhein V (2011) Communication networks and spatial ecology in nightingales. Adv Study Behav 43:239–271

Nemeth E, Kempenaers B, Matessi G, Brumm H (2012) Rock sparrow song reflects male age and reproductive success. PLoS One 7:e43259

Neudorf DL, Pitcher TE (1997) Radio transmitters do not affect nestling feeding rats by female hooded warblers. J Field Ornithol 68:64–68

Ophir AG, Galef BG Jr (2003) Female Japanese quail that "eavesdrop" on fighting males prefer losers to winners. Anim Behav 65:1–9

Otte D (1974) Effects and functions in the evolution of signaling systems. Annu Rev Ecol Syst 5:385–417

Otter K, Ratcliffe LM (2005) Enlightened decisions: female assessment and communication networks. In: McGregor PK (ed) Animal communication networks. Cambridge University Press, Cambridge, pp 133–151

Otter K, Chruszcz B, Ratcliffe L (1997) Honest advertisement and song output during the dawn chorus of black-capped chickadees. Behav Ecol 8:167–173

Otter K, McGregor PK, Terry AMR, Burford FRL, Peake TM, Dabelsteen T (1999) Do female great tits (*Parus major*) assess males by eavesdropping? A field study using interactive song playback. Proc R Soc B 266:1305–1309

Otter KA, Stewart IRK, McGregor PK, Terry AMR, Dabelsteen T, Burke T (2001) Extra-pair paternity among great tits *Parus major* following manipulation of male signals. J Avian Biol 32:338–344

Peake TM (2005) Communication networks. In: McGregor PK (ed) Animal communication networks. Cambridge University Press, Cambridge

Peake TM, Terry AMR, McGregor PK, Dabelsteen T (2002) Do great tits assess rivals by combining direct experience with information gathered by eavesdropping? Proc R Soc B 269:925–1929

Pepperberg IM (2005) Cognitive aspects of networks and avian capacities. In: McGregor PK (ed) Animal communication networks. Cambridge University Press, Cambridge, pp 568–582

Poesel A, Dabelsteen T, Pedersen SB (2004) Dawn song of male blue tits as a predictor of competitiveness in midmorning singing interactions. Acta Ethol 6:65–71

Poesel A, Dabelsteen T, Pedersen SB (2007) Implications of conspecific background noise for features of blue tit, *Cyanistes caeruleus*, communication networks at dawn. J Ornithol 148:123–128

Radesäter T, Jakobsson S, Andbjer N, Bylin A, Nyström K (1987) Song rate and pair formation in the willow warbler, *Phylloscopus trochilus*. Anim Behav 35:1645–1651

Ramsay SM, Otter KA, Ratcliffe LM (1999) Nest-site selection by female black-capped chickadees: settlement based on conspecific attraction? Auk 116:604–617

Real L (1991) Search theory and mate choice. II. Mutual interaction, assortative mating, and equilibrium variation in male and female fitness. Am Nat 138:901–917

Riebel K (2009) Song and female mate choice in zebra finches: a review. Adv Study Behav 40:197–238

Riebel K, Naguib M, Gil D (2009) Experimental manipulation of the rearing environment influences adult female zebra finch song preferences. Anim Behav 78:1397–1404
Rivera-Gutierrez HF, Pinxten R, Eens M (2010) Multiple signals for multiple messages: great tit, *Parus major*, song signals age and survival. Anim Behav 80:451–459
Rodríguez RL, Boughman JW, Gray DA, Hebets EA, Höbel G, Symes LB (2013) Diversification under sexual selection: the relative roles of mate preference strength and the degree of divergence in mate preferences. Ecol Lett 16:964–974
Rosenthal GG (2017) Mate choice: the evolution of sexual decision making from microbes to humans. Princeton University Press, Princeton, NJ
Roth T, Sprau P, Schmidt R, Naguib M, Amrhein V (2009) Sex-specific timing of mate searching and territory prospecting in the nightingale: nocturnal life of females. Proc R Soc B 276:2045–2050
Ryan MJ, Tuttle MD, Taft LK (1981) The costs and benefits of frog chorusing behaviour. Behav Ecol Sociobiol 8:273–278
Saether SA (2002) Female calls in lek-mating birds: indirect mate choice, female competition for mates or direct mate choice? Behav Ecol 13:344–352
Saino N, Galeotti P, Sacchi R, Møller AP (1997) Song and immunological condition in male barn swallows (*Hirundo rustica*). Behav Ecol 8:364–371
Schmidt R, Kunc HP, Amrhein V, Naguib M (2006) Responses to interactive playback predict future mating status in nightingales. Anim Behav 72:1355–1362
Searcy WA, Beecher MD (2009) Song as an aggressive signal in songbirds. Anim Behav 78:1281–1292
Searcy WA, Yaukawa K (1996) Song and female choice. In: Kroodsma DE, Miller EH (eds) Ecology and evolution of acoustic communication in birds. Cornell University Press, Ithaca, pp 454–473
Snijders L, van der Eijk J, van Rooij EP, de Goede P, van Oers K, Naguib M (2015a) Song trait similarity in great tits varies with social structure. PLoS One 10:e0116881
Snijders L, van Rooij E, Henskens M, van Oers K, Naguib M (2015b) Dawn song predicts behaviour during territory conflicts in personality-typed great tits. Anim Behav 109:45–52
Snijders L, van Oers K, Naguib M (2017) Sex-specific responses to territory intrusions in a communication network: evidence from radio-tagged great tits. Ecol Evol 7(3):918–927
Sprau P, Roth T, Amrhein V, Naguib M (2013) The predictive value of trill performance in a large repertoire songbird, the nightingale *Luscinia megarhynchos*. J Avian Biol 44:567–574
Tobias JA, Seddon N (2002) Female begging in European robins: do neighbors eavesdrop for extrapair copulations? Behav Ecol 13:637–642
Todt D (1981) On functions of vocal matching: effects of counter-replies on song post choice and singing. Z Tierpsychol 57:73–93
Todt D, Naguib M (2000) Vocal interactions in birds: the use of song as a model in communication. Adv Study Behav 29:247–296
Uy JAC, Safran RJ (2013) Variation in the temporal and spatial use of signals and its implications for multimodal communication. Behav Ecol Sociobiol 67:1499–1511
Uy JAC, Patricelli GL, Borgia G (2001) Complex mate searching in the satin bowerbird *Ptilonorhynchus violaceus*. Am Nat 158:530–542
Vasconcelos M (2008) Transitive inference in non-human animals: an empirical and theoretical analysis. Behav Process 78:313–334
Vehrencamp SL, Hall ML, Bohman ER, Depeine CD, Dalziell AH (2007) Song matching, overlapping, and switching in the banded wren: the sender's perspective. Behav Ecol Sociobiol 18:849–859
Vignal C, Mathevon N, Mottin S (2004) Audience drives male songbird response to partner's voice. Nature 430:448–450
Wiley RH (2015) Noise matters: the evolution of communication. Harvard University Press, Cambridge, MA
Zandberg L, Gort G, van Oers K, Hinde CA (2017) Direct fitness benefits explain mate preference, but not choice, for similarity in heterozygosity levels. Ecol Lett 20:1306–1314

Chapter 12
Interspecific Communication: Gaining Information from Heterospecific Alarm Calls

Robert D. Magrath, Tonya M. Haff, and Branislav Igic

Abstract Many birds and mammals give alarm calls when they detect predators or other threats, and these calls have been used as classic models for understanding signal design. Here we consider signal design and usage, and how individuals acquire and use information from the alarm calls of other species. Alarm calls often encode detailed information on danger, such as the type of predator, its current behavior, size, or proximity. Alarm calls are sometimes very similar among species or can share generic acoustic features, and both help to explain recognition of heterospecific alarms. However, alarm calls can vary greatly among species, and taxonomically widespread eavesdropping also requires learning the association between calls and danger. Once heterospecifics eavesdrop on alarm calls, there is potentially selection on callers to modify their alarm calls or usage. If callers benefit from eavesdroppers' responses to their alarm calls, they may be selected to enhance signal efficacy, leading to interspecific communication and mutual benefit. Alternatively, callers can be selected to manipulate eavesdroppers, using deceptive signaling, including mimicry, causing the eavesdropper to suffer a cost. If callers suffer a cost from eavesdroppers' responses, their signaling can be modified to make eavesdropping harder, leading to cue denial. Overall, alarm signals provide an insight into the evolution of signal design, and the complex flow of information within and among species in natural communities.

R. D. Magrath (✉) · T. M. Haff · B. Igic
Division of Ecology & Evolution, Research School of Biology, Australian National University, Canberra, Australia
e-mail: Robert.Magrath@anu.edu.au

T. Aubin, N. Mathevon (eds.), *Coding Strategies in Vertebrate Acoustic Communication*, Animal Signals and Communication 7,
https://doi.org/10.1007/978-3-030-39200-0_12

12.1 Alarm Calls, Heterospecific Eavesdropping, and Communication

12.1.1 Alarm Calls and Eavesdropping

Many birds and mammals give alarm calls when they detect predators or other threats. These calls have evolved to warn conspecifics of danger, or to deter predators by harassing them or informing them that they have lost the element of surprise (Klump and Shalter 1984; Caro 2005; Zuberbühler 2009). Alarm calls are of widespread ecological importance because most species are vulnerable to predators, and these calls have been used as classic models for understanding signal design and the evolution of communication (Marler 1955, 1957; Maynard Smith 1965; Sherman 1977). Furthermore, given the broad relevance of information on danger, it is not surprising that many individuals respond to the alarm calls of other species that face similar threats (Seppänen et al. 2007; Goodale et al. 2010; Schmidt et al. 2010; Magrath et al. 2015a). Here we consider signal design and usage, and how individuals acquire and use information from the alarm calls of heterospecifics.

12.1.2 Definitions and Scope of Chapter

We consider responses to heterospecific alarm calls regardless of whether they involve communication, eavesdropping, or deception. Communication entails the use of signals that have evolved because the sender benefits from the receiver's response, so that signals have an "intended" audience (Bradbury and Vehrencamp 2011). Furthermore, receivers of signals are selected to respond only if they also benefit, so that communication requires evolution of both signalers and receivers (Maynard Smith and Harper 2003). Unlike communication, eavesdropping entails reception of signals intended for others. Here we use eavesdropping to mean "interceptive eavesdropping" (Peake 2005), in which individuals respond to signals themselves, and not just communication interactions among others ("social eavesdropping"; McGregor and Dabelsteen 1996; Peake 2005; Searcy and Nowicki 2005). Eavesdroppers, like receivers in communication, will be selected to respond if they gain relevant information. In contrast to both communication and eavesdropping, senders may use signals to deceive an intended audience. Deception differs from communication because, despite the receiver being the intended recipient, it does not benefit from its response to the signal, so that there is a clear conflict of interest between sender and receiver (Searcy and Nowicki 2005). We include communication, eavesdropping, and deception because it can be difficult to know which is in play during receiver responses to alarm calls, and because there is likely to be an evolutionary continuum from eavesdropping to communication and deception (Kostan 2002; Radford et al. 2011).

In this chapter, we first consider what information is conveyed during alarm communication within species, and how it is encoded. This provides the backdrop for understanding what information could be acquired by heterospecifics. We then consider the benefits of eavesdropping on other species' calls, and how individuals are able to recognize the alarm calls of other species. We follow by considering how eavesdropping can lead to communication, deception, or suppression of information. We finish by focusing on vocal mimicry, which unites the themes of call structure, eavesdropping, communication, and deception.

12.2 Alarm Communication Within Species

12.2.1 Function of Alarm Calls

We define alarm calls broadly to include any calls given specifically when a predator is nearby, and which prompt antipredator behavior in conspecifics (Klump and Shalter 1984; Zuberbühler 2009; Magrath et al. 2015a). Calls intended to communicate with predators are sometimes considered signals of unprofitability or deterrence, rather than alarm calls (Caro 2005; Bradbury and Vehrencamp 2011), but conspecifics are also likely to respond to these calls. Furthermore, signals could evolve because of their effects on both predators and conspecifics, so that a broad definition is important for understanding alarm call evolution.

Alarm calls can be named and classified in many ways, such as according to the context of production or their acoustic structure, but many species have three main types, which are associated with different antipredator responses (Klump and Shalter 1984). (1) Flee alarm calls are given in response to immediate danger, and prompt listeners to flee or become cryptic. When given to flying predators, such as hawks, these are often called "aerial" or "hawk" alarm calls. (2) Mobbing alarm calls are given to predators not posing immediate danger, and prompt others to approach, monitor, and sometimes harass the predator. (3) Distress alarm calls are given when individuals are attacked, which might startle the predator, recruit others to help the caller, or attract secondary predators.

Alarm calls often communicate about the type of threat, the degree of danger posed, or both, with varying degrees of specificity. "Functionally referential" alarm calls communicate about the type of threat (Gill and Bierema 2013; Townsend and Manser 2013; Suzuki 2016). Vervet monkeys, *Chlorocebus pygerythrus*, for a classic example, produce acoustically distinct alarm calls to leopards, eagles, and snakes, and respond appropriately to playback of those calls, such as by looking down to "snake" alarms and up to "eagle" alarms (Seyfarth et al. 1980a, b). Similarly, Japanese great tits, *Parus minor*, have different alarm calls for different predators, and nestlings respond by fleeing the nest when warned of snakes but crouching down when warned of crows (Suzuki 2011). These responses are appropriate given that snakes can enter nests but crows cannot. Some birds even have alarm calls given specifically to brood parasites (Gill and Sealy 1996; Feeney and Langmore 2013; Wheatcroft and Price 2015). Other species have alarm calls that communicate about the predator's current behavior. For example, Siberian jays,

Perisoreus infaustus, and noisy miners, *Manorina melanocephala*, both have acoustically different calls to perched and airborne raptors (Griesser 2008; Cunningham and Magrath 2017; Farrow et al. 2017). Alarm calls can also communicate about the degree of danger, related to predator proximity or type, such as in white-browed scrubwrens, *Sericornis frontalis* (Leavesley and Magrath 2005), black-capped chickadees, *Poecile atricapillus* (Templeton et al. 2005) and New Holland honeyeaters, *Phylidonyris novaehollandiae* (McLachlan and Magrath 2020). Alarm calls may also communicate simultaneously about both the type of danger and degree of threat, as shown for meerkats, *Suricata suricatta* (Manser 2001; Manser et al. 2001, 2002). Not all alarm calls convey specific information, as some are given in a broad range of contexts, such as "general" or "variable-use" calls given to a variety of threats and in social interactions (Marler 1957; Fichtel and Kappeler 2002). In these cases, context and call variability can nonetheless guide listeners' responses (Wheatcroft 2015). Overall, different alarm calls allow listeners to take appropriate action according to the type or urgency of the threat.

12.2.2 Coding: Marler's Insight on Structure and Function

In a classic paper on signal design, Peter Marler proposed that the acoustic structure of alarm calls should reflect their function (Marler 1955). Passerine alarm calls given to flying hawks should be difficult to locate, to reduce the risk of attack, so Marler predicted that they should be high-pitched, pure tones, with a gradual onset and ending. By contrast, mobbing calls given to perched raptors should be easily locatable so that others can join in mobbing. They should therefore be abrupt, repetitive, and more broadband than calls given to flying hawks. Consistent with these predictions, many European passerines have "seeet" aerial alarm calls (Fig. 12.1a) and mobbing calls with exactly these acoustic features (Marler 1955, 1957).

Subsequent research supports Marler's main conclusions about seeet aerial alarm calls versus mobbing calls. Playback experiments show that seeet calls are harder for predators to locate than mobbing calls (Klump and Shalter 1984; Jones and Hill 2001). Seeet calls are also likely to give minimal distance cues. The long duration and gradual onset and termination will reduce the ability to detect reverberations, while the narrow frequency range stops assessment of the relative attenuation of different frequencies; both these features can be used to judge distance (Naguib and Wiley 2001). In addition to being difficult to locate, seeet calls are difficult to hear, especially for hawks. The high frequency of these calls (7–8 kHz) means they are unlikely to be heard at a great distance because of relatively rapid attenuation (Wiley and Richard 1982) and because avian hearing often becomes less sensitive above 4–5 kHz (Dooling 2004). More subtly, differences in hearing sensitivities are exploited by great tits, whose seeet call is at a frequency to which it is much more sensitive than the sparrowhawk, *Accipiter nisus*, its major predator (Klump et al. 1986). This call frequency therefore minimizes the risk of eavesdropping by hawks.

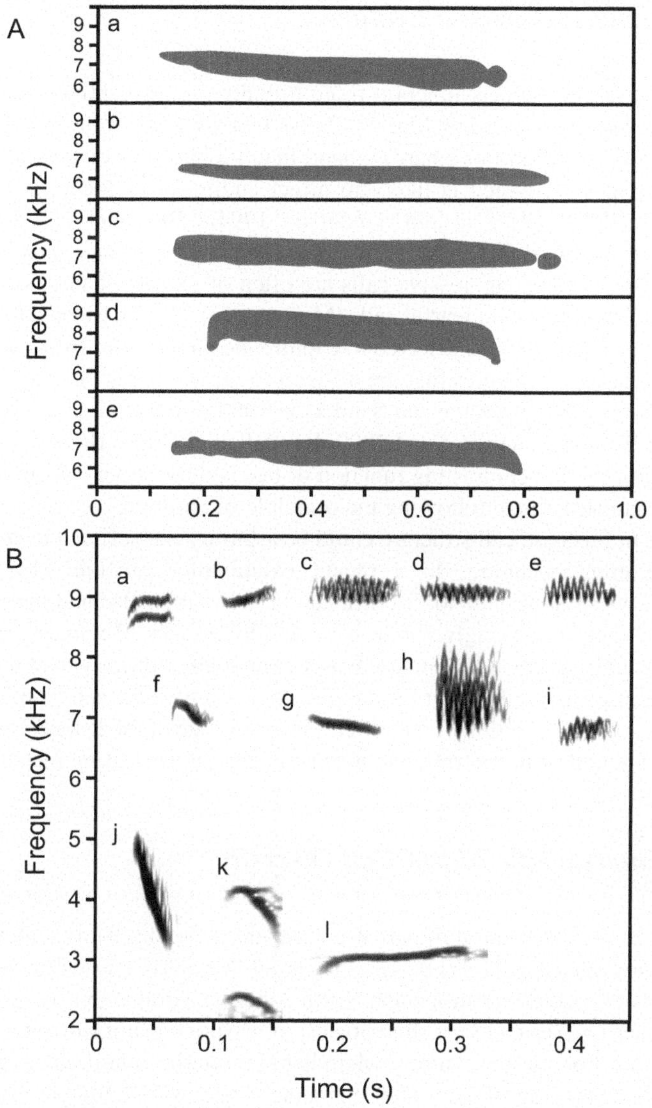

Fig. 12.1 Similarities and differences among alarm calls given by passerines to flying hawks. (**a**) Seeet alarm calls given by some European passerine species, redrawn from Marler (1957): (a) reed bunting, *Emberiza schoeniclus*; (b) blackbird, *Turdus merula*; (c) great tit, *Parus major*; (d) blue tit, *Cyanistes caeruleus*; and (e) chaffinch, *Fringilla coelebs*. (**b**) Aerial alarm calls given by Australian members of the superfamily Meliphagoidea. Only single elements are shown, but calls often have multiple similar elements. (a) Western thornbill, *Acanthiza inornata*; (b) white-winged fairy-wren, *Malurus leucopterus*; (c) superb fairy-wren, *M. cyaneus*; (d) splendid fairy-wren, *M. splendens*; (e) variegated fairy-wren, *M. lamberti*; (f) southern whiteface, *Aphelocephala leucopsis*; (g) chestnut-rumped thornbill, *Acanthiza uropygialis*; (h) white-browed scrubwren, *Sericornis frontalis*; (i) brown gerygone, *Gerygone mouki*; (j) New Holland honeyeater, *Phylidonyris novaehollandiae*; (k) white-plumed honeyeater, *Ptilotula penicillata*; (l) noisy miner, *Manorina melanocephala*; recordings by Robert Magrath and Janet Gardner. Note the different time and frequency scales in (**a**) and (**b**)

12.2.3 Generic Acoustic Features

Alarm call structure reflects function more broadly, and potentially for additional reasons than those suggested by Marler (1955). In a diversity of birds and mammals, flee alarm calls are often more narrow-band than mobbing or distress alarms given by the same species, which is likely to affect audibility or localizability (Marler 1957; Aubin 1989; Jurisevic and Sanderson 1994a; Bradbury and Vehrencamp 2011). These broad differences might also reflect constraints of motivational-structural rules, because aggressive calls are often of greater bandwidth, less tonal, and of lower frequency than fearful calls (Morton 1977, 2017). Related to this idea is that some alarm calls may be selected to capture the attention of listeners, by taking advantage of acoustic features that are difficult to ignore or easy to hear over a long distance (Aubin 1989; Jurisevic and Sanderson 1998; Fitch et al. 2002; Neudorf and Sealy 2002; Blumstein and Récapet 2009; Rendall et al. 2009; Branch and Freeberg 2012). In addition, the contrasting function of flee and mobbing calls may select for contrasting call structures, following the principle of antithesis (Darwin 1872; Hurd et al. 1995). In this case call structure could be arbitrary but selected to contrast with calls of different meaning. In a recent examination of this idea, however, superb fairy-wrens were able to discriminate between their flee and mobbing alarm calls despite their surprisingly subtle acoustic differences (Tegtman and Magrath in press). Overall, alarm call structure is not completely arbitrary, and it is likely to enhance or diminish audibility or locatability depending on the call type. In addition, alarm calls can be a reflection of the caller's internal state or the listener's perception, or enhance the ability to discriminate between calls of contrasting meaning.

12.2.4 Interspecific Alarm Call Diversity

Despite the broad association of call structure and meaning, there is huge diversity among species' alarm calls. While c. 7 kHz seeet flee alarm calls are given to flying hawks by many species in Europe and North America, comparable aerial alarm calls given by other species can vary dramatically in frequency and structure. For example, Australian honeyeater (Family Meliphagidae) aerial alarm calls are typically composed of short, rapidly repeated elements, often with frequency sweeps, and with mean frequencies of 2–4 kHz (Rooke and Knight 1977; Jurisevic and Sanderson 1994b). Furthermore, in the superfamily Meliphagoidea (including Meliphagidae, Maluridae, and Acanthizidae), aerial alarm calls vary in frequency, duration, bandwidth, and possession of rapid frequency modulation, and very few have a seeet-like structure (Jurisevic and Sanderson 1998; Fallow et al. 2011, 2013; Cunningham and Magrath 2017; Fig. 12.1b). Similarly, birds in Sri Lankan mixed flocks have a diversity of alarm calls given to raptors, many of which are of relatively low frequency and high bandwidth (Goodale and Kotagama 2005). In neither case is it clear why there is so much variation. There is also great diversity in mobbing call structure. A study of 52 species of American passerines from 12 families revealed substantial diversity among mobbing calls; only about 50% had mobbing calls with abrupt elements of broad frequency range (Ficken and Popp 1996). Similarly, calls used for mobbing vary "drastically" among 22 species of Himalayan passerines, perhaps in part because they are also used in social contexts within species (Wheatcroft and Price 2013).

12.2.5 Alarm Call Usage

In addition to encoding information within call element structure and frequency, call usage and composition can affect meaning in multiple ways. First, the number of elements included in a call can convey graded information on the degree of danger. For example, superb fairy-wrens, *Malurus cyaneus*, New Holland honeyeaters and white-browed scrubwrens include more elements in their aerial alarm calls when a flying predator is closer or more dangerous (Leavesley and Magrath 2005; Fallow and Magrath 2010; McLachlan and Magrath 2020), and black-capped chickadees include more D notes in their chick-a-dee calls when mobbing a more dangerous raptor (Templeton et al. 2005). Second, the overall rate of call delivery can change call meaning. For example, a suite of passerines from India, Japan, and Sweden have a higher call repetition rate when calling to predators than during social interactions, and playbacks show that listeners respond with antipredator behavior when call repetition rate is higher (Wheatcroft 2015). In this case, call usage changes the meaning of the same variable-use calls. Third, the relative use of different calls can convey graded and possibly categorical information about danger. For example, Arabian babblers, *Turdoides squamiceps*, include relatively more high-pitched than low-pitched calls when mobbing a live cat than an owl mount, implying that the combination of calls conveys information about the type or magnitude of danger (Naguib et al. 1999). Similarly, white-handed gibbon, *Hylobates lar*, songs contain relatively more "hoo" notes in songs prompted by predators than in duet songs with their mates (Clarke et al. 2006). Finally, call syntax can affect meaning. For example, Japanese great tits give D notes alone in the absence of predators but together with other note types when predators are near, and the response to compound calls depends on whether the D notes are before or after the other notes (Suzuki et al. 2016). In this case the order of notes within a call produces a compound meaning, rather than merely the relative abundance of different notes.

12.3 Information from Heterospecific Alarms

12.3.1 Diversity of Eavesdropping

Heterospecific alarm calls are a potentially rich source of information about danger for species facing similar threats, and over 70 species have been shown through experimental playback to respond to other species' alarm calls (Magrath et al. 2015a). Individuals usually respond to playback of heterospecific alarm calls with typical antipredator behavior, such as by becoming vigilant, fleeing to cover or freezing after flee alarm calls, or approaching and calling in response to mobbing or distress calls. Examples span a diversity of taxa, with both mammals and birds responding to alarm calls of other species of mammals and birds. There are even a few examples of lizards responding to bird alarm calls (Vitousek et al. 2007; Ito and Mori 2010; Ito et al. 2013), showing that individuals do not need to have their own alarm calls, or even to produce any vocalizations, to be responsive to the calls of other species.

12.3.2 Information Gained

Do individuals gain similarly detailed information on danger from heterospecfic and conspecific alarm calls? There has been no broad study, but in some cases they do, whereas in other cases they do not. Individuals can gain information on predator type from other species' referential alarm calls. For example, ring-tailed lemurs, *Lemur catta*, respond to Verreaux' sifaka's, *Propithecus verreauxi*, aerial alarm calls by looking up and to their terrestrial alarm calls by running to trees (Oda and Masataka 1996). Comparable patterns are found in other primates (Zuberbühler 2000b; Kirchhof and Hammerschmidt 2006). Similarly, Carolina chickadees, *Poecile carolinensis*, respond appropriately to the aerial and mobbing calls of tufted titmice, *Baeolophus bicolor* (Hetrick and Sieving 2012), Australian magpies, *Cracticus tibicen*, orient appropriately to aerial and mobbing alarm calls of noisy miners (Dawson Pell et al. 2018), and yellow-casqued hornbills, *Ceratogymna elata*, respond to the relevant "eagle" but not irrelevant "leopard" alarm calls given by Diana monkeys, *Cercopithecus diana* (Rainey et al. 2004). Individuals can also respond to graded heterospecific alarm calls. Superb fairy-wrens, white-browed scrubwrens, and red-breasted nuthatches, *Sitta canadensis*, all respond to graded heterospecific alarm calls indicating the degree of danger (Templeton and Greene 2007; Fallow and Magrath 2010). In other cases, eavesdroppers gain only partial information. For example, although banded mongooses, *Mungos mungo*, increased vigilance after playback of lapwing, *Vanellus* spp., alarm calls, they did not respond to urgency information in the alarm calls of the only species tested, the crowned plover, *V. coronatus* (Müller and Manser 2008). Overall, individuals can use at least some of the detailed information encoded in other species' alarms, but we know little about potential constraints on decoding heterospecific alarms (Murray and Magrath 2015).

12.3.3 Benefits

The ability to eavesdrop on other species' alarm calls is likely to bring many benefits, ultimately increasing survival and reproductive success. Being able to eavesdrop on heterospecifics means that there are more eyes looking for danger, but there could also be specific benefits gained from heterospecific alarm calls over conspecifics ones. For example, species on the ground or with a foraging niche that restricts vigilance may rely on alarm calls of other species that are highly vigilant, such a flycatching species that rely on vigilance to detect prey (Goodale and Kotagama 2008; Martínez and Zenil 2012). In a within-species test of this idea, New Holland honeyeaters were slower to spot gliding predator models, and more likely to flee to alarm calls, when feeding on nectar in flowers than when hawking for insects from exposed perches (McLachlan et al. 2019). Similarly, some species lack alarm calls or rarely give them, and yet respond to those of other species (Sullivan 1985; Vitousek et al. 2007; Sridhar et al. 2009; Srinivasan et al. 2010). Immediate responses to alarm calls presumably reduce the probability of detection or attack by predators, but there will also be indirect benefits. First, being attentive to heterospecific alarms can increase foraging success. Scimitarbills, *Rhinopomastus cyanomelas*, eavesdrop on alarm calls of pied babblers, *Turdoides bicolor*, and when in mixed groups they use babblers for information on danger and so reduce personal vigilance and increase foraging success (Ridley et al. 2014). Second, eavesdropping on alarm calls can also affect the choice of feeding location or habitat (Parejo et al. 2012). Third, heterospecific alarms potentially provide information on danger without resource competition, which could diminish the net benefit of comparable information from conspecifics (Seppänen and Forsman 2007). Finally, eavesdropping may facilitate associative learning about previously unrecognized predators (Curio et al. 1978; Vieth et al. 1980) or alarm calls (Potvin et al. 2018).

12.4 Mechanisms of Recognition of Heterospecific Alarm Calls

12.4.1 Learned and Unlearned Recognition

The great diversity of alarm calls within and among species raises the puzzle of how individuals recognize the alarm calls of other species. The answer is complex; animals can recognize some alarm calls on first exposure, whereas other alarm calls require learning, and both mechanisms have potential costs and benefits. Reacting to acoustic features alone allows for response to novel alarm calls on first exposure, but restricts the response to calls with a narrow range of characteristics (Hollén and Radford 2009). By contrast, learning allows individuals to gain information from alarm calls with a broad range of acoustic features, but leaves individuals more vulnerable to predators until they have learned to recognize the calls (Griffin 2004). For most species, recognition of heterospecific alarm calls probably involves both innate responses and learning, which helps explain the widespread occurrence of eavesdropping across taxa (Magrath et al. 2015a, b; Potvin et al. 2018).

12.4.2 Acoustic Similarity to Conspecific Calls

Individuals often respond to novel calls that are acoustically similar to conspecific calls. Recognition of conspecific alarm calls appears to rely on one or a few key acoustic features, with generalization of response to unfamiliar calls which share those features (Wiley and Richard 1982; Ghirlanda and Enquist 2003; ten Cate and Rowe 2007; Fallow et al. 2013; Magrath et al. 2015a; Meise et al. 2018). Indeed, the mutual response by many European songbirds to one another's high-pitched seeet flee alarms is probably due to acoustic similarity (Marler 1957; Sect. 12.2.2). More recent work has revealed some key acoustic features that are generalized to heterospecific alarms, with secondary features modifying the response (Aubin and Brémond 1989; Aubin 1991; Johnson et al. 2003; Fallow et al. 2011, 2013; Dutour et al. 2017). For example, superb fairy-wrens flee to cover to unfamiliar natural and synthetic aerial alarm calls when their peak frequencies are similar to conspecific calls (Fig. 12.2), and adjust how much time they spend in cover based on similarity to the rate of frequency modulation of conspecific calls (Fallow et al. 2011, 2013). Similarly, several European species of birds respond to synthetic distress calls that share the key property of an increasing frequency sweep, but their response is also likely to be affected by fine-scale acoustic differences (Aubin 1991). Acoustic similarities among species' alarms need not be exact to prompt response, probably because of natural variability in alarm calls and because responding to alarm calls is critical for survival. Conspecific alarm calls vary among individuals and are affected by environmental degradation during transmission, creating selection for a broad response range that is likely to mean generalization to a greater range of heterospecific alarm calls (Blumstein and Munos 2005; Leavesley and Magrath 2005; Fallow et al. 2011, 2013; Magrath et al. 2015a). In addition, the costs of not responding to an alarm call when there is a predator will select for a low threshold for signal detection (Wiley 2015), and therefore response to a greater range of both conspecific and heterospecific alarm calls.

12.4.3 Generic Acoustic Features

Although acoustic similarity to conspecific calls can allow recognition of novel alarm calls, it is not the only potential mechanism of unlearned response. The alarm calls of many species contain acoustic features that are inherently attention-grabbing or frightening, and this may also prompt immediate response to unfamiliar calls (Hirth and McCullough 1977; Morton 1977; Fitch et al. 2002; Rendall et al. 2009). Many mobbing, distress, "panic," or general alarm calls, for example, contain elements that are harsh, abrupt, broadband, or nonlinear. These acoustic features tend to be honest signals of caller distress, increase listener attention and response, and prevent habituation (Manser 2001; Blumstein and Récapet 2009; Townsend and Manser 2011; Blumstein and Chi 2012; Blesdoe and Blumstein 2014; Karp et al. 2014). For example, great-tailed grackles, *Quiscalus mexicanus*, increase vigilance after playback of novel synthetic calls that include nonlinear features such as abrupt transitions and noise (Slaughter et al. 2013), and western gray kangaroos, *Macropus fuliginosus*, do not habituate to novel sounds with nonlinear features (Biedenweg et al. 2011).

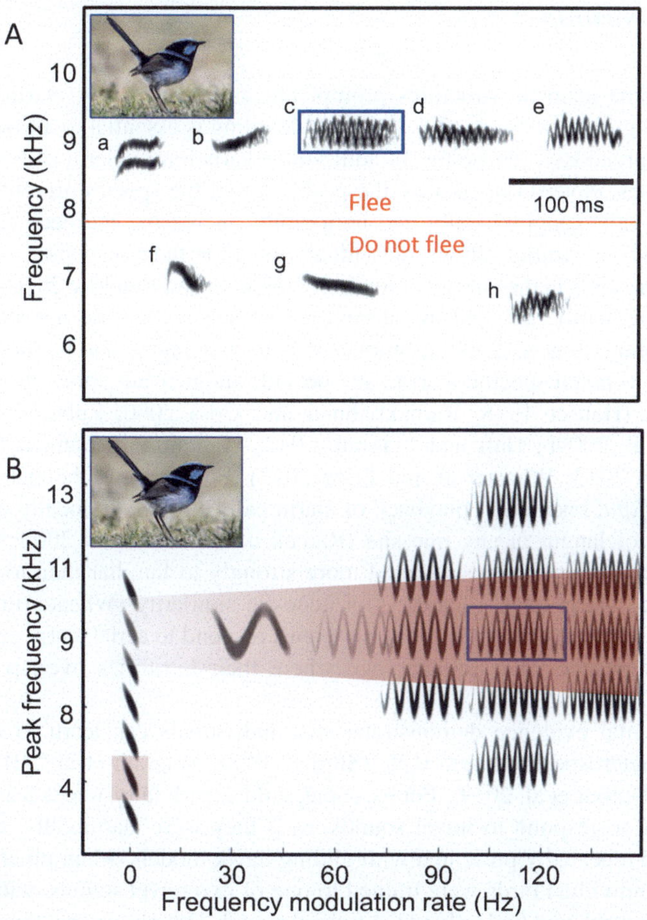

Fig. 12.2 Fairy-wrens, *Malurus cyaneus*, respond to unfamiliar aerial alarm calls that are acoustically similar in peak frequency to their own alarms. (**a**) Response of fairy-wrens (inset) to playback of conspecific and unfamiliar heterospecific aerial alarm calls from members of the Acanthizidae and other Maluridae: (a) western thornbill, *Acanthiza inornata*; (b) white-winged fairy-wren, *Malurus leucopterus*; (c) superb fairy-wren, *M. cyaneus*; (d) splendid fairy-wren, *M. splendens*; (e) variegated fairy-wren, *M. lamberti*; (f) southern whiteface, *Aphelocephala leucopsis*; (g) chestnut-rumped thornbill, *Acanthiza uropygialis*; (h) brown gerygone, *Gerygone mouki*. Results from Fallow et al. (2011). (**b**) Response by fairy-wrens (inset) to playbacks of computer-synthesized calls based on variation in the Maluridae, Acanthizidae and Meliphagidae. The red shading highlights sounds to which fairy-wrens usually fled, with a darker shade representing a higher percentage. The 4 kHz unmodulated call is similar to that of a familiar local species, the New Holland honeyeater, *Phylidonyris novaehollandiae*. Figure (**b**) modified from Fallow et al. 2013; photograph in (**a**) and (**b**) by Jessica McLachlan

12.4.4 Learning

Although some acoustic structures prompt unlearned responses, alarm calls vary greatly among species (Sect. 12.2.4), so that learning is essential to allow taxonomically widespread eavesdropping. In addition to variation in alarm calls within and among species, eavesdroppers face the problem that the species present in a community can vary geographically and temporally, increasing the need for learning (Griffin 2004). Learning allows individuals to tailor their responses to the most relevant alarm calls in the current community (Nocera and Ratcliffe 2010), as well as to ignore non-alarm calls and the alarm calls of species that do not share similar predators (Magrath et al. 2009a). Studies of both geographic and temporal patterns of response to heterospecific alarm calls provide indirect but compelling evidence for learning (Hauser 1988; Ramakrishnan and Coss 2000; Zuberbühler 2000a; Magrath et al. 2009b; Haff and Magrath 2012; Magrath and Bennett 2012; Haff and Magrath 2013; Wheatcroft and Price 2013). For example, bonnet macaques, *Macaca radiata*, respond to playback of alarm calls of only the locally common of two species of langur at any one site (Ramakrishnan and Coss 2000; Fig. 12.3); Himalayan bird communities respond more strongly to familiar than to unfamiliar heterospecific alarm calls, regardless of acoustic similarity (Wheatcroft and Price 2013); and fledgling white-browed scrubwrens, respond to aerial alarm calls of New Holland honeyeaters at a younger age where their territories overlap (Haff and Magrath 2013).

Experimental evidence demonstrates that individuals can learn to respond to unfamiliar heterospecific alarm calls (Shriner 1999; Magrath et al. 2015b; Potvin et al. 2018; Dutour et al. 2019). For example, wild superb fairy-wrens learned over a 2-day period to respond to novel sounds, as if they were alarm calls, when those sounds were repeatedly presented with gliding hawk models (Magrath et al. 2015b; Fig. 12.4). Individual birds were trained to one of two novel sounds, with the other acting as a control, and fled only to the trained sound. This shows that they learned to recognize specific sounds, and were not merely responding because they became scared of playbacks in general. Further work on this system has revealed that fairy-wrens can learn to recognize novel alarm calls through association with a chorus of known alarm calls (Potvin et al. 2018), demonstrating that learning can take place without visual pairing of predators with alarm calls or fearful responses.

Learning and acoustic attributes could simultaneously be important in alarm call recognition if some features of alarm calls facilitate learning. For example, unfamiliar calls that have "attention grabbing" features (Sect. 12.2.3) may be easier to learn about than calls that lack such features (Ghirlanda and Enquist 2003). One possible example concerns greater responses to "variable use" calls that are given at higher rates when predators are near (Sect. 12.2.5). Playback experiments show that both conspecifics and heterospecifics are more responsive to higher rate calls from their own and other species (Randler and Förschler 2011; Wheatcroft 2015), and it is plausible that high call rates promote learning about call structure. Call structure itself can also affect learning. For example, it is harder to train nestling great tits to

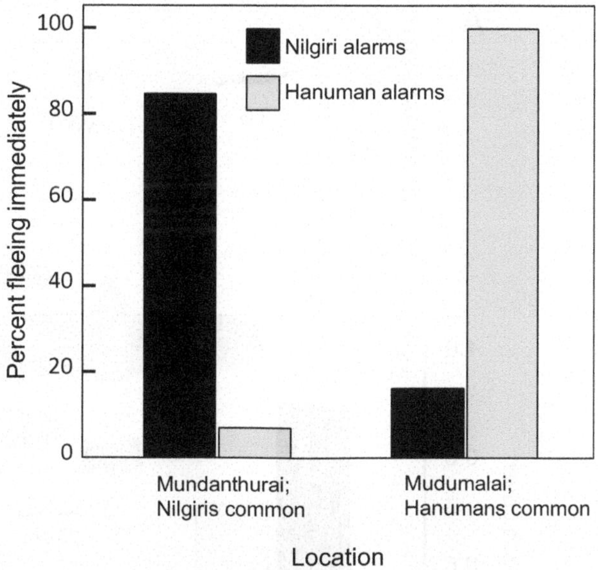

Fig. 12.3 Response of bonnet macaques, *Macaca radiata*, to playback of the alarm calls of Nilgiri langurs, *Semnopithecus johnii*, and Hanuman langurs, *S. entellus*, in two separate geographic areas. Nilgiri langurs were only common at Mundanthurai, while Hanuman langurs were only common at Mudumalai. Macaques usually only fled to playback of the familiar species' alarm calls. Redrawn from Ramakrishnan and Coss (2000)

beg to parental alarm calls than to conspecific or heterospecific songs (Rydén 1978, 1982). Similarly, young Belding's ground squirrel, *Spermophilus beldingi,* can discriminate between acoustically different parental alarm calls before emergence from their burrows, but do not treat them as alarm calls until after they have experience outside the burrow (Mateo 1996).

12.5 Eavesdropping, Communication, and Deception

12.5.1 Eavesdropping Facilitates Evolution

Eavesdropping on heterospecific alarm calls, regardless of the mechanism of recognition, sets the stage for further evolution in the structure or use of signals, and potentially reception and use by listeners (Marler 1957; Kostan 2002; Dabelsteen 2005; Searcy and Nowicki 2005). As soon as members of other species respond to alarm calls, there is potentially selection to enhance or suppress their responses by modifying alarm calls or their usage (Kostan 2002; Dabelsteen 2005). If callers are unaffected by eavesdropper responses, their signaling will be unaffected, but if callers benefit, they will be selected to enhance that benefit. By contrast, callers will be selected to make calls more difficult to overhear if eavesdropper responses are costly to callers, which has recently been dubbed "cue denial" (Searcy and Yasukawa 2017). Listeners will be selected to enhance detection and response to alarm calls if they bring benefits, but to ignore such calls if they are deceptive. Here, we consider these scenarios.

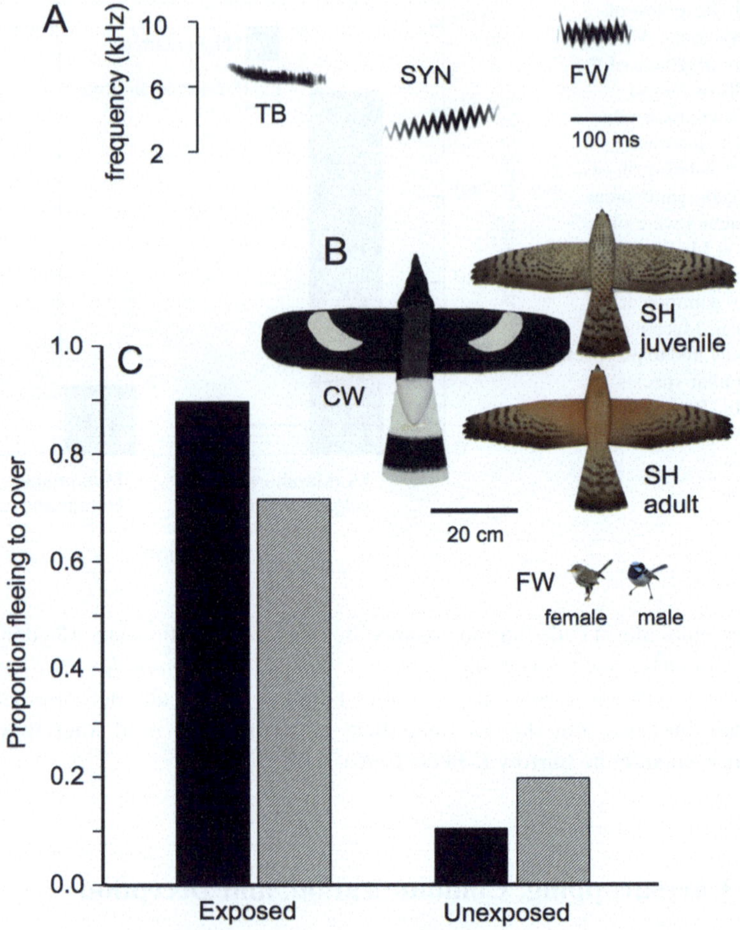

Fig. 12.4 Superb fairy-wrens, *Malurus cyaneus*, were trained to respond to unfamiliar alarm calls. (**a**) Unfamiliar calls included the aerial alarm call of the allopatric chestnut-rumped thornbill (TB), *Acanthiza uropygialis*, and a sound synthesized on computer (SYN), and both were different to conspecific calls (FW) and local heterospecific calls. (**b**) Unfamiliar calls were presented to fairy-wrens along with gliding model predators of pied currawongs (CW), *Strepera graculina*, and collared sparrowhawks (SH), *Accipiter cirrocephalus*. Each focal bird was exposed to only one of the two novel calls during training with model predators, whereas the other was unexposed and acted as a control sound. (**c**) After training, focal fairy-wrens fled only to the exposed sound and not to the unexposed novel sound. Figure reproduced from Magrath et al. (2015b)

12.5.2 Communication

Eavesdroppers and senders could both benefit, potentially leading to interspecific communication, not merely eavesdropping. If mobbing or distress calls prompt members of others species to approach and harass predators, there should be selection to call specifically when heterospecifics are nearby, or to modify calls to make them more recognizable and audible to heterospecifics. The caller gains help in mobbing, the eavesdropper gains information on danger, and collective mobbing is likely to be more effective at deterring predators (Krams and Krama 2002; Krams et al. 2006). Similarly, if heterospecifics flee or become cryptic after hearing aerial alarm calls, the caller could benefit if the predator does not detect any prey and departs the area (Maynard Smith 1965). Alternatively, the caller might reduce risk to itself by prompting all individuals to flee, thereby confusing an attacking predator, diluting the risk of attack, or reducing the risk of being singled out by fleeing alone and becoming the most conspicuous individual (Sherman 1985; Cresswell 1994; Caro 2005; De Vos and O'Riain 2010). Again, both parties are likely to benefit. In these cases, alarm calls might evolve features that make them easy to recognize or learn, which could select for similarities among species' alarm calls (Marler 1957), or acoustic attributes that prompt attention or facilitate learning (Sects. 12.2.3, 12.4.3, 12.4.4). Finally, individuals may give alarm calls specifically when other species are present, again suggesting active communication. For example, fork-tailed drongos, *Dicrurus adsimilis*, rarely give alarms to terrestrial predators when alone, but do so when near pied babblers, which are vulnerable to terrestrial threats (Ridley et al. 2007). Drongos benefit in the longer term because they can also steal food from babblers (Sect. 12.5.4).

12.5.3 Cue Denial

If callers suffer a cost from eavesdropping, signal structure and use are likely to evolve to make eavesdropping more difficult. We are not aware of examples involving eavesdropping by prey species, but seeet flee alarm calls have attributes that make them difficult for predators to locate or overhear (Sects. 12.2.2, 12.2.3). Similarly, Diana monkeys suppress calling after detecting chimpanzees, *Pan troglodytes*, because unlike other predators that abandon hunting when detected, chimpanzees can pursue monkeys through the treetops (Zuberbühler et al. 1997; Zuberbühler 2000a). These cases are the opposite of communication. In communication, signaling evolves from cues to enhance the reception by listeners (Bradbury and Vehrencamp 2011), whereas cue denial hinders reception.

12.5.4 Deception

Turning the tables, eavesdroppers are vulnerable to deception if callers can exploit their responses. Indeed, several species use alarm calls deceptively, when predators are absent, to scare members of other species and steal food (Munn 1986; Møller 1988; Goodale and Kotagama 2005; Flower 2011). In Amazonian mixed-species flocks, white-winged tanager-shrikes, *Lanio versicolor*, and bluish-slate antshrikes,

Thamnomanes schistogyns, act as sentinels in warning of danger. But they also give alarm calls deceptively when competing for a flying insect with another species, which then usually flees to cover (Munn 1986). Similarly, in addition to giving alarm calls to warn pied babblers of danger (Sect. 12.5.2), fork-tailed drongos, *Dicrurus adsimilis*, use deceptive alarm calls to steal food from both babblers and meerkats (Ridley et al. 2007; Flower 2011; Flower and Gribble 2012). There is therefore a fine balance between cooperation and deception (Ridley et al. 2007; Radford et al. 2011). Indeed, such deception is likely to be effective only if it is not too common. We consider deception and call structure further in the next section on vocal mimicry.

12.6 Vocal Mimicry

12.6.1 Avian Mimicry of Alarm Calls

About 15–20% of songbird species mimic the vocalizations of other species, commonly including their alarm calls (Marshall 1950; Baylis 1982; Kelley et al. 2008; Dalziell et al. 2015). Several species from at least 15 passerine families mimic heterospecific alarm calls (Table 12.1), and they can be given alone or in combination with nonmimetic vocalizations (Goodale and Kotagama 2006; Flower 2011; Wheatcroft and Price 2013; Igic and Magrath 2014; Dalziell and Welbergen 2016). Furthermore, heterospecific alarm calls and predator vocalizations are often mimicked together and may serve similar functions (Chu 2001b; Ratnayake et al. 2010; Kelley and Healy 2011; Goodale et al. 2014b). Mimicry can include a single type of alarm call, such as an aerial or mobbing alarm call (Flower 2011; Wheatcroft and Price 2013), or several types given in the same calling bout (Chu 2001b; Kelley and Healy 2011).

12.6.2 Mimetic Accuracy

The similarity of mimetic alarm calls to the species being copied—termed mimetic accuracy—varies among species, probably for multiple reasons. First, there are likely to be constraints on a mimic's vocal abilities. For example, superb lyrebirds, *Menura novaehollandiae*, and northern mockingbirds, *Mimus polyglottos*, are able to produce almost perfect copies of some heterospecific vocalizations but not of others, probably reflecting species-specific motor abilities (Zollinger and Suthers 2004; Zann and Dunstan 2008; Dalziell and Magrath 2012). Likewise, mimicry of alarm calls varies in accuracy (Flower 2011; Fallow et al. 2013; Igic and Magrath 2013), but it is unknown if this is related to vocal constraints. Second, mimics may only be selected to copy the key features of alarm calls that prompt heterospecific response, whereas other acoustic features might not be copied accurately. For example, brown thornbills, *Acanthiza pusilla*, produce imperfect copies of

Table 12.1 Passerines shown to mimic heterospecific alarm calls and predator vocalizations

Family	Species	Distribution	References
Menuridae	Superb lyrebird, *Menura novaehollandiae*	Australia	Dalziell and Welbergen (2016)
Ptilonorhynchidae	Spotted bowerbird, *Chlamydera maculate*	Australia	Kelley and Healy (2011)
Acanthizidae	Brown thornbill, *Acanthiza pusilla*	Australia	Igic and Magrath (2013)
Vireonidae	White-eyed vireo, *Vireo griseu*	North and Central America	Adkisson and Conner (1978)
Dicruridae	Greater racket-tailed drongo, *Dicrurus paradiseus*	Asia	Goodale and Kotagama (2006)
	Fork-tailed drongo, *D. adsimilis*	Africa	Flower (2011)
Corvidae	Blue jay, *Cyanocitta cristata*[a]	North America	Hailman (2009)
	Steller's jay, *C. steller*[a]	North America	Billings et al. (2017)
	Sri Lanka Magpie, *Urocissa ornata*[a]	Sri Lanka	Ratnayake et al. (2010)
Phylloscopidae	Greenish warbler, *Phylloscopus trochiloides*	Eurasia	Wheatcroft and Price (2013)
Acrocephalidae	Icterine warbler, *Hippolais icterina*	Europe	Riegert and Jůzlová (2018)
Alaudidae	Sabota lark, *Calendulauda sabota*	Africa	Vernon (1973)
Bombycillidae	Phainopepla, *Phainopepla nitens*	North and Central America	Chu (2001b)
Muscicapidae	Robin-chats, *Cossypha natalensis, C. heuglini, C. dichroa*	Africa	Oatley (1969)
Mimidae	Northern mockingbird, *Mimus polyglottos*	North and Central America	Gammon and Altizer (2011)
Sturnidae	Common starling, *Sturnus vulgaris*	Widespread	Hindmarsh (1984)
	Cape starling, *Lamprotornis nitens*	Africa	Vernon (1973)
Emberizidae	Eastern towhee, *Pipilo erythrophthalmus*	North America	Greenlaw et al. (1998)
Fringillidae	Thick-billed euphonia, *Euphonia laniirostris*	South America	Morton (1976)
	Violaceous euphonia, *Euphonia violacea*	South America	Snow (1974)

[a]Only mimicry of predator vocalizations reported

heterospecific alarm calls, yet retain the key acoustic features important for provoking alarm response (Fallow et al. 2013; Igic and Magrath 2013). Third, selection for acoustic mimetic accuracy may be frequency dependent, as it is in other mimetic systems (Lindström et al. 1997). This implies that accuracy of specific mimetic alarms will need to be higher when they are relatively more common compared to the heterospecific's alarm calls. Finally, because not responding to alarm calls could be lethal, listeners are likely to have low thresholds for signal recognition, and so respond to imperfect copies (Sect. 12.4.2).

12.6.3 *Function*

The function of mimetic alarm calls depends on their target audience, the type of alarm calls mimicked, and the contexts in which they are used (Dalziell et al. 2015). The intended audience of mimetic alarm calls can be conspecifics, other prey species, or predators (Flower 2011; Goodale et al. 2014a; Igic et al. 2015). Mimicry of different types of alarms calls affects function, with aerial alarms scaring or distracting listeners, and mobbing calls attracting them (Flower 2011; Goodale et al. 2014b). Furthermore, mimetic alarm calls can be used in the same context as their respective nonmimetic versions (Morton 1976; Goodale and Kotagama 2006; Wheatcroft and Price 2013), or deceptively in the absence of predators (Flower 2011).

12.6.3.1 Mobbing Predators

One proposed function of mimetic alarm calls is the attraction of heterospecifics to help mob predators. An individual mobbing a predator may benefit from attracting heterospecifics if a larger mobbing group dilutes predation risk to itself (Ficken 1989), or is more effective at causing the predator to depart (Pettifor 1990; Flasskamp 1994). That being the case, mimicry may by-pass the need for heterospecifics to learn to recognize a mimic's nonmimetic alarm calls, and so ensure response regardless of familiarity (Wheatcroft and Price 2013). Furthermore, individuals sometimes respond less intensely even to familiar heterospecific alarm calls than to conspecific alarm calls (Wheatcroft and Price 2013; Ridley et al. 2014). In such instances, alarm mimicry may act to deceive heterospecifics about the identity of the caller and so increase their mobbing. For example, orange-billed babblers, *Turdoides rufescens*, and ashy-headed laughing-thrushes, *Garrulax cinereifrons*, are more likely to mob after hearing mimicry of their own mobbing alarm calls by greater racket-tailed drongos, *Dicurus paradiseus*, than after hearing the drongo's own mobbing alarm calls (Goodale et al. 2014b). Mimicry might also be more effective in attracting heterospecific mobbers when a mixture of nonmimetic and mimetic calls evokes stronger mobbing than nonmimetic calls alone (Chu 2001a). More generally, it is possible that mimicry of a multispecies mobbing chorus is more evocative than a single-species mobbing choruses.

12.6.3.2 Deceiving Predators

Mimetic alarm calls can be used deceptively to trick predators. For example, brown thornbills scare a common nest predator by using their own aerial alarm calls in combination with mimicry of heterospecific aerial alarm calls when their nestlings are attacked (Igic and Magrath 2014; Igic et al. 2015; Fig. 12.5). Pied currawongs, *Strepera graculina*, their major nest predator, respond to playback of these calls as if they themselves are under threat from a flying predator, either scanning the sky for danger or fleeing to a safe location, which could provide the thornbill's nestlings with an opportunity to escape (Igic et al. 2015). The inclusion of mimetic calls increases the effectiveness of deception, as currawongs respond more strongly to this mixture of calls than to the thornbill's own alarms. As another example in the same context, female superb lyrebirds mimic both predator calls and alarm calls of local heterospecifics during nest defense (Dalziell and Welbergen 2016). Outside the context of nest defense, mimicry of alarm calls when captured by a predator might startle predators into releasing the mimic (Chu 2001b; Igic and Magrath 2013), a potential function of distress calls (Wise et al. 1999; Neudorf and Sealy 2002).

12.6.3.3 Deception and Kleptoparasitism

Mimetic alarm calls can also be used deceptively to increase the success of stealing food from heterospecifics (see also Sect. 12.5.4). In the best-studied example, fork-tailed drongos deceptively use both their own and mimetic alarm calls to steal food from meerkats and pied babblers (Flower 2011). When a target individual is handling food, drongos produce nonmimetic alarm calls, mimetic alarm calls, or a mixture of the two, despite the absence of a predator (Flower 2011). The victim often drops the food and flees to cover, enabling the drongo to steal it (Flower 2011). Mimicry increases the success of deception by allowing the drongo to switch to a different alarm call when a particular alarm call becomes ineffective at deceiving the target (Flower et al. 2014). Alarm mimicry therefore helps the drongo to avoid frequency-dependent constraints on deception. As noted in Sect. 12.5.4, drongos also increase the efficiency of their deception by acting as sentinels for their targets and producing "true" alarm calls in the presence of predators.

12.6.3.4 Intraspecific Communication and Teaching

Mimicking heterospecific alarm calls and predator vocalizations may help mimics communicate with conspecifics, or teach young about sounds indicating danger. Alarm calls encode general information about predators, such as their type, size, location, and behavior, thereby enabling listeners to choose appropriate responses (Sect. 12.3.2). Mimicry of predator calls could potentially convey even more detail, including the species of predator (Hailman 2009; Billings et al. 2017), which might help listeners locate predators more quickly and respond more effectively. Furthermore, mimicry might help young learn to recognize heterospecific alarm calls or the calls of predators. During nest defense, the parents of some species produce mimetic alarm calls and predator calls in combination with nonmimetic alarm calls (Dalziell et al. 2015), and this may promote learning by offspring to recognize heterospecific calls of danger (Oatley 1969; Goodale et al. 2014a; Potvin et al. 2018). Young could associate these "dangerous" sounds with nonmimetic alarm calls, which young often recognize innately (Magrath et al. 2010).

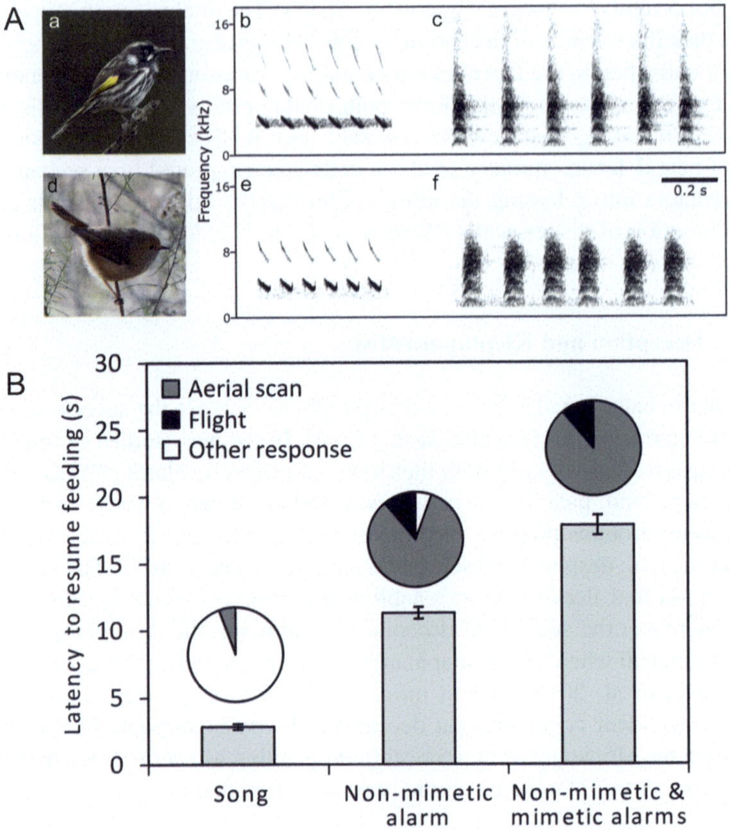

Fig. 12.5 Brown thornbills, *Acanthiza pusilla*, use vocal mimicry of alarm calls to deceive predators. (**a**) (a) New Holland honeyeaters, *Phylidonyris novaehollandiae*, produce (b) aerial and (c) mobbing calls, and (d) brown thornbills, mimic both the honeyeater's (e) aerial and (f) mobbing alarm calls. Photographs by Jessica McLachlan. (**b**) Playback of thornbill alarm call mimicry at artificial thornbill nests scares their major nest predator, the pied currawong, *Strepera graculina*. Columns show the mean ± standard error delay to resume feeding after playback, and pie charts show the immediate response of currawongs to playback. Modified from Igic et al. (2015)

12.7 Conclusions and Opportunities

Communication is often envisaged as an interaction between two individuals, usually members of the same species. One individual produces a signal that is directed at the other individual and both benefit. The signal has evolved specifically because the sender benefits from the receiver's response, and the receiver benefits from the information gained from the signal and its subsequent response (Maynard Smith and Harper 2003). But life is more complex, as exemplified by the production and use of alarm calls in natural communities. Alarm calls can be directed at multiple individuals, including conspecifics and heterospecifics; listeners can be the intended recipients or eavesdroppers that listen in on signals intended for others; callers usually benefit from listeners' responses but may be indifferent or suffer a cost; and listeners usually benefit from their responses to alarm calls, but may be deceived into costly responses.

In this chapter we have summarized research on the information encoded in alarm calls and the responses by heterospecifics. The alarm calls of both conspecifics and heterospecifics are a rich source of information on danger, providing immediate and longer term benefits. Alarm calls are sometimes very similar among species and can share generic acoustic features, and both help to explain recognition by heterospecifics. However, alarm calls can also vary greatly among species, and are often not recognized without learning their association with danger. Individuals are likely to treat heterospecific alarm calls like any other cue of danger but, once they start to respond, there is potentially selection on callers to modify their alarm calls or usage. If callers benefit from eavesdroppers' responses, they may be selected to enhance signal efficacy, leading to interspecific communication and mutual benefit. Alternatively, callers can be selected to manipulate eavesdroppers, using deceptive signaling and even mimicry, causing the eavesdropper to suffer a cost. If callers suffer a cost from eavesdropper's responses, their signaling can be modified to make eavesdropping harder, leading to cue denial.

We see many opportunities for research. Despite Marler's (1955) observations on acoustic form and function in alarm calls, we know little about the evolutionary history of alarm call structure, or the combined importance of acoustic structure and learning in the development of responses to heterospecific alarm calls. Again, despite the celebrated early study of vervet monkey alarm communication and subsequent work on other species (Sect. 12.3.2), we know little about the detail of information gained by heterospecifics. Furthermore, we are only beginning to understand the roles that eavesdropping, communication, and deception play during interspecific interactions. The study of avian vocal mimicry has already provided insights into signal design, and the balance between communication and deception, and will repay further work. Overall, alarm calls provide an excellent model to study the complex web of interspecific interactions.

Acknowledgments We thank all our collaborators in our work on communication and eavesdropping in Australian birds. In recent years these have included Lauren Ascah, Tom Bennett, Andrew Cockburn, Sean Cunningham, Francesca Dawson Pell, Anastasia Dalziell, Pam Fallow, Janet Gardner, Esteban Fernández Juricic, Jessica McLachlan, Trevor Murray, Helen Osmond, Dominique Potvin, Andy Radford, Chaminda Ratnayake, Tom Rowell, Natalie Tegtman, You Zhou, and others who have helped in various ways. We are also grateful for financial support from the Australian Research Council and the Research School of Biology, ANU, and to the Australian Bird and Bat Banding Scheme, Australian National Botanic Gardens, and ANU Ethics Committee permission to carry out the work.

References

Adkisson CS, Conner RN (1978) Interspecific vocal imitation in white-eyed vireos. Auk 95:602–606

Aubin T (1989) The role of frequency modulation in the process of distress call recognition by the starling (*Sturnus vulgaris*). Behaviour 108:57–72

Aubin T (1991) Why do distress calls evoke interspecific responses? An experimental study applied to some species of birds. Behav Process 23:103–111

Aubin T, Brémond JC (1989) Parameters used for recognition of distress calls in two species: *Larus argentatus* and *Sturnus vulgaris*. Bioacoustics 2:22–33

Baylis JR (1982) Avian vocal mimicry: its function and evolution. In: Kroodsma DE, Miller EH, Ouellet H (eds) Acoustic communication in birds. Academic, New York, pp 51–83

Biedenweg TA, Parsons MH, Fleming PA, Blumstein DT (2011) Sounds scary? Lack of habituation following the presentation of novel sounds. PLoS One 6:e14549

Billings AC, Greene E, MacArthur-Waltz D (2017) Steller's jays assess and communicate about predator risk using detection cues and identity. Behav Ecol 28:776–783

Blesdoe EK, Blumstein DT (2014) What is the sound of fear? Behavioral responses of white-crowned sparrows *Zonotrichia leucophrys* to synthesized nonlinear acoustic phenomena. Curr Zool 60:534–541

Blumstein DT, Chi YY (2012) Scared and less noisy: glucocorticoids are associated with alarm call entropy. Biol Lett 8:189–192

Blumstein DT, Munos O (2005) Individual, age and sex-specific information is contained in yellow-bellied marmot alarm calls. Anim Behav 69:353–361

Blumstein DT, Récapet C (2009) The sound of arousal: the addition of novel non-linearities increases responsiveness in marmot alarm calls. Ethology 115:1074–1081

Bradbury JW, Vehrencamp SL (2011) Principles of animal communication, 2nd edn. Sinaur, Sunderland, MA

Branch CL, Freeberg TM (2012) Distress calls in tufted titmice (*Baeolophus bicolor*): are conspecifics or predators the target? Behav Ecol 23:854–862

Caro TM (2005) Antipredator defenses in birds and mammals. University of Chicago Press, Chicago

Chu M (2001a) Heterospecific responses to scream calls and vocal mimicry by Phainopeplas (*Phainopepla nitens*) in distress. Behaviour 138:775–787

Chu M (2001b) Vocal mimicry in distress calls of Phainopeplas. Condor 103:389–395

Clarke E, Reichard UH, Zuberbühler K (2006) The syntax and meaning of wild gibbon songs. PLoS One 1:e73

Cresswell W (1994) The function of alarm calls in redshanks, *Tringa totanus*. Anim Behav 47:736–738

Cunningham S, Magrath RD (2017) Functionally referential alarm calls in noisy miners communicate about predator behaviour. Anim Behav 129:171–179

Curio E, Ernst U, Vieth W (1978) Adaptive significance of avian mobbing II. Cultural transmission of enemy recognition in blackbirds-effectiveness and some constraints. Z Tierpsychol 48:184–202

Dabelsteen T (2005) Public, private or anonymous? Facilitating and countering eavesdropping. In: McGregor PK (ed) Animal communication networks. Cambridge University Press, Melbourne, pp 38–62

Dalziell AH, Magrath RD (2012) Fooling the experts: accurate vocal mimicry in the song of the superb lyrebird, *Menura novaehollandiae*. Anim Behav 83:1401–1410

Dalziell AH, Welbergen JA (2016) Elaborate mimetic vocal displays by female Superb lyrebirds. Front Ecol Evol 4:34

Dalziell AH, Welbergen JA, Igic B, Magrath RD (2015) Avian vocal mimicry: a unified conceptual framework. Biol Rev 90:643–658

Darwin C (1872) The expression of emotions in man and animals, University of Chicago Press, 1965 edn. Murray, London

Dawson Pell FSE, Potvin DA, Ratnayake CP, Fernández-Juricic E, Magrath RD, Radford AN (2018) Birds orient their heads appropriately in response to functionally referential alarm calls of heterospecifics. Anim Behav 140:109–118

De Vos A, O'Riain J (2010) Sharks shape the geometry of a selfish seal herd: experimental evidence from seal decoys. Biol Lett 6:48–50

Dooling R (2004) Audition: can birds hear everything they sing? In: Marler P, Slabbekoorn H (eds) Nature's music: the science of birdsong. Elsevier, San Diego, pp 206–225

Dutour M, Léna J-P, Lengagne T (2017) Mobbing calls: a signal transcending species boundaries. Anim Behav 131:3–11

Dutour M, Léna J-P, Dumet A, Gardette V, Mondy N, Lengagne T (2019) The role of associative learning process on the response of fledgling great tits (*Parus major*) to mobbing calls. Anim Cogn 22:1095–1103

Fallow PM, Magrath RD (2010) Eavesdropping on other species: mutual interspecific understanding of urgency information in avian alarm calls. Anim Behav 79:411–417

Fallow PM, Gardner JL, Magrath RD (2011) Sound familiar? Acoustic similarity provokes responses to unfamiliar heterospecific alarm calls. Behav Ecol 22:401–410

Fallow PM, Pitcher BJ, Magrath RD (2013) Alarming features: birds use specific acoustic properties to identify heterospecific alarm calls. Proc R Soc B 280:20122539

Farrow LF, Doohan SJ, McDonald PG (2017) Alarm calls of a cooperative bird are referential and elicit context-specific antipredator behavior. Behav Ecol 28:724–731

Feeney WE, Langmore NE (2013) Social learning of a brood parasite by its host. Biol Lett 9:20130443

Fichtel C, Kappeler PM (2002) Anti-predator behavior of group-living Malagasy primates: mixed evidence for a referential alarm call system. Behav Ecol Sociobiol 51:262–275

Ficken MS (1989) Are mobbing calls of Steller's jays a "confusion chorus"? J Field Ornithol 60:52–55

Ficken MS, Popp J (1996) A comparative analysis of passerine mobbing calls. Auk 113:370–380

Fitch WT, Neubauer J, Herzel H (2002) Calls out of chaos: the adaptive significance of nonlinear phenomena in mammalian vocal production. Anim Behav 63:407–418

Flasskamp A (1994) The adaptive significance of avian mobbing V. An experimental test of the 'move on' hypothesis. Ethology 96:322–333

Flower TP (2011) Fork-tailed drongos use deceptive mimicked alarm calls to steal food. Proc R Soc B 278:1548–1555

Flower TP, Gribble M (2012) Kleptoparasitism by attacks versus false alarm calls in fork-tailed drongos. Anim Behav 83:403–410

Flower TP, Gribble M, Ridley AR (2014) Deception by flexible alarm mimicry in an African bird. Science 344:513–516

Gammon DE, Altizer CE (2011) Northern mockingbirds produce syntactical patterns of vocal mimicry that reflect taxonomy of imitated species. J Field Ornithol 82:158–164

Ghirlanda S, Enquist M (2003) A century of generalization. Anim Behav 66:15–36

Gill SA, Bierema AKM (2013) On the meaning of alarm calls: a review of functional reference in avian alarm calling. Ethology 119:449–461

Gill SA, Sealy SG (1996) Nest defence by yellow warblers: recognition of a brood parasite and an avian nest predator. Behaviour 133:263–282

Goodale E, Kotagama SW (2005) Alarm calling in Sri Lankan mixed-species bird flocks. Auk 122:108–120

Goodale E, Kotagama SW (2006) Context-dependent vocal mimicry in a passerine bird. Proc R Soc B 273:875–880

Goodale E, Kotagama SW (2008) Response to conspecific and heterospecific alarm calls in mixed-species bird flocks of a Sri Lankan rainforest. Behav Ecol 19:887–894

Goodale E, Beauchamp G, Magrath RD, Nieh JC, Ruxton GD (2010) Interspecific information transfer influences animal community structure. Trends Ecol Evol 25:354–361

Goodale E, Ratnayake CP, Kotagama SW (2014a) The frequency of vocal mimicry associated with danger varies due to proximity to nest and nesting stage in a passerine bird. Behaviour 151:73–88

Goodale E, Ratnayake CP, Kotagama SW (2014b) Vocal mimicry of alarm-associated sounds by a drongo elicits flee and mobbing responses from other species that participate in mixed-species bird flocks. Ethology 120:266–274

Greenlaw JS, Shackelford CE, Brown RE (1998) Call mimicry by eastern towhees and its significance in relation to auditory learning. Wilson Bull 110:431–434

Griesser M (2008) Referential calls signal predator behavior in a group-living bird species. Curr Biol 18:69–73

Griffin AS (2004) Social learning about predators: a review and prospectus. Learn Behav 32:131–140

Haff TM, Magrath RD (2012) Learning to listen? Nestling response to heterospecific alarm calls. Anim Behav 84:1401–1410

Haff TM, Magrath RD (2013) Eavesdropping on the neighbours: fledgling response to heterospecific alarm calls. Anim Behav 85:411–418

Hailman JP (2009) Context of blue jay (*Cyanocitta cristata*) mimicking Cooper's Hawk (*Accipiter cooperii*) cackle. Fla Field Nat 37:94–94

Hauser MD (1988) How infant Vervet monkeys learn to recognize starling alarm calls: the role of experience. Behaviour 105:187–201

Hetrick SA, Sieving KE (2012) Antipredator calls of tufted titmice and interspecific transfer of encoded threat information. Behav Ecol 23:83–92

Hindmarsh AM (1984) Vocal mimicry in starlings. Behaviour 90:302–324

Hirth DH, McCullough DR (1977) The evolution of alarm signals in ungulates with special reference to the white-tailed deer. Am Nat 111:31–42

Hollén LI, Radford AN (2009) The development of alarm call behaviour in mammals and birds. Anim Behav 78:791–800

Hurd PL, Wachtmeister C-A, Enquist M (1995) Darwin's principle of antithesis revisted: a role for perceptual biases in the evolution of intraspecific signals. Proc R Soc B 259:201–205

Igic B, Magrath RD (2013) Fidelity of vocal mimicry: identification and accuracy of mimicry of heterospecific alarm calls by the brown thornbill. Anim Behav 85:593–603

Igic B, Magrath RD (2014) A songbird mimics different heterospecific alarm calls in response to different types of threat. Behav Ecol 25:538–548

Igic B, McLachlam J, Lehtinen I, Magrath RD (2015) Crying wolf to a predator: deceptive vocal mimicry by a bird protecting young. Proc R Soc B 282:20150798

Ito R, Mori A (2010) Vigilance against predators induced by eavesdropping on heterospecific alarm calls in a non-vocal lizard *Oplurus cuvieri cuvieri* (Reptilia: Iguania). Proc R Soc B 277:1275–1280

Ito R, Ikeuchi I, Mori A (2013) A day gecko darkens its body color in response to avian alarm calls. Curr Herpetol 32:26–33

Johnson FR, McNaughton EJ, Shelly CD, Blumstein DT (2003) Mechanisms of heterospecific recognition in avian mobbing calls. Aust J Zool 51:577–585

Jones KJ, Hill WL (2001) Auditory perception of hawks and owls for passerine alarm calls. Ethology 107:717–726

Jurisevic MA, Sanderson KJ (1994a) Alarm vocalisations in Australian birds: convergent characteristics and phylogenetic differences. Emu 94:69–77

Jurisevic MA, Sanderson KJ (1994b) The vocal repertoires of six honeyeater (Meliphagidae) species from Adelaide, South Australia. Emu 94:141–148

Jurisevic MA, Sanderson KJ (1998) A comparative analysis of distress call structure in Australian passerine and non-passerine species: influence of size and phylogeny. J Avian Biol 29:61–71

Karp D, Manser M, Wiley EM, Townsend SW (2014) Nonlinearities in meerkat alarm calls prevent receivers from habituating. Ethology 120:189–196

Kelley LA, Healy SD (2011) The mimetic repertoire of the spotted bowerbird *Ptilonorhynchus maculatus*. Naturwissenschaften 98:501–507
Kelley LA, Coe RL, Madden JR, Healy SD (2008) Vocal mimicry in songbirds. Anim Behav 76:521–528
Kirchhof J, Hammerschmidt K (2006) Functionally referential alarm calls in tamarins (*Saguinus fuscicollis* and *Saguinus mystax*) – evidence from playback experiments. Ethology 112:346–354
Klump GM, Shalter MD (1984) Acoustic behavior of birds and mammals in the predator context. 1. Factors affecting the structure of alarm signals. 2. The functional-significance and evolution of alarm signals. Z Tierpsychol 66:189–226
Klump GM, Kretzschmar E, Curio E (1986) The hearing of an avian predator and its avian prey. Behav Ecol Sociobiol 18:317–324
Kostan KM (2002) The evolution of mutualistic interspecific communication: assessment and management across species. J Comp Psychol 116:206–209
Krams I, Krama T (2002) Interspecific reciprocity explains mobbing behaviour of the breeding chaffinches, *Fringilla coelebs*. Proc R Soc B 269:2345–2350
Krams I, Krama T, Igaune K (2006) Mobbing behaviour: reciprocity-based co-operation in breeding pied flycatchers *Ficedula hypoleuca*. Ibis 148:50–54
Leavesley AJ, Magrath RD (2005) Communicating about danger: urgency alarm calling in a bird. Anim Behav 70:365–373
Lindström L, Alatalo RV, Mappes J (1997) Imperfect Batesian mimicry—the effects of the frequency and the distastefulness of the model. Proc R Soc B 264:149–153
Magrath RD, Bennett T (2012) A micro-geography of fear: learning to eavesdrop on alarm calls of neighbouring heterospecifics. Proc R Soc B 279:902–909
Magrath RD, Pitcher BJ, Gardner JL (2009a) An avian eavesdropping network: alarm signal reliability and heterospecific response. Behav Ecol 20:745–752
Magrath RD, Pitcher BJ, Gardner JL (2009b) Recognition of other species' aerial alarm calls: speaking the same language or learning another? Proc R Soc B 276:769–774
Magrath RD, Haff TM, Horn A, Leonard ML (2010) Calling in the face of danger: predation risk and acoustic communication by parent birds and their offspring. Adv Study Behav 41:187–253
Magrath RD, Haff TM, Fallow PM, Radford AN (2015a) Eavesdropping on heterospecific alarm calls: from mechanisms to consequences. Biol Rev 90:560–586
Magrath RD, Haff TM, McLachlan JR, Igic B (2015b) Wild birds learn to eavesdrop on heterospecific alarm calls. Curr Biol 25:1–4
Manser MB (2001) The acoustic structure of suricates' alarm calls varies with predator type and the level of response urgency. Proc R Soc B 268:2315–2324
Manser MB, Bell MB, Fletcher LB (2001) The information that receivers extract from alarm calls in suricates. Proc R Soc B 268:2485–2491
Manser MB, Seyfarth RM, Cheney DL (2002) Suricate alarm calls signal predator class and urgency. Trends Cogn Sci 6:55–57
Marler P (1955) Characteristics of some animal calls. Nature 176:6–8
Marler P (1957) Specific distinctiveness in the communication signals of birds. Behaviour 11:13–37
Marshall AJ (1950) The function of vocal mimicry in birds. Emu 50:5–16
Martínez AE, Zenil RT (2012) Foraging guild influences dependence on heterospecific alarm calls in Amazonian bird flocks. Behav Ecol 23:544–550
Mateo JM (1996) Early auditory experience and the ontogeny of alarm-call discrimination in Belding's ground squirrels (*Spermophilus beldingi*). J Comp Psychol 110:115–124
Maynard Smith J (1965) The evolution of alarm calls. Am Nat 99:59–63
Maynard Smith J, Harper D (2003) Animal signals. Oxford University Press, Oxford
McGregor PK, Dabelsteen T (1996) Communication networks. In: Kroodsma DE, Miller EH (eds) Ecology and evolution of acoustic communication in birds. Cornell University press, Ithaca, pp 409–425
McLachlan JR, Ratnayake CP, Magrath RD (2019) Personal information about danger trumps social information from avian alarm calls. Proc R Soc B 286:20182945

McLachlan JR, Magrath RD (2020) Speedy revelations: how alarm calls can convey rapid, reliable information about urgent danger. Proc R Soc B. (in press)

Meise K, Franks DW, Bro-Jørgensen J (2018) Multiple adaptive and non-adaptive processes determine responsiveness to heterospecific alarm calls in African savannah herbivores. Proc R Soc B 285:20172676

Møller AP (1988) False alarm calls as a means of resource usurpation in the great tit *Parus major*. Ethology 79:25–30

Morton ES (1976) Vocal mimicry in the thick-billed Euphonia. Wilson Bull 88:485–486

Morton ES (1977) On the occurrence and significance of motivation-structural rules in some bird and mammal sounds. Am Nat 111:855–869

Morton ES (2017) Animal vocal communication: assessment and management roles. Cambridge University Press, Cambridge

Müller CA, Manser MB (2008) The information banded mongooses extract from heterospecific alarms. Anim Behav 75:897–904

Munn CA (1986) Birds that 'cry wolf'. Nature 319:143–145

Murray TG, Magrath RD (2015) Does signal deterioration compromise eavesdropping on other species' alarm calls? Anim Behav 108:33–41

Naguib M, Wiley RH (2001) Estimating the distance to a source of sound: mechanisms and adaptations for long-range communication. Anim Behav 62:825–837

Naguib M, Mundry R, Ostreiher R, Hultsch H, Schrader L, Todt D (1999) Cooperatively breeding Arabian babblers call differently when mobbing in different predator-induced situations. Behav Ecol 10:636–640

Neudorf DL, Sealy SG (2002) Distress calls of birds in a neotropical cloud forest. Biotropica 34:118–126

Nocera JJ, Ratcliffe LM (2010) Migrant and resident birds adjust antipredator behavior in response to social information accuracy. Behav Ecol 21:121–128

Oatley T (1969) The functions of vocal imitation by African Cossyphas. Ostrich 40:85–89

Oda R, Masataka N (1996) Interspecific responses of ringtailed lemurs to playback of antipredator alarm calls given by Verreaux's sifakas. Ethology 102:441–453

Parejo D, Avilés JM, Rodríguez J (2012) Alarm calls modulate the spatial structure of a breeding owl community. Proc R Soc B 279:2135–2141

Peake TM (2005) Eavesdropping in communication networks. In: McGregor PK (ed) Animal communication networks. Cambridge University Press, Cambridge, pp 13–37

Pettifor RA (1990) The effects of avian mobbing on a potential predator, the European kestrel, *Falco tinnunculus*. Anim Behav 39:821–827

Potvin DA, Ratnayake CP, Radford AN, Magrath RD (2018) Birds learn socially to recognize heterospecific alarm calls by acoustic association. Curr Biol 28:2632–2637

Radford AN, Bell MBV, Hollen LI, Ridley AR (2011) Singing for your supper: sentinel calling by kleptoparasites can mitigate the cost to victims. Evolution 65:900–906

Rainey HJ, Züberbuhler K, Slater PJB (2004) Hornbills can distinguish between primate alarm calls. Proc R Soc B 271:755–759

Ramakrishnan U, Coss RG (2000) Recognition of heterospecific alarm vocalizations by bonnet macaques (*Macaca radiata*). J Comp Psychol 114:3–12

Randler C, Förschler MI (2011) Heterospecifics do not respond to subtle differences in chaffinch mobbing calls: message is encoded in number of elements. Anim Behav 82:725–730

Ratnayake CP, Goodale E, Kotagama SW (2010) Two sympatric species of passerine birds imitate the same raptor calls in alarm contexts. Naturwissenschaften 97:103–108

Rendall D, Owren MJ, Ryan MJ (2009) What do animal signals mean? Anim Behav 78:233–240

Ridley AR, Child MF, Bell MBV (2007) Interspecific audience effects on the alarm-calling behaviour of a kleptoparasitic bird. Biol Lett 3:589–591

Ridley AR, Wiley EM, Thompson AM (2014) The ecological benefits of interceptive eavesdropping. Funct Ecol 28:197–205

Riegert J, Jůzlová Z (2018) Vocal mimicry in the song of Icterine warblers (*Hippolais icterina*): possible functions and sources of variability. Ethol Ecol Evol 30:430–446

Rooke IJ, Knight TA (1977) Alarm calls of honeyeaters with reference to locating sources of sound. Emu 77:193–198

Rydén O (1978) The significance of antecedent auditory experiences on later reactions to the 'seeet' alarm-call in great tit nestlings *Parus major*. Z Tierpsychol 47:396–409

Rydén O (1982) Selective resistance to approach: a precursor to fear responses to an alarm call in great tit nestlings *Parus major*. Dev Psychobiol 15:113–120

Schmidt KA, Dall SRX, van Gils JA (2010) The ecology of information: an overview on the ecological significance of making informed decisions. Oikos 119:304–316

Searcy WA, Nowicki S (2005) The evolution of animal communication: reliability and deception in signalling systems. Princeton University Press, Princeton, NJ

Searcy WA, Yasukawa K (2017) Eavesdropping and cue denial in avian acoustic signals. Anim Behav 124:273–282

Seppänen A, Forsman JT (2007) Interspecific social learning: novel preference can be acquired from a competing species. Curr Biol 17:1248–1252

Seppänen JT, Forsman JT, Monkkonen M, Thomson RL (2007) Social information use is a process across time, space, and ecology, reaching heterospecifics. Ecology 88:1622–1633

Seyfarth RM, Cheney DL, Marler P (1980a) Monkey responses to three different alarm calls: evidence of predator classification and semantic communication. Science 210:801–803

Seyfarth RM, Cheney DL, Marler P (1980b) Vervet monkey alarm calls: semantic communication in a free-ranging primate. Anim Behav 28:1070–1094

Sherman PW (1977) Nepotism and the evolution of alarm calls. Science 197:1246–1253

Sherman PW (1985) Alarm calls of Belding's ground squirrels to aerial predators: nepotism or self-preservation? Behav Ecol Sociobiol 17:313–323

Shriner WM (1999) Antipredator responses to a previously neutral sound by free-living adult golden-mantled ground squirrels, *Spermophilus lateralis* (Sciuridae). Ethology 105:747–757

Slaughter EI, Berlin ER, Bower JT, Blumstein DT (2013) A test of the nonlinearity hypothesis in great-tailed grackles (*Quiscalus mexicanus*). Ethology 119:309–315

Snow BK (1974) Vocal mimicry in violaceous Euphonia, *Euphonia violacea*. Wilson Bull 86:179–180

Sridhar H, Beauchamp G, Shanker K (2009) Why do birds participate in mixed-species foraging flocks? A large-scale synthesis. Anim Behav 78:337–347

Srinivasan U, Raza RH, Quadar S (2010) The nuclear question: rethinking species importance in multi-species animal groups. J Anim Ecol 79:948–954

Sullivan KA (1985) Selective alarm calling by downy woodpeckers in mixed-species flocks. Auk 102:184–187

Suzuki TN (2011) Parental alarm calls warn nestlings about different predatory threats. Curr Biol 21:R15–R16

Suzuki TN (2016) Semantic communication in birds: evidence from field research over the past two decades. Ecol Res 31:307–319

Suzuki TN, Wheatcroft D, Griesser M (2016) Experimental evidence for compositional syntax in bird calls. Nat Commun 7:10986

Tegtman N, Magrath RD (in press) Discriminating between similar alarm calls of contrasting function. Philos Trans R Soc B.

Templeton CN, Greene E (2007) Nuthatches eavesdrop on variations in heterospecific chickadee mobbing alarm calls. Proc Natl Acad Sci USA 104:5479–5482

Templeton CN, Greene E, Davis K (2005) Allometry of alarm calls: black-capped chickadees encode information about predator size. Science 308:1934–1937

ten Cate C, Rowe C (2007) Biases in signal evolution: learning makes a difference. Trends Ecol Evol 22:380–387

Townsend SW, Manser M (2011) The function of nonlinear phenomena in meerkat alarm calls. Biol Lett 7:47–49

Townsend SW, Manser MB (2013) Functionally referential communication in mammals: the past, present and the future. Ethology 119:1–11

Vernon CJ (1973) Vocal imitation by South African birds. Ostrich 44:23–30

Vieth W, Curio E, Ernst U (1980) The adaptive significance of avian mobbing III. Cultural transmission of enemy recognition in blackbirds - cross-species tutoring and properties of learning. Anim Behav 28:1217–1229

Vitousek MN, Adelman JS, Gregory NC, St Clair JJH (2007) Heterospecific alarm call recognition in a non-vocal reptile. Biol Lett 3:632–634

Wheatcroft D (2015) Repetition rate of calls used in multiple contexts communicates presence of predators to nestlings and adult bird. Anim Behav 103:35–44

Wheatcroft D, Price T (2013) Learning and signal copying facilitate communication among bird species. Proc R Soc B 280:20123070

Wheatcroft D, Price TD (2015) Rates of signal evolution are associated with the nature of interspecific communication. Behav Ecol 26:83–90

Wiley RH (2015) Noise matters: the evolution of communication. Harvard, Cambridge, MA

Wiley RH, Richard DG (1982) Adaptations for acoustic communication in birds: sound transmission and signal detection. In: Kroodsma DE, Miller EH, Ouellet H (eds) Acoustic communication in birds, Production, perception, and design features of sound, vol 1. Academic, New York, pp 131–181

Wise KK, Conover MR, Knowlton FF (1999) Response of coyotes to avian distress calls: testing the startle-predator and predator-attraction hypotheses. Behaviour 136:935–949

Zann R, Dunstan E (2008) Mimetic song in superb lyrebirds: species mimicked and mimetic accuracy in different populations and age classes. Anim Behav 76:1043–1054

Zollinger SA, Suthers RA (2004) Motor mechanisms of a vocal mimic: implications for birdsong production. Proc R Soc B 271:483–491

Zuberbühler K (2000a) Causal knowledge of predators' behaviour in wild Diana monkeys. Anim Behav 59:209–220

Zuberbühler K (2000b) Interspecies semantic communication in two forest primates. Proc R Soc B 267:713–718

Zuberbühler K (2009) Survivor signals: the biology and psychology of animal alarm calling. Adv Study Behav 40:277–322

Zuberbühler K, Noë R, Seyfarth RM (1997) Diana monkey long-distance calls: messages for conspecifics and predators. Anim Behav 53:589–604

Chapter 13
A Framework to Understand Interspecific Multimodal Signaling Systems

Alexis C. Billings and Daniel T. Blumstein

Abstract Continued interest in multimodal signaling systems has resulted in new frameworks to understand the evolution and use of these complex signals. Most of these studies have focused on multimodal communication within a species (sexual and agonistic signaling), but members of different species also benefit by communicating through both eavesdropping and evolved signals. Here we develop a framework to understand interspecific multimodal signaling systems that asks three questions: (1) Is there an ecological incentive to communicate? (2) Is interspecific communication mechanistically possible? and (3) Is there a fitness consequence to this communication? Many aspects of multimodal signaling systems are expected to be similar within and across species, and signal reliability underlies all signaling. However, we identify unique constraints that apply to interspecific signaling systems: the need for overlapping sensory systems, sensory thresholds and cognitive abilities between the two species. This new framework should help identify the processes shaping multimodal signaling evolution in interspecific signaling systems.

13.1 Introduction

Multimodal signaling occurs when signals consist of components from two or more sensory modalities. Multimodal signals are common within most animal signaling systems, and perhaps are the norm (Hebets and Papaj 2004; Partan and Marler 2005). For instance, the black-tailed prairie dog's (*Cynomys ludovicianus*) multifunction, contagious jump–yip contains a visual component (the jump) and an acoustic component (the yip) (Hare et al. 2014). However, given the inherent costs of producing and receiving signals, a fundamental question is how have these complex

A. C. Billings (✉)
Division of Biological Sciences, The University of Montana, Missoula, MT, USA

D. T. Blumstein
Department of Ecology and Evolutionary Biology, University of California, Los Angeles, CA, USA

T. Aubin, N. Mathevon (eds.), *Coding Strategies in Vertebrate Acoustic Communication*, Animal Signals and Communication 7,
https://doi.org/10.1007/978-3-030-39200-0_13

signals evolved (Bro-Jørgensen 2010)? Previous research has focused on trying to understand the benefits of multimodal communication; however, this research has focused predominantly on intraspecific signaling systems, specifically sexual and agonistic signals (Bro-Jørgensen 2010; Bro-Jørgensen and Dabelsteen 2008; Candolin 2003). Here we present a framework to explain the conditions under which multimodal signaling has evolved in *inter*specific signaling systems.

A number of hypotheses have been developed to explain the evolution of multimodal signals (reviewed in Bro-Jørgensen 2010). Many of the adaptive explanations for multimodal signaling in intraspecific signaling systems are likely to apply to interspecific signaling systems because there will be similar selection for increased robustness, content, and/or reliability. However, there are likely notable differences between intra- and interspecific multimodal signaling systems because of differences in ecology, sensory systems, and cognition that exist between different species.

13.2 A Brief Background of Multimodal Signaling

Multimodal signals are often classified as redundant or nonredundant depending on the information contained in the components of the signal (Partan and Marler 2005). Furthermore, the evolution of multimodal signals can be explained using two main mechanisms: content-driven selection for increased information (i.e., the multiple messages hypothesis; Johnstone 1996) and efficacy-driven selection for increased robustness (i.e., the backup signals hypothesis; Johnstone 1996; Hebets and Papaj 2004). To study receiver responses, each component of a multimodal signal is tested separately and then together to understand how the combination of the components alters response (Partan and Marler 2005). For example, male fruit flies (*Drosophila melanogaster*) that use both an acoustic and a chemical component in their female courtship display have more successful matings than males that use only acoustic or chemical components (Rybak et al. 2002). This illustrates redundant enhancement (Partan and Marler 2005), where both components provide information to assess male suitability (redundant), but when combined males have significantly more matings than from either component presented alone (enhancement).

Prior work and existing frameworks fail to distinguish between intraspecific and interspecific signaling systems. The majority of empirical examples of multimodal signaling have focused on intraspecific signaling systems (Bro-Jørgensen and Dabelsteen 2008; Candolin 2003; Partan and Marler 2005; Wilkins et al. 2015). Recently, however, frameworks and hypotheses that enable the analysis of a signaling system as a whole have been proposed (Bro-Jørgensen 2010; Hebets et al. 2016; Wilkins et al. 2015). These frameworks use network and systems approaches that account for dynamic selection and consider the possibility of interactions between components across contexts (i.e., inter-signal interaction; Hebets and Papaj 2004). Although, these new frameworks and hypotheses aim to better understand a signaling system as a whole, again no distinction has been formally made between intraspecific and interspecific signaling systems.

13.3 Interspecific Signaling and Communication

Kostan (2002) developed a framework for the evolution of interspecific communication that exists along a gradient of reciprocity: eavesdropping by one species, both species eavesdropping on one another, asymmetrical communication where one species is intentionally signaling to the other, or mutualistic communication where both species produce signals that alter the behavior of the other (Kostan 2002). Interspecific communicative interactions can occur within any of these categories and in a variety of contexts that include (but are not limited to) predator–prey interactions, habitat selection, resource acquisition, and species recognition.

Regardless of the category of communication, it is important to understand the costs and benefits of the exchange for both a signaler and a receiver (Westrip and Bell 2015). In eavesdropping situations, the receiver benefits from the information in the signal and the sender can either be negatively affected (sender −, receiver +; as seen when a predator eavesdrops on prey; Rhebergen et al. 2015) or not affected at all (sender 0, receiver +; as seen when one species eavesdrops on the alarm calls of another species; Fallow and Magrath 2010). In mutualistic asymmetrical communication, both the sender and the receiver benefit (sender +, receiver +; as seen when flowers signal to their pollinators). The fitness benefits of both the sender and the receiver are important in order to understand the category of the signaling system and the mechanisms that maintain it.

We develop an integrative framework that specifies the conditions under which we expect to find multimodal signaling systems among species that is based on three broad questions: (1) Is there an ecological incentive for interspecific communication? (2) Is interspecific multimodal communication mechanistically possible? and (3) Is there a fitness consequence to multimodal communication?

13.4 An Interspecific Multimodal Framework (Fig. 13.1)

13.4.1 Is There an Ecological Incentive to Communicate?

In order for interspecific communication to evolve, at least one species in a pair should gain fitness benefits from communicating with the other (e.g., two species share predators, share food, or there is a predator–prey relationship; Murray and Magrath 2015). Some species have more opportunities for interactions than others. For instance, increased ecological similarity between species should lead to greater potential for interactions and perhaps stronger selection for communication to evolve.

If there is an ecological incentive to communicate, then are there aspects of each species' ecology that favor multimodal signals over unimodal or multicomponent signals? In intraspecific systems, it is hypothesized that selection for increased content, reliability, or robustness (i.e., content- and efficacy-driven selection) leads

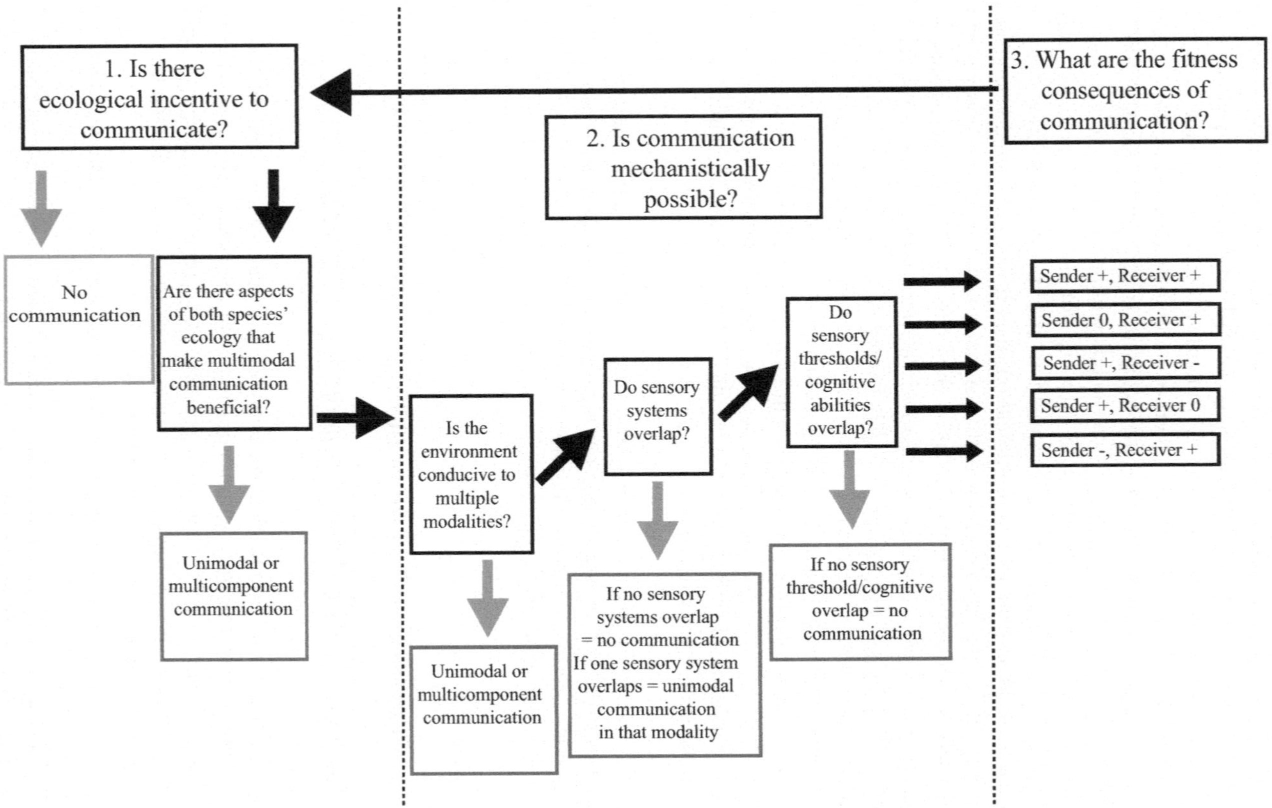

Fig. 13.1 Conceptual framework for understanding how and why multimodal signaling systems evolve in interspecific signaling systems. Black arrows indicate a "yes" response to the question posed in the boxes, where a gray arrow indicates a "no" response

to multimodal signals over unimodal or multicomponent signals, and similar selection may exist for interspecific signaling systems. For example, aposematic signals are often multimodal and combine visual, acoustic, and often olfactory components to deter predators. One hypothesis for the use of multimodal signals rather than unimodal or multicomponent signals is that the multimodal signals aid in learning and associating a defended prey with unpalatability (Rowe and Halpin 2013).

13.4.2 Is Interspecific Multimodal Communication Mechanistically Possible?

Sensory drive suggests that the relationship between the environmental conditions, the sensory systems, and signals together drive the evolution of signaling systems (Endler 1992; Tobias et al. 2010). Following this, we ask three main questions to identify the mechanisms behind interspecific multimodal signaling: (1) Do the environmental conditions favor specific modalities? (2) Do the species' sensory modalities overlap? and (3) Do the species' sensory thresholds and cognitive abilities overlap? The answers to these questions highlight the main differences between intraspecific and interspecific communication.

Do the environmental conditions favor specific modalities? Environments influence communication signals in two ways: the environment can influence the transmission and diffusion properties of a potential signal, and the environment can influence the ability of the receiver to detect the signal above the background noise (Bradbury and Vehrencamp 2011; Endler 1992). Environmental influences differ across signal modalities because of inherently different transmission and diffusion rates. For example, both acoustic and chemical signals can be used for short- and long-distance communication, but acoustic signals are generally short-term signals, while chemical signals can persist for a longer time (Weissburg et al. 2014). However, the environment can influence transmission and diffusion rates in other ways. For example, an acoustic signal travels further in water than in air (Bradbury and Vehrencamp 2011). Furthermore, habitats are not always stable (Bro-Jørgensen 2010). For example, there may be seasonal variation in the transmission properties of a habitat type. Great tit (*Parus major*) song transmits differently in a deciduous forest habitat before and after foliation (Blumenrath and Dabelsteen 2004). In addition, within a habitat type, variation in sounds produced by other species, abiotic features, and anthropogenic sources will further modify or mask signal transmission. For example, Uy and Safran (2013) found that the habitat density influences the use of the components of a multimodal signal used for species recognition. A subspecies of *Monarcha* flycatcher found in dense habitats used the acoustic and visual components sequentially, with the acoustic signal acting as a long-range signal and the visual signal acting as a close-range signal. Whereas another subspecies found in more open habitats used both acoustic and visual signals simultaneously (Uy and Safran 2013).

Do the sensory modalities overlap? Receivers can influence signal evolution through how they perceive and process signals (Rowe 1999). Therefore, the receiver's sensory system can be a selective force on the evolution of signals, including multimodal signals. In interspecific signaling systems, this means that the species must have overlapping sensory modalities: the sender needs to produce signal components in modalities the receiver possesses. For example, California ground squirrels (*Spermophilus beecheyi*) produce a multimodal signal to one predator and not another because of the predator's sensory systems (Rundus et al. 2007). Squirrels augment their antipredator tail waving behavior with infrared radiation when confronted with infrared-sensitive rattlesnakes (*Crotalus oreganus*), but do not when confronted with infrared insensitive gopher snakes (*Pituophis melanoleucus*). Furthermore, when rattlesnakes are exposed to tail waving and infrared radiation signals they shift from predatory to defensive behavior more often than when confronted with tail waving alone. The rattlesnake pit organs, which are responsible for detecting infrared radiation, make the rattlesnake an exceptional rodent predator; however, the California ground squirrel has exploited this adaptation to put rattlesnakes on the defensive.

Do the sensory thresholds and cognitive abilities overlap? Beyond possessing overlapping sensory systems, the signals must also be within the receiver's detection thresholds and cognitive processing abilities (Murray and Magrath 2015). For example, Murray and Magrath (2015) found responses to conspecific and heterospecific mobbing calls in superb fairy-wrens (*Malurus cyaneus*) and white-browed scrubwrens (*Sericornis frontalis*) differed because of reduced recognition of heterospecific calls. They concluded that there may be constraints on eavesdropping, such as the lack of perceptual specializations to detect heterospecific mobbing calls despite a large ecological overlap between the two species (Murray and Magrath 2015). This suggests that ecological overlap alone does not guarantee communication, but that sensory thresholds are important in signal use.

The opportunity for overlapping sensory modalities, sensory thresholds, and cognitive abilities is the key difference between intraspecific and interspecific communication systems. In general, conspecifics share similar sensory systems and thresholds (but see Gall and Lucas 2010). However, different species may not necessarily have the same sensory systems, the same sensory sensitivity or thresholds within a given sensory system, the same cognitive abilities, or the same information processing abilities. Therefore, it is imperative when investigating the possibility of interspecific multimodal communication to account for the sensory system as a whole (modalities, thresholds, and cognitive abilities). Overlap can be accomplished through the coevolution of sender and receiver, such as seen in plant–pollinator signaling systems (e.g., sensory drive; Endler 1992) or through sensory exploitation of sensory systems evolved for another purpose, such as seen in California ground squirrels exploiting the rattlesnakes' infrared sensory system to deter predation (Rundus et al. 2007; Ryan 1998).

Asking whether multimodal communication is mechanistically possible between species reveals important and unique constraints on the evolution of such communication because of the need for overlapping sensory modalities, sensory thresholds, and cognitive abilities between species. Although receiver psychology (including sensory thresholds and cognitive abilities) has been acknowledged as important in intraspecific signaling systems (Rowe 1999), it may be even more crucial in interspecific signaling systems because of the greater chance of a mismatch between sensory modalities, sensory thresholds, or cognitive/processing abilities. Therefore, future research into interspecific multimodal signaling systems needs to address the sensory systems of receivers in order to completely understand how the signaling system functions.

13.4.3 Is There a Fitness Consequence to Interspecific Multimodal Communication?

Like intraspecific signaling systems, the fitness consequences of signaling are dictated by the costs and benefits of signaling for both senders and receivers. These costs and benefits may also be driven by selection for increased robustness (efficacy-driven selection) or increased information or reliability (content-driven selection). Furthermore, there may be an adaptive reason to signal or the signaling system may be a by-product of another process (e.g., sensory exploitation). Finally, the costs and benefits of signaling between a sender and receiver can act as an ecological incentive to interact. Below, we present examples for three of the cost/benefit situations between sender and receiver using our framework: both sender and receiver benefit (sender +, receiver +), sender benefits (sender +, receiver −), or receiver benefits (sender −, receiver +).

Signaler +, Receiver +: Plant–Pollinator Signaling Systems
Most plant–pollinator signaling systems are asymmetrical communication systems (Kostan 2002) that involve an olfactory component and a visual component. The ecological incentive to communicate is the mutual benefit to both sender and receiver: plants get pollinated (sender +) and pollinators get an energy reward (receiver +). Selection for increased information and reliability about the nutritional reward (content-driven selection; Leonard et al. 2011) and robustness against a noisy background with multiple olfactory and visual signals bombarding pollinators (efficacy-driven selection; Leonard and Masek 2014) may drive the need for multimodal signals over unimodal signals. Communication is mechanistically possible because these systems are thought to coevolve, with plant multimodal signals coevolving with the sensory and perceptual systems of their pollinators (Haverkamp et al. 2016; Leonard and Masek 2014). For example, the evening primrose (*Oenthera neomexicana*) and hawkmoth (*Manduca sexta*) plant–pollinator system uses both olfactory and visual signals (Raguso and Willis 2002). Hawkmoths require both a visual component and an olfactory component from evening primrose in order to elicit feeding behavior (i.e., proboscis extension). Decoupling the visual and olfactory stimulus will elicit approach but not feeding.

Signaler +, Receiver −: Predators Manipulating Their Prey

Some asymmetrical signaling systems (Kostan 2002) are deceptive or manipulative where the signaler benefits and the receiver does not. For example, painted redstarts (*Myioborus pictus*) are flush-pursuing birds (Jabłoński and Lee 2006). Redstarts use visual signals accompanied by substrate vibrations to exploit their insect prey escape responses so they can pursue them in aerial chases. The ecological incentive to signal is a benefit to the sender in a predator–prey context. The use of a multimodal signal (visual + substrate vibrations) over a unimodal signal may result from sensory exploitation where the two components together flush prey more often or successfully than a unimodal signal. Another possibility is the signals are linked through morphology and one cannot be produced without the other. If sensory exploitation is involved in this system, then the signal will be in the modalities, thresholds, and cognitive abilities that the prey evolved for other purposes. This is conjecture because the necessary experiments have not been completed. Yet, sensory exploitation seems reasonable to expect since the visual stimulus alone has been linked to sensory exploitation (Jablonski 2001).

Signaler −, Receiver +: Predators Eavesdropping on Their Prey

Some signaling systems are characterized by eavesdropping receivers that benefit at the cost to the sender. Predators often eavesdrop on their prey's signals to locate them (Halfwerk et al. 2014; Rhebergen et al. 2015; Roberts et al. 2007). For example, male Túngara frogs (*Physalaemus pustulosus*) produce acoustic signals to attract females and compete with other males. A by-product of these acoustic signals is the visual signal of the inflating vocal sac. Females in this system prefer males with a linked acoustic and visual signal (Taylor and Ryan 2013; Taylor et al. 2011). Fringe-lipped bats (*Trachops cirrhosus*) also prefer the acoustic and visual components to aid in localizing their prey (Halfwerk et al. 2014; Rhebergen et al. 2015). Therefore, in this system, the multimodal signal evolved as a sexual signal, but is being eavesdropped on by predators. Male frogs use a multimodal signal over a unimodal signal because of intraspecific sexual selection (Taylor et al. 2011; Taylor and Ryan 2013). For the predators, the multimodal signal improves prey localization under various acoustic environmental conditions, suggesting the environment that these signals are produced in favors multimodal over unimodal because Túngara frogs call in choruses (Rhebergen et al. 2015). Finally, the bats have the capacity to hear the acoustic signals and use echolocation on the vocal sac to aid in localizing prey (Rhebergen et al. 2015), which is interesting because the bats are not using their visual sensory system for the visual component, but instead are using their unique sensory system (echolocation).

13.5 Conclusions

The literature on intraspecific multimodal signaling and communication is rapidly expanding. This has been driven by clearly articulated frameworks that help us understand the multimodal signaling system as a whole. We hope that by developing a similar framework for interspecific multimodal signaling systems researchers will better understand the costs and benefits that influence the adaptive value of interspecific multimodal signaling systems. Potential information is everywhere and for many species, particularly those with overlapping sensory and cognitive systems, there may be clear benefits from acquiring and using information produced by other species. And, there is often strong selection on species to exploit other species' sensory abilities for their own benefit. We hope that this framework provides structure to help understand empirical examples and provides predictions that can be empirically tested in future work. Finally, understanding interspecific multimodal signaling puts us in a better position to assess and understand how anthropogenic changes that effect multimodal signaling systems will influence these important interspecific relationships (Halfwerk and Slabbekoorn 2015; Partan 2017).

Acknowledgments We thank Robert Magrath, Erick Greene, Mike Webster, Ray Callaway, and two anonymous reviewers for many constructive comments on previous versions of this manuscript, Nicole Munoz for discussions about multimodal assessment, and Pauhana for other support. A.B. was supported by the Dan Pletcher Memorial Scholarship and the Bertha Morton Scholarship from the University of Montana, and from a N.S.F. East Asia and Pacific Summer Institute Fellowship (EAPSI). D.T.B. is supported by the N.S.F.

References

Blumenrath SH, Dabelsteen T (2004) Degradation of great tit (*Parus major*) song before and after foliation: implications for vocal communication in a deciduous forest. Behaviour 141 (8):935–958. https://doi.org/10.2307/4536177

Bradbury JW, Vehrencamp SL (2011) Principles of animal communication, 2nd edn. Sinauer Associates Incorporated, Sunderland, MA

Bro-Jørgensen J (2010) Dynamics of multiple signalling systems: animal communication in a world in flux. Trends Ecol Evol 25(5):292–300. https://doi.org/10.1016/j.tree.2009.11.003

Bro-Jørgensen J, Dabelsteen T (2008) Knee-clicks and visual traits indicate fighting ability in eland antelopes: multiple messages and back-up signals. BMC Biol 6:47–55. https://doi.org/10.1186/1741-7007-6-47

Candolin U (2003) The use of multiple cues in mate choice. Biol Rev 78:575–595

Endler JA (1992) Signals, signal conditions, and the direction of evolution. Am Nat 139:S125–S153. https://doi.org/10.2307/2462431

Fallow PM, Magrath RD (2010) Eavesdropping on other species: mutual interspecific understanding of urgency information in avian alarm calls. Anim Behav 79(2):411–417. https://doi.org/10.1016/j.anbehav.2009.11.018

Gall MD, Lucas JR (2010) Sex differences in auditory filters of brown-headed cowbirds (Molothrus ater). J Comp Physiol A 196(8):559–567. https://doi.org/10.1007/s00359-010-0543-3

Halfwerk W, Slabbekoorn H (2015) Pollution going multimodal: the complex impact of the human-altered sensory environment on animal perception and performance. Biol Lett 11(4):1–7. https://doi.org/10.1098/rsbl.2014.1051

Halfwerk W, Dixon MM, Ottens KJ, Taylor RC, Ryan MJ, Page RA, Jones PL (2014) Risks of multimodal signaling: bat predators attend to dynamic motion in frog sexual displays. J Exp Biol 217(17):3038–3044. https://doi.org/10.1242/jeb.107482

Hare JF, Campbell KL, Senkiw RW (2014) Catch the wave: prairie dogs assess neighbours' awareness using contagious displays. Proc R Soc Lond Ser B Biol Sci 281(1777):20132153. https://doi.org/10.1098/rspb.2013.2153

Haverkamp A, Bing J, Badeke E, Hansson BS, Knaden M (2016) Innate olfactory preferences for flowers matching proboscis length ensure optimal energy gain in a hawkmoth. Nat Commun 7:1–9. https://doi.org/10.1038/ncomms11644

Hebets EA, Papaj DR (2004) Complex signal function: developing a framework of testable hypotheses. Behav Ecol Sociobiol 57(3):197–214. https://doi.org/10.1007/s00265-004-0865-7

Hebets EA, Barron AB, Balakrishnan CN, Hauber ME, Mason PH, Hoke KL (2016) A systems approach to animal communication. Proc R Soc Lond B 283(1826):1–10. https://doi.org/10.1098/rspb.2015.2889

Jablonski PG (2001) Sensory exploitation of prey: manipulation of the initial direction of prey escapes by a conspicuous "rare enemy". Proc R Soc B 268(1471):1017–1022. https://doi.org/10.1098/rspb.2001.1623

Jabłoński PG, Lee SD (2006) Effects of visual stimuli, substrate-borne vibrations and air current stimuli on escape reactions in insect prey of flush-pursuing birds and their implications for evolution of flush-pursuers. Behaviour 143(3):303–324. https://doi.org/10.2307/4536346

Johnstone RA (1996) Multiple displays in animal communication: "backup signals" and "multiple messages". Phil Trans Biol Sci 351(1337):329–338. https://doi.org/10.2307/56390

Kostan KM (2002) The evolution of mutualistic interspecific communication: assessment and management across species. J Comp Psychol 116(2):206–209. https://doi.org/10.1037//0735-7036.116.2.206

Leonard AS, Masek P (2014) Multisensory integration of colors and scents: insights from bees and flowers. J Comp Physiol A 200(6):463–474. https://doi.org/10.1007/s00359-014-0904-4

Leonard AS, Dornhaus A, Papaj DR (2011) Flowers help bees cope with uncertainty: signal detection and the function of floral complexity. J Exp Biol 214(1):113–121. https://doi.org/10.1242/jeb.047407

Murray TG, Magrath RD (2015) Does signal deterioration compromise eavesdropping on other species' alarm calls? Anim Behav 108(108):33–41. https://doi.org/10.1016/j.anbehav.2015.07.015

Partan SR (2017) Multimodal shifts in noise: switching channels to communicate through rapid environmental change. Anim Behav 124:325–337. https://doi.org/10.1016/j.anbehav.2016.08.003

Partan SR, Marler P (2005) Issues in the classification of multimodal communication signals. Am Nat 166(2):231–245. https://doi.org/10.1086/431246

Raguso RA, Willis MA (2002) Synergy between visual and olfactory cues in nectar feeding by naïve hawkmoths, *Manduca sexta*. Anim Behav 64(5):685–695. https://doi.org/10.1006/anbe.2002.4010

Rhebergen F, Taylor RC, Ryan MJ, Page RA, Halfwerk W (2015) Multimodal cues improve prey localization under complex environmental conditions. Proc R Soc B 282(1814):1–8. https://doi.org/10.1098/rspb.2015.1403

Roberts JA, Taylor PW, Uetz GW (2007) Consequences of complex signaling: predator detection of multimodal cues. Behav Ecol 18(1):236–240. https://doi.org/10.1093/beheco/arl079

Rowe C (1999) Receiver psychology and the evolution of multicomponent signals. Anim Behav 58 (5):921–931. https://doi.org/10.1006/anbe.1999.1242

Rowe C, Halpin C (2013) Why are warning displays multimodal? Behav Ecol Sociobiol 67 (9):1425–1439. https://doi.org/10.1007/s00265-013-1515-8

Rundus AS, Owings DH, Joshi SS, Chinn E, Giannini N (2007) Ground squirrels use an infrared signal to deter rattlesnake predation. Proc Natl Acad Sci U S A 104(36):14372–14376. https://doi.org/10.1073/pnas.0702599104

Ryan MJ (1998) Sexual selection, receiver biases, and the evolution of sex differences. Science 281 (5385):1999–2003. https://doi.org/10.1126/science.281.5385.1999

Rybak F, Sureau G, Aubin T (2002) Functional coupling of acoustic and chemical signals in the courtship of the male *Drosophila melanogaster*. Proc R Soc B 269:695–701

Taylor RC, Ryan MJ (2013) Interactions of multisensory components perceptually rescue Túngara frog mating signals. Science 341(6143):273–274. https://doi.org/10.1126/science.1237113

Taylor RC, Klein BA, Stein J, Ryan MJ (2011) Multimodal signal variation in space and time: how important is matching a signal with its signaler? J Exp Biol 214(5):815–820. https://doi.org/10.1242/jeb.043638

Tobias JA, Aben J, Brumfield RT, Derryberry EP, Halfwerk W, Slabbekoorn H, Seddon N (2010) Song divergence by sensory drive in Amazonian birds. Evol Int J Org Evol 64(10):2820–2839. https://doi.org/10.1111/j.1558-5646.2010.01067.x

Uy JAC, Safran RJ (2013) Variation in the temporal and spatial use of signals and its implications for multimodal communication. Behav Ecol Sociobiol 67(9):1499–1511. https://doi.org/10.1007/s00265-013-1492-y

Weissburg M, Smee DL, Ferner MC (2014) The sensory ecology of nonconsumptive predator effects. Am Nat 2:141–157. https://doi.org/10.1086/676644?ref=no-x

Westrip JRS, Bell MBV (2015) Breaking down the species boundaries: selective pressures behind interspecific communication in vertebrates. Ethology 121(8):725–732. https://doi.org/10.1111/eth.12379

Wilkins MR, Shizuka D, Joseph MB, Hubbard JK, Safran RJ (2015) Multimodal signalling in the North American barn swallow: a phenotype network approach. Proc R Soc B 282(1816):1–10. https://doi.org/10.1098/rspb.2015.1574

The manufacturer's authorised representative in the EU is Springer Nature Customer Service Centre GmbH, Europaplatz 3, 69115 Heidelberg, Germany. If you have any concerns regarding our products, please contact ProductSafety@springernature.com

Printed and bound by CPI Group (UK) Ltd, Croydon, CR0 4YY

30/09/2026

02225507-0014